THE SACRED BEETLE

AND OTHER GREAT ESSAYS IN SCIENCE

The Sacred Beetle

and other great essays in science

Chosen and introduced by
MARTIN GARDNER

OXFORD UNIVERSITY PRESS

Oxford University Press, Walton Street, Oxford OX2 6DP
Oxford New York Toronto
Delhi Bombay Calcutta Madras Karachi
Petaling Jaya Singapore Hong Kong Tokyo
Nairobi Dar es Salaam Cape Town
Melbourne Auckland
and associated companies in
Berlin Ibadan

Oxford is a trade mark of Oxford University Press

First edition published 1957 in the USA as Great Essays in Science *by Washington Square Press Inc.*
Revised edition first published 1984 in the USA as The Sacred Beetle and other great essays in science *by Prometheus Books*
First published in the UK 1985 as an Oxford University Press paperback
Reprinted (twice) 1990

British Library Cataloguing in Publication Data
The sacred beetle and other great essays in science.
—Rev. ed.
1. Science
I. Title II. Gardner, Martin, 1914–
III. Gardner, Marting, 1914– Great essays in science
500 Q158.5
ISBN 0–19–286047–X

Printed in Great Britain by
Biddles Ltd.
Guildford and King's Lynn

For
Charlotte

Acknowledgements

For copyrighted material the editor is indebted to the following authors, publishers and agents:

Isaac Asimov, 'Science and Beauty', from *The Roving Mind* (Prometheus Books, 1983): the author.

John Burroughs, 'Science and Literature', from *Indoor Studies* (1889): Houghton Mifflin Co.

Rachel Carson, 'The Sunless Sea', from *The Sea Around Us*, copyright 1951 by Rachel L. Carson: Staples Press Ltd. and Oxford University Press Inc.

G. K. Chesterton, 'The Logic of Elfland', from 'The Ethics of Elfland', in *Orthodoxy*, copyright 1908, 1935, by G. K. Chesterton: John Lane The Bodley Head Ltd. and Dodd, Mead & Co. Inc.

Charles Darwin, 'Recapitulation and Conclusion', from *The Origin of Species* (1859): Appleton-Century-Crofts Inc.

John Dewey, 'The Influence of Darwinism on Philosophy', from *The Influence of Darwinism on Philosophy and Other Essays* (1910), reprinted from *Popular Science Monthly* (July 1909): Henry Holt & Co.

John Dos Passos, 'Proteus', from *The 42nd Parallel*, part of the trilogy *U.S.A.*, copyright 1930 by John Dos Passos: the author.

Arthur Stanley Eddington, 'The Decline of Determinism', Chapter IV of *New Pathways in Science* (1934): Cambridge University Press.

Albert Einstein, '$E = mc^2$', from *Out of My Later Years* (1950), reprinted from *Science Illustrated* (April 1946), copyright McGraw-Hill Publishing Co. Inc.: Philosophical Library Inc.

Havelock Ellis, 'What Makes a Woman Beautiful', from 'Sexual Selection in Man', in *Studies in the Psychology of Sex*, copyright 1905 by F. A. Davis Co., copyright 1933 by Havelock Ellis: Random House Inc.

J. Henri Fabre, 'The Sacred Beetle', from *The Sacred Beetle and Others*, copyright 1918 by Dodd, Mead & Co., Inc.: Harold Ober Associates and Dodd, Mead & Co. Inc.

Laura Fermi, 'Success', from *Atoms in the Family*, copyright 1954 by the University of Chicago: The University Of Chicago Press.

Sigmund Freud, 'Dreams of the Death of Beloved Persons', originally titled 'Dreams of the Death of Persons of Whom the Dreamer is Fond', from *The Interpretation of Dreams* (original German publication 1900), translated by James Strachey (1953): George Allen & Unwin Ltd. and Basic Books Inc.

Samuel A. Goudsmit, 'The Gestapo in Science', from *Alsos*, copyright 1947: Abelard-Schuman Inc.

Stephen Jay Gould, 'Nonmoral Nature', from *Natural History*, Vol. 91, No. 2, copyright © 1982 by the American Museum of Natural History: *Natural History*.

Aldous Huxley, 'Science in the Brave New World', Chapter XVI of *Brave New World*, copyright 1932 by Aldous Huxley: Chatto & Windus Ltd. and Harper & Bros.

Julian Huxley, 'An Essay on Bird-Mind', from *Essays of a Biologist* (1923): Chatto & Windus and Alfred A. Knopf Inc.

T. H. Huxley, 'Science and Culture', from *Science and Education* (1893): Appleton-Century-Crofts Inc.

William James, 'The Problem of Being', from *Some Problems of Philosophy*, copyright 1911 by Henry James Jr.: Paul R. Reynolds & Son.

Joseph Wood Krutch, 'The Colloid and the Crystal' (copyright 1950 by Street & Smith Publications Inc., reprinted by permission of *Mademoiselle*), from *The Best of Two Worlds*, copyright 1950 by Joseph Wood Krutch: William Sloane Associates Inc.

Jonathan Norton Leonard, 'Other-Worldly Life', from *Flight into Space*, copyright 1953 by Jonathan Norton Leonard: Sidgwick & Jackson Ltd. and Random House Inc.

Maurice Maeterlinck, 'The Nuptial Flight', from *The Life of the Bee* (1901), translated by Alfred Sutro: Countess Maurice Maeterlinck.

Ernest Nagel, 'Automation', originally titled 'Self-Regulation', copyright 1955 by Scientific American Inc.: Simon & Schuster Inc.

J. Robert Oppenheimer, 'Physics in the Contemporary World', copyright 1955 by J. Robert Oppenheimer: Dr. J. Robert Oppenheimer and the A. D. Little Lecture Committee, and Simon & Schuster Inc.

José Ortega y Gasset, 'The Barbarism of "Specialization"', from *The Revolt of the Masses*, copyright 1932 by W. W. Norton & Co. Inc.: George Allen & Unwin Ltd. and W. W. Norton & Co. Inc.

Bertrand Russell, 'The Science to Save Us from Science', from the 19 March 1950 issue of *The New York Times Magazine*, and 'The Greatness of Albert Einstein', from the 30 May 1955 issue of *The New Leader*: the author.

Carl Sagan, 'Can We Know the Universe? Reflections on a Grain of Salt', from *Broca's Brain*, copyright 1979 by Random House: the author and Scott Meredith Literary Agency Inc.

Lewis Thomas, 'Seven Wonders', from *Late Night Thoughts*, copyright © 1983 by Lewis Thomas: Oxford University Press and Viking Penguin Inc.

H. G. Wells, 'The New Source of Energy', Chapter I, Sections 1 and 2, of *The World Set Free*, copyright 1914, 1942, by Herbert George Wells: E. P. Dutton & Co. Inc. and the executors of the estate of H. G. Wells; 'Science and Ultimate Truth', Section 4 of Chapter II of *The Work, Wealth and Happiness of Mankind*, copyright 1931 by Herbert George Wells: the executors of the estate of H. G. Wells.

A. N. Whitehead, 'Religion and Science', Chapter XII of *Science and the Modern World* (1925): Cambridge University Press and The Macmillan Co.

Francis Bacon, 'The Sphinx', is taken from his *De Sapienta Veterum* (On the Wisdom of the Ancients) of 1609; Robert Louis Stevenson's 'Pan's Pipes' was first published in *London Magazine*, 4 May 1876, and later in his book *Virginibus Puerisque and other papers* (Kegan Paul, 1881).

The details given above, and the dates in the list of contents, are those of first publication, wherever these are known to the editor. In one or two cases the original details have not been traced, but the dates given are unlikely to be misleadingly late.

Contents

Preface xiii

Prologue
FRANCIS BACON
The Sphinx (1609) 1

CHARLES DARWIN
Recapitulation and Conclusion (1859) 5

JOHN DEWEY
The Influence of Darwinism on Philosophy (1909) 18

STEPHEN JAY GOULD
Nonmoral Nature (1982) 32

WILLIAM JAMES
The Problem of Being (1911) 46

HAVELOCK ELLIS
What Makes a Woman Beautiful? (1905) 53

JEAN HENRI FABRE
The Sacred Beetle (1918) 72

GILBERT KEITH CHESTERTON
The Logic of Elfland (1908) 95

CARL SAGAN
Can We Know the Universe? Reflections on a Grain of Salt (1979) 102

JOSEPH WOOD KRUTCH
The Colloid and the Crystal (1950) 110

JOSÉ ORTEGA Y GASSET
The Barbarism of "Specialization" (1932) 121

THOMAS HENRY HUXLEY
Science and Culture (1893) 130

JOHN BURROUGHS
Science and Literature (1889) 149

ISAAC ASIMOV
Science and Beauty (1983) 167

ERNEST NAGEL
Automation (1955) 173

JONATHAN NORTON LEONARD
Other-Worldly Life (1953) 181

J. ROBERT OPPENHEIMER
Physics in the Contemporary World (1955) 196

ALFRED NORTH WHITEHEAD
Religion and Science (1925) 215

JOHN DOS PASSOS
Proteus (1930) 229

JULIAN HUXLEY
An Essay on Bird-Mind (1923) 233

ARTHUR STANLEY EDDINGTON
The Decline of Determinism (1934) 251

ALDOUS HUXLEY
Science in the Brave New World (1932) 273

RACHEL CARSON
The Sunless Sea (1951) 286

MAURICE MAETERLINCK
The Nuptial Flight (1901) 305

H. G. WELLS
The New Source of Energy (1914) 321
Science and Ultimate Truth (1931) 330

LAURA FERMI
Success (1954) 336

SAMUEL GOUDSMIT
The Gestapo in Science (1947) 349

ROBERT LOUIS STEVENSON
Pan's Pipes (1876) 368

SIGMUND FREUD
Dreams of the Death of Beloved Persons (1900) 374

BERTRAND RUSSELL
The Science to Save Us from Science (1950) 397
The Greatness of Albert Einstein (1955) 408

ALBERT EINSTEIN
$E = mc^2$ (1946) 413

LEWIS THOMAS
Seven Wonders (1983) 419

Index 428

Preface

IN 1955 Herbert Alexander, then president of Pocket Books, asked me to edit an anthology to be titled *Great Essays in Science.* The paperback edition (it never saw hard covers) came out in 1957. This expanded and revised edition, with its new title, brings the book back in print after a lapse of more than a decade.

Most of the small changes made in the introductions have been occasioned by the deaths of contributors. The computer revolution has moved much faster than anyone anticipated when Ernest Nagel wrote about automation, but I am allowing that essay to remain, with my introduction unaltered, because the coming of the industrial robots has made Nagel's piece as timely and as accurate as it was in 1957.

The only selection I removed was a long chapter on the moon that first appeared in Sir Robert Ball's classic work of 1885, *The Story of the Heavens.* Although this chapter was written more than seventy years before I put it in my anthology, it contained surprisingly little that was then out of date, and indeed one can still learn much from it. "Significant new knowledge," I wrote in my introduction, "will have to wait upon the first lunar explorers." I was right in predicting that "soon the artificial satellites will be whirling around the earth" and that the moon is "certain to be the first spot visited by our spacemen." Unfortunately, I added another guess that was far too cautious. "Our children," I wrote, "may well live to see a rocket ship circle or land on the moon and return."

Four new essays have been added. I wish there could have been more, but the size of a book is limited and I decided with reluctance to limit my choices to essays by four Americans who began their distinguished science-writing careers after the date of the first edition of this collection: Isaac Asimov, Stephen Jay Gould, Carl Sagan, and Lewis Thomas. By an astonishing coincidence, all four of these men grew up in Brooklyn! I like to think of the Brooklyn Bridge as a giant symbol of the joining of C. P. Snow's "two cultures." It is a bridge that all four writers travel back and forth in their books as easily and often as they must have once traveled back and forth between Manhattan and Brooklyn.

As I said in my original preface, I had hoped that every major branch of science might be represented in this anthology, but it was not possible. Medicine and the social sciences, for example, are not here, although many of the book's essays raise momentous political questions.

A limit also had to be placed on the age of the selections. For many reasons I decided to include nothing published earlier than 1859, the pivotal year of Darwin's *Origin of Species*. (The prologue by Francis Bacon is the sole exception.) Only the piece by Havelock Ellis has been cut. The others are reprinted in full, with no textual alterations.

As a literary form the essay has always had irresponsible boundaries, and in this collection its definition becomes no clearer. There are "essays" here that are chapters, or segments from chapters, not written to be read apart from the book in which they first appeared. Several were originally lectures. Two are from works of fiction. Some are brief enough to be called sketches, others long enough to be called treatises. Some are heavy with scientific erudition, others glance at science casually over a shoulder. Some are wandering, informal expressions of opinion; others labor a thesis with the systematic vigor of a medieval schoolman. Even the cardinal rule that a great essay must be beautifully written has been violated in one or two instances.

Nothing could be further from my intent than to hand the reader a volume designed primarily to teach him science or bring him up to date on the latest trends and discoveries. There is no end to the making of such anthologies and even the best of them have a distressing way of becoming out of date before the pages are bound. Rather, the purpose of this book is to spread before the reader, whether his or her interest in science be passionate or mild, a sumptuous feast of great writing—absorbing, thought-disturbing pieces that have something important to say about science and say it forcibly and well.

Martin Gardner

Prologue

FRANCIS BACON

The Sphinx

SPHINX, says the story, was a monster combining many shapes in one. She had the face and voice of a virgin, the wings of a bird, the claws of a griffin. She dwelt on the ridge of a mountain near Thebes and infested the roads, lying in ambush for travellers, whom she would suddenly attack and lay hold of; and when she had mastered them, she propounded to them certain dark and perplexing riddles, which she was thought to have obtained from the Muses. And if the wretched captives could not at once solve and interpret the same, as they stood hesitating and confused she cruelly tore them to pieces. Time bringing no abatement of the calamity, the Thebans offered to any man who should expound the Sphinx's riddles (for this was the only way to subdue her) the sovereignty of Thebes as his reward. The greatness of the prize induced Œdipus, a man of wisdom and penetration, but lame from wounds in his feet, to accept the condition and make the trial: who presenting himself full of confidence and alacrity before the Sphinx, and being asked what kind of animal it was, which was born four-footed, afterwards became two-footed, then three-footed, and at last four-footed again, answered readily that it was man; who at his birth and during his infancy sprawls on all fours, hardly attempting to creep; in a little while walks upright on two feet; in later years leans on a walking-stick and so goes as it were on three; and at last in extreme age and decrepitude, his sinews all failing, sinks into a quadruped again, and keeps his bed. This was the right answer and gave him the victory; whereupon he slew the Sphinx; whose body was put on the back of an ass and carried about in triumph; while himself was made according to compact King of Thebes.

The fable is an elegant and a wise one, invented apparently in allusion to Science; especially in its application to practical life. Science, being the wonder of the ignorant and unskilful, may be not absurdly called a monster. In figure and aspect it is represented as many-shaped, in allusion to the immense variety of matter with which it deals. It is said to have the face and voice of a woman, in respect of its beauty and facility of utterance. Wings are added because the sciences and the discoveries of science spread and fly abroad in an instant; the communication of knowledge being like that of one candle with another, which lights up at once. Claws, sharp and hooked, are ascribed to it with great elegance, because the axioms and arguments of science penetrate and hold fast the mind, so that it has no means of evasion or escape; a point which the sacred philosopher also noted: *The words of the wise are as goads, and as nails driven deep in.* Again, all knowledge may be regarded as having its station on the heights of mountains; for it is deservedly esteemed a thing sublime and lofty, which looks down upon ignorance as from an eminence, and has moreover a spacious prospect on every side, such as we find on hill-tops. It is described as infesting the roads, because at every turn in the journey or pilgrimage of human life, matter and occasion for study assails and encounters us. Again Sphinx proposes to men a variety of hard questions and riddles which she received from the Muses. In these, while they remain with the Muses, there is probably no cruelty; for so long as the object of meditation and inquiry is merely to know, the understanding is not oppressed or straitened by it, but is free to wander and expatiate, and finds in the very uncertainty of conclusion and variety of choice a certain pleasure and delight; but when they pass from the Muses to Sphinx, that is from contemplation to practice, whereby there is necessity for present action, choice, and decision, then they begin to be painful and cruel; and unless they be solved and disposed of they strangely torment and worry the mind, pulling it first this way and then that, and fairly tearing it to pieces. Moreover the riddles of the Sphinx have always a twofold condition attached to them; distraction and laceration of mind, if you fail to solve them; if you succeed, a kingdom. For he who understands his subject is master of his end; and every workman is king over his work.

Now of the Sphinx's riddles there are in all two kinds; one concerning the nature of things, another concerning the nature of man; and in like manner there are two kinds of kingdom offered as the reward of solving them; one over nature, and the other over man. For the command over things natural,—over bodies, medicines, mechanical powers, and infinite other of the kind—is the one proper and ultimate end of true natural philosophy; however the philosophy of the School, content with what it finds, and swelling with talk, may neglect or spurn the search after realities and works. But the riddle proposed to Œdipus, by the solution of which he became King of Thebes, related to the nature of man; for whoever has a thorough insight into the nature of man may shape his fortune almost as he will, and is born for empire; as was well declared concerning the arts of the Romans,—

> Be thine the art,
> O Rome, with government to rule the nations,
> And to know whom to spare and whom to abate,
> And settle the condition of the world.

And therefore it fell out happily that Augustus Cæsar,

whether on purpose or by chance, used a Sphinx for his seal. For he certainly excelled in the art of politics if ever man did; and succeeded in the course of his life in solving most happily a great many new riddles concerning the nature of man, which if he had not dexterously and readily answered he would many times have been in imminent danger of destruction. The fable adds very prettily that when the Sphinx was subdued, her body was laid on the back of an ass: for there is nothing so subtle and abstruse, but when it is once thoroughly understood and published to the world, even a dull wit can carry it. Nor is that other point to be passed over, that the Sphinx was subdued by a lame man with club feet; for men generally proceed too fast and in too great a hurry to the solution of the Sphinx's riddles; whence it follows that the Sphinx has the better of them, and instead of obtaining the sovereignty by works and effects, they only distract and worry their minds with disputations.

CHARLES DARWIN

"AT LAST gleams of light have come," Darwin wrote in 1844 to a friend, "and I am almost convinced (quite contrary to the opinion I started with) that species are not (it is like confessing a murder) immutable."

Charles Darwin (1809-1882) was a man of extraordinary patience and humility. "Almost convinced," he wrote, and this after more than ten years of painstaking labor in the gathering of relevant facts, and fourteen years before he felt compelled to publish his views! The theory of evolution had been propounded before; but not until Darwin issued his *Origin of Species,* in 1859, had such a mountainous mass of evidence been brought together into one orderly, irrefutable argument. In truth, a murder of a sort *had* been committed. The book dealt a mortal blow to prevailing interpretations of the opening chapters of *Genesis,* and Christian orthodoxy was never the same again.

Darwin himself, as a young biologist aboard H.M.S. *Beagle,* was so thoroughly orthodox that the ship's officers laughed at his propensity for quoting Scripture. Then "disbelief crept over me at a very slow rate," he recalled, "but was at last complete. The rate was so slow that I felt no distress." The phrase "by the creator," in the final sentence of the selection chosen here, did not appear in the first edition of *Origin of Species.* It was added to the second edition to conciliate angry clerics. Darwin later wrote, "I have long since regretted that I truckled to public opinion and used the Pentateuchal term of creation, by which I really meant 'appeared' by some wholly unknown process."

Darwin knew nothing, of course, of modern mutation theory. He thought environment could modify an individual organism and that these modifications could be communicated through the blood stream to the germ plasm and so passed on to the next generation. This Lamarckian aspect of his views has long been discarded, but natural selection by survival of the fittest remains the indispensable cornerstone of the evolutionary process. "There is grandeur in this view of life," Darwin writes. And there is grandeur also in the unpretentious sentences of this great and unassuming man.

CHARLES DARWIN

Recapitulation and Conclusion

I HAVE now recapitulated the facts and considerations which have thoroughly convinced me that species have been modified, during a long course of descent. This has been effected chiefly through the natural selection of numerous successive, slight, favourable variations; aided in an important manner by the inherited effects of the use and disuse of parts; and in an unimportant manner, that is in relation to adaptive structures, whether past or present, by the direct action of external conditions, and by variations which seem to us in our ignorance to arise spontaneously. It appears that I formerly underrated the frequency and value of these latter forms of variation, as leading to permanent modifications of structure independently of natural selection. But as my conclusions have lately been much misrepresented, and it has been stated that I attribute the modification of species exclusively to natural selection, I may be permitted to remark that in the first edition of this work, and subsequently, I placed in a most conspicuous position—namely, at the close of the Introduction —the following words: "I am convinced that natural selection has been the main but not the exclusive means of modification." This has been of no avail. Great is the power of steady misrepresentation; but the history of science shows that fortunately this power does not long endure.

It can hardly be supposed that a false theory would explain, in so satisfactory a manner as does the theory of natural selection, the several large classes of facts above specified. It has recently been objected that this is an unsafe method of arguing; but it is a method used in judging of the common events of life, and has often been used by the greatest natural philosophers. The undulatory theory of light has thus been arrived at; and the belief in the revolution of the earth on its own axis was until lately supported by hardly any direct evidence. It is no valid objection that science as yet throws no light on the far higher problem of the essence or origin of life. Who can explain what is the essence of the attraction of gravity? No one now objects to following out the results consequent on this unknown element of attraction; notwithstanding that Leibnitz formerly accused Newton of introducing "occult qualities and miracles into philosophy."

I see no good reason why the views given in this volume should shock the religious feelings of any one. It is satisfactory, as showing how transient such impressions are, to remember that the greatest discovery ever made by man, namely, the law of the attraction of gravity, was also attacked by Leibnitz, "as subversive of natural, and inferentially of revealed, religion." A celebrated author and divine has written to me that "he has gradually learnt to see that it is just as noble a conception of the Deity to believe that He created a few original forms capable of self-development into other and needful forms, as to believe that He required a fresh act of creation to supply the voids caused by the action of His laws."

Why, it may be asked, until recently did nearly all the most eminent living naturalists and geologists disbelieve in the mutability of species? It cannot be asserted that organic beings in a state of nature are subject to no variation; it cannot be proved that the amount of variation in the course of long ages is a limited quantity; no clear distinction has been, or can be, drawn between species and well-marked varieties. It cannot be maintained that species when intercrossed are invariably sterile, and varieties invariably fertile; or that sterility is a special endowment and sign of creation. The belief that species were immutable productions was almost unavoidable as long as the history of the world was thought to be of short duration; and now that we have acquired some idea of the lapse of time, we are too apt to assume, without proof, that the geological record is so perfect that it would have afforded us plain evidence of the mutation of species, if they had undergone mutation.

But the chief cause of our natural unwillingness to admit that one species has given birth to other and distinct species, is that we are always slow in admitting great changes of which we do not see the steps. The difficulty is the same as that felt by so many geologists, when Lyell first insisted that long lines of inland cliffs had been formed, and great valleys excavated, by the agencies which we see still at work. The mind cannot possibly grasp the full meaning of the term of even a million years; it cannot add up and perceive the full effects of many slight variations, accumulated during an almost infinite number of generations.

Although I am fully convinced of the truth of the views given in this volume under the form of an abstract, I by no means expect to convince experienced naturalists whose minds are stocked with a multitude of facts all viewed, during a long course of years, from a point of view directly opposite to mine. It is so easy to hide our ignorance under such expressions as the "plan of creation," "unity of design," &c., and to think that we give an explanation when we only re-state a fact. Any one whose disposition leads him to attach more weight to unexplained difficulties than to the explanation of a certain number of facts will certainly reject the theory. A few naturalists, endowed with much flexibility of mind, and who have already begun to doubt the immutability of species, may be influenced by this volume; but I look with confidence to the future,—to young and rising naturalists, who will be able to view both sides of the question with impartiality. Whoever is led to believe that species are mutable will do good service by conscientiously expressing his conviction; for thus only can the load of prejudice by which this subject is overwhelmed be removed.

Several eminent naturalists have of late published their belief that a multitude of reputed species in each genus are not real species; but that other species are real, that is, have been independently created. This seems to me a strange conclusion to arrive at. They admit that a multitude of forms, which till lately they themselves thought were special creations, and which are still thus looked at by the majority of naturalists, and which consequently have all the external characteristic features of true species,—they admit that these have been produced by variation, but they refuse to extend the same view to other and slightly different forms. Nevertheless they do not pretend that they can define, or even conjecture, which are the created forms of life, and which are those produced by secondary laws. They admit variation as a *vera causa* in one case, they arbitrarily reject it in another, without assigning any distinction in the two cases. The day will come when this will be given as a curious illustration of the blindness of preconceived opinion. These authors seem no more startled at a miraculous act of creation than at an ordinary birth. But do they really believe that at innumerable periods in the earth's history certain elemental atoms have been commanded suddenly to flash into living tissues? Do they believe that at each supposed act of creation one individual or many were produced? Were all the infinitely numerous kinds of animals and plants created as eggs or seed, or as full grown? And in the case of mammals, were they created bearing the false marks of nourishment from the mother's womb? Undoubtedly some of these same questions cannot be answered by those who believe in the appearance or creation of only a few forms of life, or of some one form alone. It has been maintained by several authors that it is as easy to believe in the creation of a million beings as of one; but Maupertuis' philosophical axiom "of least action" leads the mind more willingly to admit the smaller number; and certainly we ought not to believe that innumerable beings within each great class have been created with plain, but deceptive, marks of descent from a single parent.

As a record of a former state of things, I have retained in the foregoing paragraphs, and elsewhere, several sentences which imply that naturalists believe in the separate creation of each species; and I have been much censured for having thus expressed myself. But undoubtedly this was the general belief when the first edition of the present work appeared. I formerly spoke to very many naturalists on the subject of evolution, and never once met with any sympathetic agreement. It is probable that some did then believe in evolution, but they were either silent, or expressed themselves so ambiguously that it was not easy to understand their meaning. Now things are wholly changed, and almost every naturalist admits the great principle of evolution. There are, however, some who still think that species have suddenly given birth, through quite unexplained means, to new and totally different forms: but, as I have attempted to show, weighty evidence can be opposed to the admission of great and abrupt modifications. Under a scientific point of view, and as leading to further investigation, but little advantage is gained by believing that new forms are suddenly developed in an inexplicable manner from old and widely different forms, over the old belief in the creation of species from the dust of the earth.

It may be asked how far I extend the doctrine of the modification of species. The question is difficult to answer, because the more distinct the forms are which we consider, by so much the arguments in favour of community of descent become fewer in number and less in force. But some arguments of the greatest weight extend very far. All the members of whole classes are connected together by a chain of affinities, and all can be classed on the same principle, in groups subordinate to groups. Fossil remains sometimes tend to fill up very wide intervals between existing orders.

Organs in a rudimentary condition plainly show that an early progenitor had the organ in a fully developed condition; and this in some cases implies an enormous amount of modification in the descendants. Throughout whole classes various structures are formed on the same pattern, and at a very early age the embryos closely resemble each other. Therefore I cannot doubt that the theory of descent with modification embraces all the members of the same great class or kingdom. I believe that animals are descended from at most only four or five progenitors, and plants from an equal or lesser number.

Analogy would lead me one step farther, namely, to the belief that all animals and plants are descended from some one prototype. But analogy may be a deceitful guide. Nevertheless all living things have much in common, in their chemical composition, their cellular structure, their laws of growth, and their liability to injurious influences. We see this even in so trifling a fact as that the same poison often similarly affects plants and animals; or that the poison secreted by the gall-fly produces monstrous growths on the wild rose or oak-tree. With all organic beings, excepting perhaps some of the very lowest, sexual reproduction seems to be essentially similar. With all, as far as is at present known, the germinal vesicle is the same; so that all organisms start from a common origin. If we look even to the two main divisions—namely, to the animal and vegetable kingdoms—certain low forms are so far intermediate in character that naturalists have disputed to which kingdom they should be referred. As Professor Asa Gray has remarked, "the spores and other reproductive bodies of many of the lower algæ may claim to have first a characteristically animal, and then an unequivocally vegetable existence." Therefore, on the principle of natural selection with divergence of character, it does not seem incredible that, from some such low and intermediate form, both animals and plants may have been developed; and, if we admit this, we must likewise admit that all the organic beings which have ever lived on this earth may be descended from some one primordial form. But this inference is chiefly grounded on analogy, and it is immaterial whether or not it be accepted. No doubt it is possible, as Mr. G. H. Lewes has urged, that at the first commencement of life many different forms were evolved; but if so, we may conclude that only a very few have left modified descendants. For, as I have recently remarked in regard to the members of each great kingdom, such as the Vertebrata, Articulata, &c., we have distinct evidence in their embryological, homologous, and rudimentary structures, that within each kingdom all the members are descended from a single progenitor.

When the views advanced by me in this volume, and by Mr. Wallace, or when analogous views on the origin of species are generally admitted, we can dimly foresee that there will be a considerable revolution in natural history. Systematists will be able to pursue their labours as at present; but they will not be incessantly haunted by the shadowy doubt whether this or that form be a true species. This, I feel sure and I speak after experience, will be no slight relief. The endless disputes whether or not some fifty species of British brambles are good species will cease. Systematists will have only to decide (not that this will be easy) whether any form be sufficiently constant and distinct from other forms, to be capable of definition; and if definable, whether the differences be sufficiently important to deserve a specific name. This latter point will become a far more essential consideration than it is at present; for differences, however slight, between any two forms, if not blended by intermediate gradations, are looked at by most naturalists as sufficient to raise both forms to the rank of species.

Hereafter we shall be compelled to acknowledge that the only distinction between species and well-marked varieties is, that the latter are known, or believed to be connected at the present day by intermediate gradations whereas species were formerly thus connected. Hence, without rejecting the consideration of the present existence of intermediate gradations between any two forms, we shall be led to weigh more carefully and to value higher the actual amount of difference between them. It is quite possible that forms now generally acknowledged to be merely varieties may hereafter be thought worthy of specific names; and in this case scientific and common language will come into accordance. In short, we shall have to treat species in the same manner as those naturalists treat genera, who admit that genera are merely artificial combinations made for convenience. This may not be a cheering prospect; but we shall at least be freed from the vain search for the undiscovered and undiscoverable essence of the term species.

The other and more general departments of natural history will rise greatly in interest. The terms used by naturalists, of affinity, relationship, community of type, paternity, morphology, adaptive characters, rudimentary and aborted organs, &c., will cease to be metaphorical, and will have a plain signification. When we no longer look at an organic being as a savage looks at a ship, as something wholly beyond his comprehension; when we regard every production of nature as one which has had a long history; when we contemplate every complex structure and instinct as the summing up of many contrivances, each useful to the possessor, in the same way as any great mechanical invention is the summing up of the labour, the experience, the reason, and even the blunders of numerous workmen; when we thus view each organic being, how far more interesting— I speak from experience—does the study of natural history become!

A grand and almost untrodden field of inquiry will be opened, on the causes and laws of variation, on correlation, on the effects of use and disuse, on the direct action of external conditions, and so forth. The study of domestic productions will rise immensely in value. A new variety raised by man will be a more important and interesting subject for study than one more species added to the infinitude of already recorded species. Our classifications will come to be, as far as they can be so made, genealogies; and will then truly give what may be called the plan of creation. The rules for classifying will no doubt become simpler when we have a definite object in view. We possess no pedigrees or armorial bearings; and we have to discover and trace the many diverging lines of descent in our natural genealogies, by characters of any kind which have long been inherited. Rudimentary organs will speak infallibly with respect to the nature of long-lost structures. Species and groups of species which are called aberrant, and which may fancifully be called living fossils, will aid us in forming a picture of the ancient forms of life. Embryology will often reveal to us the structure, in some degree obscured, of the prototypes of each great class.

When we can feel assured that all the individuals of the same species, and all the closely allied species of most genera, have within a not very remote period descended from one parent, and have migrated from some one birth-place; and when we better know the many means of migration, then, by the light which geology now throws, and will continue to throw, on former changes of climate and of the level of the land, we shall surely be enabled to trace in an admirable manner the former migrations of the inhabitants of the whole world. Even at present, by comparing the differences between the inhabitants of the sea on the opposite sides of a continent, and the nature of the various inhabitants on that continent in relation to their apparent means of immigration, some light can be thrown on ancient geography.

The noble science of Geology loses glory from the extreme imperfection of the record. The crust of the earth with its imbedded remains must not be looked at as a well-filled museum, but as a poor collection made at hazard and at rare intervals. The accumulation of each great fossiliferous formation will be recognised as having depended on an unusual concurrence of favourable circumstances, and the blank intervals between the successive stages as having been of vast duration. But we shall be able to gauge with some security the duration of these intervals by a comparison of the preceding and succeeding organic forms. We must be cautious in attempting to correlate as strictly contemporaneous two formations, which do not include many identical species, by the general succession of the forms of life. As species are produced and exterminated by slowly acting and still existing causes, and not by miraculous acts of creation; and as the most important of all causes of organic change is one which is almost independent of altered and perhaps suddenly altered physical conditions, namely, the mutual relation of organism to organism,—the improvement of one organism entailing the improvement or the extermination of others; it follows, that the amount of organic change in the fossils of consecutive formations probably serves as a fair measure of the relative, though not actual lapse of time. A number of species, however, keeping in a body might remain for a long period unchanged, whilst within the same period several of these species by migrating into new countries and coming into competition with foreign associates, might become modified; so that we must not overrate the accuracy of organic change as a measure of time.

In the future I see open fields for far more important researches. Psychology will be securely based on the foundation already well laid by Mr. Herbert Spencer, that of the necessary acquirement of each mental power and capacity by gradation. Much light will be thrown on the origin of man and his history.

Authors of the highest eminence seem to be fully satisfied with the view that each species has been independently created. To my mind it accords better with what we know of the laws impressed on matter by the Creator, that the production and extinction of the past and present inhabitants of the world should have been due to secondary causes, like those determining the birth and death of the individual. When I view all beings not as special creations, but as the lineal descendants of some few beings which lived long before the first bed of the Cambrian system was deposited, they seem to me to become ennobled. Judging from the past, we may safely infer that not one living species will transmit its unaltered likeness to a distant futurity. And of the species now living very few will transmit progeny of any kind to a far distant futurity; for the manner in which all organic beings are grouped, shows that the greater number of species in each genus, and all the species in many genera, have left no descendants, but have become utterly extinct. We can so far take a prophetic glance into futurity as to foretell that it will be the common and widely-spread species, belonging to the larger and dominant groups within each class, which will ultimately prevail and procreate new and dominant species. As all the living forms of life are the lineal descendants of those which lived long before the Cambrian epoch, we may feel certain that the ordinary succession by generation has never once been broken, and that no cataclysm has desolated the whole world. Hence we may look with some confidence to a secure future of great length. And as natural selection works solely by and for the good of each being, all corporeal and mental endowments will tend to progress towards perfection.

It is interesting to contemplate a tangled bank, clothed with many plants of many kinds, with birds singing on the bushes, with various insects flitting about, and with worms crawling through the damp earth, and to reflect that these elaborately constructed forms, so different from each other, and dependent upon each other in so complex a manner, have all been produced by laws acting around us. These laws, taken in the largest sense, being Growth with Reproduction; Inheritance which is almost implied by reproduction: Variability from the indirect and direct action of the conditions of life, and from use and disuse: a Ratio of Increase so high as to lead to a Struggle for Life, and as a consequence to Natural Selection, entailing Divergence of Character and the Extinction of less-improved forms. Thus, from the war of nature, from famine and death, the most exalted object which we are capable of conceiving, namely, the production of the higher animals, directly follows. There is grandeur in this view of life, with its several powers, having been originally breathed by the Creator into a few forms or into one; and that, whilst this planet has gone cycling on according to the fixed law of gravity, from so simple a beginning endless forms most beautiful and most wonderful have been, and are being evolved.

JOHN DEWEY

Most people, when they look at a spectrum, see a series of distinct colors side by side. When John Dewey looked at a spectrum he saw a continuum—a shading of one color into another, with no boundaries to indicate precisely where one color ends and another begins. Mind fades into matter, subject into object, means into ends. The individual merges with the social, liberal education mixes with vocational; science itself is part of a spectrum of things people do, like plowing the earth and sailing ships. There are no eternal essences with fixed outlines. The species move. "Truth" is simply that plastic, growing body of knowledge which serves as a tool in man's struggle to perpetuate his species.

No philosopher wasted less time brooding in metaphysical towers than this absent-minded, carelessly dressed pedagogue with the rimless glasses and Vermont drawl. He was active in hundreds of liberal organizations and causes. His influence on political thought, spelling out the meaning of such terms as "freedom" and "democracy," has been immense. He was never afraid to take partisan positions even when they were unfashionable; for instance, his vigorous condemnation of Stalin's purge trials at a time when most liberals tried to look the other way. Perhaps his greatest influence was in the field of elementary education. The old-fashioned bolted-down desk symbolized for him the old restraints. He wanted to unbolt them. He wanted to unbolt the mind. It could be done, he believed, only by extending the scientific attitude into every phase of human activity.

By a pleasant coincidence, John Dewey (1859-1952) was born the same year that *The Origin of Species* appeared. For Dewey, evolution was the great dissolver of fusty absolutisms, and in the selection chosen here he gives his reasons for thinking so. Originally a lecture delivered in 1909, it has become one of his best known, most influential essays, going to the very heart of his pragmatic philosophy. In later and more technical writings his style was often involved and dull, a fact which led Max Eastman to observe that if Dewey ever wrote a quotable sentence it had become permanently lost in the pile of his 36 books and 815 magazine articles. Perhaps we can recover such a sentence in this essay. "We do not solve them," Dewey concludes, concerning those great, burning questions of history that seem to demand exclusive either/or alternatives, "we get over them."

JOHN DEWEY

The Influence of Darwinism on Philosophy

I

THAT THE publication of the "Origin of Species" marked an epoch in the development of the natural sciences is well known to the layman. That the combination of the very words origin and species embodied an intellectual revolt and introduced a new intellectual temper is easily overlooked by the expert. The conceptions that had reigned in the philosophy of nature and knowledge for two thousand years, the conceptions that had become the familiar furniture of the mind, rested on the assumption of the superiority of the fixed and final; they rested upon treating change and origin as signs of defect and unreality. In laying hands upon the sacred ark of absolute permanency, in treating the forms that had been regarded as types of fixity and perfection as originating and passing away, the "Origin of Species" introduced a mode of thinking that in the end was bound to transform the logic of knowledge, and hence the treatment of morals, politics, and religion.

No wonder, then, that the publication of Darwin's book, a half century ago, precipitated a crisis. The true nature of the controversy is easily concealed from us, however, by the theological clamor that attended it. The vivid and popular features of the anti-Darwinian row tended to leave the impression that the issue was between science on one side and theology on the other. Such was not the case—the issue lay primarily within science itself, as Darwin himself early recognized. The theological outcry he discounted from the start, hardly noticing it save as it bore upon the "feelings of his female relatives." But for two decades before final publication he contemplated the possibility of being put down by his scientific peers as a fool or as crazy; and he set, as the measure of his success, the degree in which he should affect three men of science: Lyell in geology, Hooker in botany, and Huxley in zoology.

Religious considerations lent fervor to the controversy, but they did not provoke it. Intellectually, religious emotions are not creative but conservative. They attach themselves readily to the current view of the world and consecrate it. They steep and dye intellectual fabrics in the seething vat of emotions; they do not form their warp and woof. There is not, I think, an instance of any large idea about the world being independently generated by religion. Although the ideas that rose up like armed men against Darwinism owed their intensity to religious associations, their origin and meaning are to be sought in science and philosophy, not in religion.

II

Few words in our language foreshorten intellectual history as much as does the word species. The Greeks, in initiating the intellectual life of Europe, were impressed by characteristic traits of the life of plants and animals; so impressed indeed that they made these traits the key to defining nature and to explaining mind and society. And truly, life is so wonderful that a seemingly successful reading of its mystery might well lead men to believe that the key to the secrets of heaven and earth was in their hands. The Greek rendering of this mystery, the Greek formulation of the aim and standard of knowledge, was in the course of time embodied in the word species, and it controlled philosophy for two thousand years. To understand the intellectual face-about expressed in the phrase "Origin of Species" we must, then, understand the long dominant idea against which it is a protest.

Consider how men were impressed by the facts of life. Their eyes fell upon certain things slight in bulk, and frail in structure. To every appearance, these perceived things were inert and passive. Suddenly, under certain circumstances, these things—henceforth known as seeds or eggs or germs—begin to change, to change rapidly in size, form, and qualities. Rapid and extensive changes occur, however, in many things—as when wood is touched by fire. But the changes in the living thing are orderly; they are cumulative; they tend constantly in one direction; they do not, like other changes, destroy or consume, or pass fruitless into wandering flux; they realize and fulfil. Each successive stage, no matter how unlike its predecessor, preserves its net effect and also prepares the way for a fuller activity on the part of its successor. In living beings, changes do not happen as they seem to happen elsewhere, any which way; the earlier changes are regulated in view of later results. This progressive organization does not cease till there is achieved a true final term, a τέλος, a completed, perfected end. This final form exercises in turn a plenitude of functions, not the least noteworthy of which is production of germs like those from which it took its own origin, germs capable of the same cycle of self-fulfilling activity.

But the whole miraculous tale is not yet told. The same drama is enacted to the same destiny in countless myriads of individuals so sundered in time, so severed in space, that they have no opportunity for mutual consultation and no means of interaction. As an old writer quaintly said, "things of the same kind go through the same formalities"—celebrate, as it were, the same ceremonial rites.

This formal activity which operates throughout a series of changes and holds them to a single course; which subordinates their aimless flux to its own perfect manifestation; which, leaping the boundaries of space and time, keeps individuals distant in space and remote in time to a uniform type of structure and function: this principle seemed to give insight into the very nature of reality itself. To it Aristotle gave the name, εἶδος. This term the scholastics translated as *species*.

The force of this term was deepened by its application to everything in the universe that observes order in flux and manifests constancy through change. From the casual drift of daily weather, through the uneven recurrence of seasons and unequal return of seed time and harvest, up to the majestic sweep of the heavens—the image of eternity in time—and from this to the unchanging pure and contemplative intelligence beyond nature lies one unbroken fulfilment of ends. Nature as a whole is a progressive realization of purpose strictly comparable to the realization of purpose in any single plant or animal.

The conception of εἶδος, species, a fixed form and final cause, was the central principle of knowledge as well as of nature. Upon it rested the logic of science. Change as change is mere flux and lapse; it insults intelligence. Genuinely to know is to grasp a permanent end that realizes itself through changes, holding them thereby within the metes and bounds of fixed truth. Completely to know is to relate all special forms to their one single end and good: pure contemplative intelligence. Since, however, the scene of nature which directly confronts us is in change, nature as directly and practically experienced does not satisfy the conditions of knowledge. Human experience is in flux, and hence the instrumentalities of sense-perception and of inference based upon observation are condemned in advance. Science is compelled to aim at realities lying behind and beyond the processes of nature, and to carry on its search for these realities by means of rational forms transcending ordinary modes of perception and inference.

There are, indeed, but two alternative courses. We must either find the appropriate objects and organs of knowledge in the mutual interactions of changing things; or else, to escape the infection of change, we *must* seek them in some transcendent and supernal region. The human mind, deliberately as it were, exhausted the logic of the changeless, the final, and the transcendent, before it essayed adventure on the pathless wastes of generation and transformation. We dispose all too easily of the efforts of the schoolmen to interpret nature and mind in terms of real essences, hidden forms, and occult faculties, forgetful of the seriousness and dignity of the ideas that lay behind. We dispose of them by laughing at the famous gentleman who accounted for the fact that opium put people to sleep on the ground it had a dormitive faculty. But the doctrine, held in our own day, that knowledge of the plant that yields the poppy consists in referring the peculiarities of an individual to a type, to a universal form, a doctrine so firmly established that any other method of knowing was conceived to be unphilosophical and unscientific, is a survival of precisely the same logic. This identity of conception in the scholastic and anti-Darwinian theory may well suggest greater sympathy for what has become unfamiliar as well as greater humility regarding the further unfamiliarities that history has in store.

Darwin was not, of course, the first to question the classic philosophy of nature and of knowledge. The beginnings of the revolution are in the physical science of the sixteenth and seventeenth centuries. When Galileo said: "It is my opinion that the earth is very noble and admirable by reason of so many and so different alterations and generations which are incessantly made therein," he expressed the changed temper that was coming over the world; the transfer of interest from the permanent to the changing. When Descartes said: "The nature of physical things is much more easily conceived when they are beheld coming gradually into existence, than when they are only considered as produced at once in a finished and perfect state," the modern world became self-conscious of the logic that was henceforth to control it, the logic of which Darwin's "Origin of Species" is the latest scientific achievement. Without the methods of Copernicus, Kepler, Galileo, and their successors in astronomy, physics, and chemistry, Darwin would have been helpless in the organic sciences. But prior to Darwin the impact of the new scientific method upon life, mind, and politics, had been arrested, because between these ideal or moral interests and the inorganic world intervened the kingdom of plants and animals. The gates of the garden of life were barred to the new ideas; and only through this garden was there access to mind and politics. The influence of Darwin upon philosophy resides in his having conquered the phenomena of life for the principle of transition, and thereby freed the new logic for application to mind and morals and life. When he said of species what Galileo had said of the earth, *e pur se muove,* he emancipated, once for all, genetic and experimental ideas as an organon of asking questions and looking for explanations.

III

The exact bearings upon philosophy of the new logical outlook are, of course, as yet, uncertain and inchoate. We live in the twilight of intellectual transition. One must add the rashness of the prophet to the stubbornness of the partizan to venture a systematic exposition of the influence upon philosophy of the Darwinian method. At best, we can but inquire as to its general bearing—the effect upon mental temper and complexion, upon that body of half-conscious, half-instinctive intellectual aversions and preferences which determine, after all, our more deliberate intellectual enterprises. In this vague inquiry there happens to exist as a kind of touchstone a problem of long historic currency that has also been much discussed in Darwinian literature. I refer to the old problem of design *versus* chance, mind *versus* matter, as the causal explanation, first or final, of things.

As we have already seen, the classic notion of species carried with it the idea of purpose. In all living forms, a specific type is present directing the earlier stages of growth to the realization of its own perfection. Since this purposive regulative principle is not visible to the senses, it follows that it must be an ideal or rational force. Since, however, the perfect form is gradually approximated through the sensible changes, it also follows that in and through a sensible realm a rational ideal force is working out its own ultimate manifestation. These inferences were extended to nature: (*a*) She does nothing in vain; but all for an ulterior purpose. (*b*) Within natural sensible events there is therefore contained a spiritual causal force, which as spiritual escapes perception, but is apprehended by an enlightened reason. (*c*) The manifestation of this principle brings about a subordination of matter and sense to its own realization, and this ultimate fulfilment is the goal of nature and of man. The design argument thus operated in two directions. Purposefulness accounted for the intelligibility of nature and the possibility of science, while the absolute or cosmic character of this purposefulness gave sanction and worth to the moral and religious endeavors of man. Science was underpinned and morals authorized by one and the same principle, and their mutual agreement was eternally guaranteed.

This philosophy remained, in spite of sceptical and polemic outbursts, the official and the regnant philosophy of Europe for over two thousand years. The expulsion of fixed first and final causes from astronomy, physics, and chemistry had indeed given the doctrine something of a shock. But, on the other hand, increased acquaintance with the details of plant and animal life operated as a counterbalance and perhaps even strengthened the argument from design. The marvelous adaptations of organisms to their environment, of organs to the organism, of unlike parts of a complex organ—like the eye—to the organ itself; the foreshadowing by lower forms of the higher; the preparation in earlier stages of growth for organs that only later had their functioning—these things were increasingly recognized with the progress of botany, zoology, paleontology, and embryology. Together, they added such prestige to the design argument that by the late eighteenth century it was, as approved by the sciences of organic life, the central point of theistic and idealistic philosophy.

The Darwinian principle of natural selection cut straight under this philosophy. If all organic adaptations are due simply to constant variation and the elimination of those variations which are harmful in the struggle for existence that is brought about by excessive reproduction, there is no call for a prior intelligent causal force to plan and preordain them.

Hostile critics charged Darwin with materialism and with making chance the cause of the universe.

Some naturalists, like Asa Gray, favored the Darwinian principle and attempted to reconcile it with design. Gray held to what may be called design on the installment plan. If we conceive the "stream of variations" to be itself intended, we may suppose that each successive variation was designed from the first to be selected. In that case, variation, struggle, and selection simply define the mechanism of "secondary causes" through which the "first cause" acts; and the doctrine of design is none the worse off because we know more of its *modus operandi*.

Darwin could not accept this mediating proposal. He admits or rather he asserts that it is "impossible to conceive this immense and wonderful universe including man with his capacity of looking far backwards and far into futurity as the result of blind chance or necessity." But nevertheless he holds that since variations are in useless as well as useful directions, and since the latter are sifted out simply by the stress of the conditions of struggle for existence, the design argument as applied to living beings is unjustifiable; and its lack of support there deprives it of scientific value as applied to nature in general. If the variations of the pigeon, which under artificial selection give the pouter pigeon, are not preordained for the sake of the breeder, by what logic do we argue that variations resulting in natural species are pre-designed?

IV

So much for some of the more obvious facts of the discussion of design *versus* chance, as causal principles of nature and of life as a whole. We brought up this discussion, you recall, as a crucial instance. What does our touchstone indicate as to the bearing of Darwinian ideas upon philosophy? In the first place, the new logic outlaws, flanks, dismisses—what you will—one type of problems and substitutes for it another type. Philosophy forswears inquiry after absolute origins and absolute finalities in order to explore specific values and the specific conditions that generate them.

Darwin concluded that the impossibility of assigning the world to chance as a whole and to design in its parts indicated the insolubility of the question. Two radically different reasons, however, may be given as to why a problem is insoluble. One reason is that the problem is too high for intelligence; the other is that the question in its very asking makes assumptions that render the question meaningless. The latter alternative is unerringly pointed to in the celebrated case of design *versus* chance. Once admit that the sole verifiable or fruitful object of knowledge is the particular set of changes that generate the object of study together with the consequences that then flow from it, and no intelligible question can be asked about what, by assumption, lies outside. To assert—as is often asserted—that specific values of particular truth, social bonds and forms of beauty, if they can be shown to be generated by concretely knowable conditions, are meaningless and in vain; to assert that they are justified only when they and their particular causes and effects have all at once been gathered up into some inclusive first cause and some exhaustive final goal, is intellectual atavism. Such argumentation is reversion to the logic that explained the extinction of fire by water through the formal essence of aqueousness and the quenching of thirst by water through the final cause of aqueousness. Whether used in the case of the special event or that of life as a whole, such logic only abstracts some aspect of the existing course of events in order to reduplicate it as a petrified eternal principle by which to explain the very changes of which it is the formalization.

When Henry Sidgwick casually remarked in a letter that as he grew older his interest in what or who made the world was altered into interest in what kind of a world it is anyway, his voicing of a common experience of our own day illustrates also the nature of that intellectual transformation effected by the Darwinian logic. Interest shifts from the wholesale essence back of special changes to the question of how special changes serve and defeat concrete purposes; shifts from an intelligence that shaped things once for all to the particular intelligences which things are even now shaping; shifts from an ultimate goal of good to the direct increments of justice and happiness that intelligent administration of existent conditions may beget and that present carelessness or stupidity will destroy or forego.

In the second place, the classic type of logic inevitably set philosophy upon proving that life *must* have certain qualities and values—no matter how experience presents the matter—because of some remote cause and eventual goal. The duty of wholesale justification inevitably accompanies all thinking that makes the meaning of special occurrences depend upon something that once and for all lies behind them. The habit of derogating from present meanings and uses prevents our looking the facts of experience in the face; it prevents serious acknowledgment of the evils they present and serious concern with the goods they promise but do not as yet fulfil. It turns thought to the business of finding a wholesale transcendent remedy for the one and guarantee for the other. One is reminded of the way many moralists and theologians greeted Herbert Spencer's recognition of an unknowable energy from which welled up the phenomenal physical processes without and the conscious operations within. Merely because Spencer labeled his unknowable energy "God," this faded piece of metaphysical goods was greeted as an important and grateful concession to the reality of the spiritual realm. Were it not for the deep hold of the habit of seeking justification for ideal values in the remote and transcendent, surely this reference of them to an unknowable absolute would be despised in comparison with the demonstrations of experience that knowable energies are daily generating about us precious values.

The displacing of this wholesale type of philosophy will doubtless not arrive by sheer logical disproof, but rather by growing recognition of its futility. Were it a thousand times true that opium produces sleep because of its dormitive energy, yet the inducing of sleep in the tired, and the recovery to waking life of the poisoned, would not be thereby one least step forwarded. And were it a thousand times dialectically demonstrated that life as a whole is regulated by a transcendent principle to a final inclusive goal, none the less truth and error, health and disease, good and evil, hope and fear in the concrete, would remain just what and where they now are. To improve our education, to ameliorate our manners, to advance our politics, we must have recourse to specific conditions of generation.

Finally, the new logic introduces responsibility into the intellectual life. To idealize and rationalize the universe at large is after all a confession of inability to master the courses of things that specifically concern us. As long as mankind suffered from this impotency, it naturally shifted a burden of responsibility that it could not carry over to the more competent shoulders of the transcendent cause. But if insight into specific conditions of value and into specific consequences of ideas is possible, philosophy must in time become a method of locating and interpreting the more serious of the conflicts that occur in life, and a method of projecting ways for dealing with them: a method of moral and political diagnosis and prognosis.

The claim to formulate *a priori* the legislative constitution of the universe is by its nature a claim that may lead to elaborate dialectic developments. But it is also one that removes these very conclusions from subjection to experimental test, for, by definition, these results make no differences in the detailed course of events. But a philosophy that humbles its pretensions to the work of projecting hypotheses for the education and conduct of mind, individual and social, is thereby subjected to test by the way in which the ideas it propounds work out in practice. In having modesty forced upon it, philosophy also acquires responsibility.

Doubtless I seem to have violated the implied promise of my earlier remarks and to have turned both prophet and partizan. But in anticipating the direction of the transformations in philosophy to be wrought by the Darwinian genetic and experimental logic, I do not profess to speak for any save those who yield themselves consciously or unconsciously to this logic. No one can fairly deny that at present there are two effects of the Darwinian mode of thinking. On the one hand, they are making many sincere and vital efforts to revise our traditional philosophic conceptions in accordance with its demands. On the other hand, there is as definitely a recrudescence of absolutistic philosophies; an assertion of a type of philosophic knowing distinct from that of the sciences, one which opens to us another kind of reality from that to which the sciences give access; an appeal through experience to something that essentially goes beyond experience. This reaction affects popular creeds and religious movements as well as technical philosophies. The very conquest of the biological sciences by the new ideas has led many to proclaim an explicit and rigid separation of philosophy from science.

Old ideas give way slowly; for they are more than abstract logical forms and categories. They are habits, predispositions, deeply engrained attitudes of aversion and preference. Moreover, the conviction persists—though history shows it to be a hallucination—that all the questions that the human mind has asked are questions that can be answered in terms of the alternatives that the questions themselves present. But in fact intellectual progress usually occurs through sheer abandonment of questions together with both of the alternatives they assume—an abandonment that results from their decreasing vitality and a change of urgent interest. We do not solve them: we get over them. Old questions are solved by disappearing, evaporating, while new questions corresponding to the changed attitude of endeavor and preference take their place. Doubtless the greatest dissolvent in contemporary thought of old questions, the greatest precipitant of new methods, new intentions, new problems, is the one effected by the scientific revolution that found its climax in the "Origin of Species."

STEPHEN JAY GOULD

No scientist today who writes both for the public and for the technical journals more deserves to wear the mantle of Thomas Huxley than the Harvard paleontologist Stephen Jay Gould. Let me cite some of the resemblances.

Like Huxley, Gould has never hesitated to engage in vigorous battle with ignorant creationists, suddenly as noisy and combative in the United States as they were in England when Huxley debated Bishop ("Soapy Sam") Wilberforce and clashed with the crude Biblicism of Prime Minister William Gladstone. Like Huxley, Gould not only is thoroughly familiar with the geology and biology of his time, but has made significant contributions to both those fields and to the theory of evolution. Gould is a leading advocate of "punctuated equilibrium"—or "punk eke," as it is sometimes called—a view also held by Huxley. "Nature does make jumps now and then," Huxley wrote, "and a recognition of this is of no small importance in disposing of minor objections to the doctrine of transmutation [of species]."

Like Huxley, Gould's writing is clear, forceful, brilliant, revealing a knowledge of philosophy, literature, and the arts that far exceeds the knowledge of science on the part of most professors in the humanities. Like Huxley, he is a metaphysical agnostic—a term that Huxley coined. Like Huxley, he has little patience with scientific know-nothingism. He is unashamed to call a crank a crank, or to engage in what he likes to call the "debunking" of bogus science, even though, as he once confessed, such activity "has no intellectual content whatever. You don't learn anything."

My selection is from *Hen's Teeth and Horse's Toes,* an anthology of essays, most of which first appeared in Gould's monthly column in *Natural History Magazine.* In addition to informing us about an incredible adaptation in the insect world, the essay plunges into the most disturbing of all difficulties that confront those who believe in a benevolent deity. Why is there so much evil in the world? How can a loving God be reconciled with the existence of such seemingly vile creatures as the parasitic wasps?

Gould's thesis is simple and irrefutable. Mother Nature has no morals. She is silent on the problem of evil, providing not even a hint of a solution. "The cosmic process has no sort of relation to moral ends" was the way Huxley expressed it. "The ethical progress of society depends, not on imitating the cosmic process, still less in running away from it, but in combating it."

STEPHEN JAY GOULD

Nonmoral Nature

WHEN THE Right Honorable and Reverend Francis Henry, earl of Bridgewater, died in February, 1829, he left £8,000 to support a series of books "on the power, wisdom and goodness of God, as manifested in the creation." William Buckland, England's first official academic geologist and later dean of Westminster, was invited to compose one of the nine Bridgewater Treatises. In it he discussed the most pressing problem of natural theology: If God is benevolent and the Creation displays his "power, wisdom and goodness," then why are we surrounded with pain, suffering, and apparently senseless cruelty in the animal world?

Buckland considered the depredation of "carnivorous races" as the primary challenge to an idealized world in which the lion might dwell with the lamb. He resolved the issue to his satisfaction by arguing that carnivores actually increase "the aggregate of animal enjoyment" and "diminish that of pain." The death of victims, after all, is swift and relatively painless, victims are spared the ravages of decrepitude and senility, and populations do not outrun their food supply to the greater sorrow of all. God knew what he was doing when he made lions. Buckland concluded in hardly concealed rapture:

> The appointment of death by the agency of carnivora, as the ordinary termination of animal existence, appears therefore in its main results to be a dispensation of benevolence; it deducts much from the aggregate amount of the pain of universal death; it abridges, and almost annihilates,

> throughout the brute creation, the misery of disease, and accidental injuries, and lingering decay; and imposes such salutary restraint upon excessive increase of numbers, that the supply of food maintains perpetually a due ratio to the demand. The result is, that the surface of the land and depths of the waters are ever crowded with myriads of animated beings, the pleasures of whose life are co-extensive with its duration; and which throughout the little day of existence that is allotted to them, fulfill with joy the functions for which they were created.

We may find a certain amusing charm in Buckland's vision today, but such arguments did begin to address "the problem of evil" for many of Buckland's contemporaries—how could a benevolent God create such a world of carnage and bloodshed? Yet these claims could not abolish the problem of evil entirely, for nature includes many phenomena far more horrible in our eyes than simple predation. I suspect that nothing evokes greater disgust in most of us than slow destruction of a host by an internal parasite—slow ingestion, bit by bit, from the inside. In no other way can I explain why *Alien,* an uninspired, grade-C, formula horror film, should have won such a following. That single scene of Mr. Alien, popping forth as a baby parasite from the body of a human host, was both sickening and stunning. Our nineteenth-century forebears maintained similar feelings. Their greatest challenge to the concept of a benevolent deity was not simple predation—for one can admire quick and efficient butcheries, especially since we strive to construct them ourselves—but slow death by parasitic ingestion. The classic case, treated at length by all the great naturalists, involved the so-called ichneumon fly. Buckland had sidestepped the major issue.

The ichneumon fly, which provoked such concern among natural theologians, was a composite creature representing the habits of an enormous tribe. The Ichneumonoidea are a group of wasps, not flies, that include more species than all the vertebrates combined (wasps, with ants and bees, constitute the order Hymenoptera; flies, with their two wings—wasps have four—form the order Diptera). In addition, many related wasps of similar habits were often cited for the same grisly details. Thus, the famous story did not merely implicate a single aberrant species (perhaps a perverse leakage from Satan's realm), but perhaps hundreds of thousands of them—a large chunk of what could only be God's creation.

The ichneumons, like most wasps, generally live freely as adults but pass their larval life as parasites feeding on the bodies of other animals, almost invariably members of their own phylum, Arthropoda. The most common victims are caterpillars (butterfly and moth larvae), but some ichneumons prefer aphids and others attack spiders. Most hosts are parasitized as larvae, but some adults are attacked, and many tiny ichneumons inject their brood directly into the egg of their host.

The free-flying females locate an appropriate host and then convert it to a food factory for their own young. Parasitologists speak of ectoparasitism when the uninvited guest lives on the surface of its host, and endoparasitism when the parasite dwells within. Among endoparasitic ichneumons, adult females pierce the host with their ovipositor and deposit eggs within it. (The ovipositor, a thin tube extending backward from the wasp's rear end, may be many times as long as the body itself.) Usually, the host is not otherwise inconvenienced for the moment, at least until the eggs hatch and the ichneumon larvae begin their grim work of interior excavation. Among ectoparasites, however, many females lay their eggs directly upon the host's body. Since an active host would easily dislodge the egg, the ichneumon mother often simultaneously injects a toxin that paralyzes the caterpillar or other victim. The paralysis may be permanent, and the caterpillar lies, alive but immobile, with the agent of its future destruction secure on its belly. The egg hatches, the helpless caterpillar twitches, the wasp larva pierces and begins its grisly feast.

Since a dead and decaying caterpillar will do the wasp larva no good, it eats in a pattern that cannot help but recall, in our inappropriate, anthropocentric interpretation, the ancient English penalty for treason—drawing and quartering, with its explicit object of extracting as much torment as possible by keeping the victim alive and sentient. As the king's executioner drew out and burned his client's entrails, so does the ichneumon larva eat fat bodies and digestive organs first, keeping the caterpillar alive by preserving intact the essential heart and central nervous system. Finally, the larva completes its work and kills its victim, leaving behind the caterpillar's empty shell. Is it any wonder that ichneumons, not snakes or lions, stood as the paramount challenge to God's benevolence during the heyday of natural theology?

As I read through the nineteenth- and twentieth-century literature on ichneumons, nothing amused me more than the tension between an intellectual knowledge that wasps should not be described in human terms and a literary or emotional inability to avoid the familiar categories of epic and narrative, pain and destruction, victim and vanquisher. We seem to be caught in the mythic structures of our own cultural sagas, quite unable, even in our basic descriptions, to use any other language than the metaphors of battle and conquest. We cannot render this corner of natural history as anything but story, combining the themes of grim horror and fascination and usually ending not so much with pity for the caterpillar as with admiration for the efficiency of the ichneumon.

I detect two basic themes in most epic descriptions: the struggles of prey and the ruthless efficiency of parasites. Although we acknowledge that we witness little more than automatic instinct or physiological reaction, still we describe the defenses of hosts as though they represented conscious struggles. Thus, aphids kick and caterpillars may wriggle violently as wasps attempt to insert their ovipositors. The pupa of the tortoise-shell butterfly (usually considered an inert creature silently awaiting its conversion from duckling to swan) may contort its abdominal region so sharply that attacking wasps are thrown into the air. The caterpillars of *Hapalia,* when attacked by the wasp *Apanteles machaeralis,* drop suddenly from their leaves and suspend themselves in air by a silken thread. But the wasp may run down the thread and insert its eggs nonetheless. Some hosts can encapsulate the injected egg with blood cells that aggregate and harden, thus suffocating the parasite.

J. H. Fabre, the great nineteenth-century French entomologist, who remains to this day the preeminently literate natural historian of insects, made a special study of parasitic wasps and wrote with an unabashed anthropocentrism about the struggles of paralyzed victims (see his books *Insect Life* and *The Wonders of Instinct*). He describes some imperfectly paralyzed caterpillars that struggle so violently every time a parasite approaches that the wasp larvae must feed with unusual caution. They attach themselves to a silken strand from the roof of their burrow and descend upon a safe and exposed part of the caterpillar:

> The grub is at dinner: head downwards, it is digging into the limp belly of one of the caterpillars. . . . At the least sign of danger in the heap of caterpillars, the larva retreats . . . and climbs back to the ceiling, where the swarming rabble cannot reach it. When peace is restored, it slides down [its silken cord] and returns to table, with its head over the viands and its rear upturned and ready to withdraw in case of need.

In another chapter, he describes the fate of a paralyzed cricket:

> One may see the cricket, bitten to the quick, vainly move its antennae and abdominal styles, open and close its empty jaws, and even move a foot, but the larva is safe and searches its vitals with impunity. What an awful nightmare for the paralyzed cricket!

Fabre even learned to feed some paralyzed victims by placing a syrup of sugar and water on their mouthparts—thus showing that they remained alive, sentient, and (by implication) grateful for any palliation of their inevitable fate. If Jesus, immobile and thirsting on the cross, received only vinegar from his tormentors, Fabre at least could make an ending bittersweet.

The second theme, ruthless efficiency of the parasites, leads to the opposite conclusion—grudging admiration for the victors. We learn of their skill in capturing dangerous hosts often many times larger than themselves. Caterpillars may be easy game, but the psammocharid wasps prefer spiders. They must insert their ovipositors in a safe and precise spot. Some leave a paralyzed spider in its own burrow. *Planiceps hirsutus,* for example, parasitizes a California trapdoor spider. It searches for spider tubes on sand dunes, then digs into nearby sand to disturb the spider's home and drive it out. When the spider emerges, the wasp attacks, paralyzes its victim, drags it back into its own tube, shuts and fastens the trapdoor, and deposits a single egg upon the spider's abdomen. Other psammocharids will drag a heavy spider back to a previously prepared cluster of clay or mud cells. Some amputate a spider's legs to make the passage easier. Others fly back over water, skimming a buoyant spider along the surface.

Some wasps must battle with other parasites over a host's body. *Rhyssella curvipes* can detect the larvae of wood wasps deep within alder wood and drill down to its potential victims with its sharply ridged ovipositor. *Pseudorhyssa alpestris,* a related parasite, cannot drill directly into wood since its slender ovipositor bears only rudimentary cutting ridges. It locates the holes made by *Rhyssella,* inserts its ovipositor, and lays an egg on the host (already conveniently paralyzed by *Rhyssella*), right next to the egg deposited by its relative. The two eggs hatch at about the same time, but the larva of *Pseudorhyssa* has a bigger head bearing much larger mandibles. *Pseudorhyssa* seizes the smaller *Rhyssella* larva, destroys it, and proceeds to feast upon a banquet already well prepared.

Other praises for the efficiency of mothers invoke the themes of early, quick, and often. Many ichneumons don't even wait for their hosts to develop into larvae, but parasitize the egg directly (larval wasps may then either drain the egg itself or enter the developing host larva). Others simply move fast. *Apanteles militaris* can deposit up to seventy-two eggs in a single second. Still others are doggedly persistent. *Aphidius gomezi* females produce up to 1,500 eggs and can parasitize as many as 600 aphids in a single working day. In a bizarre twist upon "often," some wasps indulge in polyembryony, a kind of iterated supertwinning. A single egg divides into cells that aggregate into as many as 500 individuals. Since some polyembryonic wasps parasitize caterpillars much larger than themselves and may lay up to six eggs in each, as many as 3,000 larvae may develop within, and feed upon, a single host. These wasps are endoparasites and do not paralyze their victims. The caterpillars writhe back and forth, not (one suspects) from pain, but merely in response to the commotion induced by thousands of wasp larvae feeding within.

The efficiency of mothers is matched by their larval offspring. I have already mentioned the pattern of eating less essential parts first, thus keeping the host alive and fresh to its final and merciful dispatch. After the larva digests every edible morsel of its victim (if only to prevent later fouling of its abode by decaying tissue), it may still use the outer shell of its host. One aphid parasite cuts a hole in the belly of its victim's shell, glues the skeleton to a leaf by sticky secretions from its salivary gland, and then spins a cocoon to pupate within the aphid's shell.

In using inappropriate anthropocentric language in this romp through the natural history of ichneumons, I have tried to emphasize just why these wasps became a preeminent challenge to natural theology—the antiquated doctrine that attempted to infer God's essence from the products of his creation. I have used twentieth-century examples for the most part, but all themes were known and stressed by the great nineteenth-century natural theologians. How then did they square the habits of these wasps with the goodness of God? How did they extract themselves from this dilemma of their own making?

The strategies were as varied as the practitioners; they shared only the theme of special pleading for an a priori doctrine—they knew that God's benevolence was lurking somewhere behind all these tales of apparent horror. Charles Lyell, for example, in the first edition of his epochal *Principles of Geology* (1830-1833), decided that caterpillars posed such a threat to vegetation that any natural checks upon them could only reflect well upon a creating deity, for caterpillars would destroy human agriculture "did not Providence put causes in operation to keep them in due bounds."

The Reverend William Kirby, rector of Barham and Britain's

foremost entomologist, chose to ignore the plight of caterpillars and focused instead upon the virtue of mother love displayed by wasps in provisioning their young with such care.

> The great object of the female is to discover a proper nidus for her eggs. In search of this she is in constant motion. Is the caterpillar of a butterfly or moth the appropriate food for her young? You see her alight upon the plants where they are most usually to be met with, run quickly over them, carefully examining every leaf, and, having found the unfortunate object of her search, insert her sting into its flesh, and there deposit an egg.... The active Ichneumon braves every danger, and does not desist until her courage and address have insured subsistence for one of her future progeny.

Kirby found this solicitude all the more remarkable because the female wasp will never see her child and enjoy the pleasures of parenthood. Yet her love compels her to danger nonetheless:

> A very large proportion of them are doomed to die before their young come into existence. But in these the passion is not extinguished.... When you witness the solicitude with which they provide for the security and sustenance of their future young, you can scarcely deny to them love for a progeny they are never destined to behold.

Kirby also put in a good word for the marauding larvae, praising them for their forbearance in eating selectively to keep their caterpillar prey alive. Would we all husband our resources with such care!

> In this strange and apparently cruel operation one circumstance is truly remarkable. The larva of the Ichneumon, though every day, perhaps for months, it gnaws the inside of the caterpillar, and though at last it has devoured almost every part of it except the skin and intestines, carefully all this time it avoids injuring the vital organs, as if aware that its own existence depends on that of the insect upon which it preys!... What would be the impression which a

> similar instance amongst the race of quadrupeds would make upon us? If, for example, an animal . . . should be found to feed upon the inside of a dog, devouring only those parts not essential to life, while it cautiously left uninjured the heart, arteries, lungs, and intestines—should we not regard such an instance as a perfect prodigy, as an example of instinctive forebearance almost miraculous? [The last three quotes come from the 1856, and last pre-Darwinian, edition of Kirby and Spence's *Introduction to Entomology*.]

This tradition of attempting to read moral meaning from nature did not cease with the triumph of evolutionary theory after Darwin published *On the Origin of Species* in 1859—for evolution could be read as God's chosen method of peopling our planet, and ethical messages might still populate nature. Thus, St. George Mivart, one of Darwin's most effective evolutionary critics and a devout Catholic, argued that "many amiable and excellent people" had been misled by the apparent suffering of animals for two reasons. First, however much it might hurt, "physical suffering and moral evil are simply incommensurable." Since beasts are not moral agents, their feelings cannot bear any ethical message. But secondly, lest our visceral sensitivities still be aroused, Mivart assures us that animals must feel little, if any, pain. Using a favorite racist argument of the time—that "primitive" people suffer far less than advanced and cultured people—Mivart extrapolated further down the ladder of life into a realm of very limited pain indeed: Physical suffering, he argued,

> depends greatly upon the mental condition of the sufferer. Only during consciousness does it exist, and only in the most highly organized men does it reach its acme. The author has been assured that lower races of men appear less keenly sensitive to physical suffering than do more cultivated and refined human beings. Thus only in man can there really be any intense degree of suffering, because only in him is there that intellectual recollection of past moments and that anticipation of future ones, which constitute in great part the bitterness of suffering. The momentary

> pang, the present pain, which beasts endure, though real enough, is yet, doubtless, not to be compared as to its intensity with the suffering which is produced in man through his high prerogative of self-consciousness [from *Genesis of Species,* 1871].

It took Darwin himself to derail this ancient tradition—in that gentle way so characteristic of his radical intellectual approach to nearly everything. The ichneumons also troubled Darwin greatly and he wrote of them to Asa Gray in 1860:

> I own that I cannot see as plainly as others do, and as I should wish to do, evidence of design and beneficence on all sides of us. There seems to me too much misery in the world. I cannot persuade myself that a beneficent and omnipotent God would have designedly created the Ichneumonidae with the express intention of their feeding within the living bodies of Caterpillars, or that a cat should play with mice.

Indeed, he had written with more passion to Joseph Hooker in 1856: "What a book a devil's chaplain might write on the clumsy, wasteful, blundering, low, and horribly cruel works of nature!"

This honest admission—that nature is often (by our standards) cruel and that all previous attempts to find a lurking goodness behind everything represent just so much absurd special pleading—can lead in two directions. One might retain the principle that nature holds moral messages for humans, but reverse the usual perspective and claim that morality consists in understanding the ways of nature and doing the opposite. Thomas Henry Huxley advanced this argument in his famous essay on *Evolution and Ethics* (1893):

> The practice of that which is ethically best—what we call goodness or virtue—involves a course of conduct which, in all respects, is opposed to that which leads to success in the cosmic struggle for existence. In place of ruthless self-assertion it demands self-restraint; in place of thrusting aside, or treading down, all competitors, it requires that the

> individual shall not merely respect, but shall help his fellows. . . . It repudiates the gladiatorial theory of existence. . . . Laws and moral precepts are directed to the end of curbing the cosmic process.

The other argument, more radical in Darwin's day but common now, holds that nature simply is as we find it. Our failure to discern the universal good we once expected does not record our lack of insight or ingenuity but merely demonstrates that nature contains no moral messages framed in human terms. Morality is a subject for philosophers, theologians, students of the humanities, indeed for all thinking people. The answers will not be read passively from nature; they do not, and cannot, arise from the data of science. The factual state of the world does not teach us how we, with our powers for good and evil, should alter or preserve it in the most ethical manner.

Darwin himself tended toward this view, although he could not, as a man of his time, thoroughly abandon the idea that laws of nature might reflect some higher purpose. He clearly recognized that the specific manifestations of those laws—cats playing with mice, and ichneumon larvae eating caterpillars—could not embody ethical messages, but he somehow hoped that unknown higher laws might exist "with the details, whether good or bad, left to the working out of what we may call chance."

Since ichneumons are a detail, and since natural selection is a law regulating details, the answer to the ancient dilemma of why such cruelty (in our terms) exists in nature can only be that there isn't any answer—and that the framing of the question "in our terms" is thoroughly inappropriate in a natural world neither made for us nor ruled by us. It just plain happens. It is a strategy that works for ichneumons and that natural selection has programmed into their behavioral repertoire. Caterpillars are not suffering to teach us something; they have simply been outmaneuvered, for now, in the evolutionary game. Perhaps they will evolve a set of adequate defenses sometime in the future, thus sealing the fate of ichneumons. And perhaps, indeed probably, they will not.

Another Huxley, Thomas's grandson Julian, spoke for this position, using as an example—yes, you guessed it—the ubiquitous ichneumons:

> Natural selection, in fact, though like the mills of God in grinding slowly and grinding small, has few other attributes that a civilized religion would call divine. . . . Its products are just as likely to be aesthetically, morally, or intellectually repulsive to us as they are to be attractive. We need only think of the ugliness of *Sacculina* or a bladder-worm, the stupidity of a rhinoceros or a stegosaur, the horror of a female mantis devouring its mate or a brood of ichneumon flies slowly eating out a caterpillar.

It is amusing in this context, or rather ironic since it is too serious to be amusing, that modern creationists accuse evolutionists of preaching a specific ethical doctrine called secular humanism and thereby demand equal time for their unscientific and discredited views. If nature is nonmoral, then evolution cannot teach any ethical theory at all. The assumption that it can has abetted a panoply of social evils that ideologues falsely read into nature from their beliefs—eugenics and (misnamed) social Darwinism prominently among them. Not only did Darwin eschew any attempt to discover an antireligious ethic in nature, he also expressly stated his personal bewilderment about such deep issues as the problem of evil. Just a few sentences after invoking the ichneumons, and in words that express both the modesty of this splendid man and the compatibility, through lack of contact, between science and true religion, Darwin wrote to Asa Gray,

> I feel most deeply that the whole subject is too profound for the human intellect. A dog might as well speculate on the mind of Newton. Let each man hope and believe what he can.

WILLIAM JAMES

To THAT STRANGE, disquieting question, "Why does anything exist?" science can never hope to provide an answer. The reason is simple. Science can only answer a "why?" by placing an event within the framework of a more general descriptive law. Why does the apple fall? Because of the law of gravitation. Why the law of gravitation? Because of certain equations that are part of the theory of relativity. Should physicists succeed some day in writing one ultimate equation from which all physical laws can be derived, one could still ask, "Why *that* equation?" If physicists reduce all existence to a finite number of particles or waves, one can always ask, "Why *those* particles?" or "Why *those* waves?" There necessarily must remain a basic substratum—a "dark abyss," as Santayana once described it, "before which intelligence must be silent for fear of going mad." It is the Unknowable of Spencer, the Noumena of Kant, the transcendent "Wholly Other" world of Plato and Christianity and all the great religions. It is the Tao that cannot be seen or heard or named because if it could be seen or heard or named it would not be Tao.

Though reason must be silent, the emotions need not be, and it is hard to conceive of a physicist with soul so dead, who never to himself hath said, "This is my own, my native hand!" "The man who cannot wonder," wrote Carlyle, ". . . were he President of innumerable Royal Societies, and carried the whole *Mécanique Céleste* and *Hegel's Philosophy*, and the epitome of all Laboratories and Observatories with their results, in his single head,—is but a Pair of Spectacles behind which there is no Eye. Let those who have Eyes look through him, then he may be useful."

In recent years it is the existentialists who have been most preoccupied with the emotional consequences of meditating on the absurdity of being. Jean Paul Sartre's remarkable novel *Nausea* is one long monolog of a man obsessed with this absurdity. He stares at his hand, at his reflection in a mirror, at the gnarled root of a chestnut tree, until he is overcome by nausea—a sickening awareness of the soft, sticky, bloated, obscene, all-pervading jelly of existence. He is the exact opposite of the central character in another but lesser known existentialist novel, *Manalive,* by G. K. Chesterton. Chesterton's protagonist is equally startled by the quaint discovery that he is alive, but he finds this such an exhilarating fact that he constantly invents ingenious methods of shocking himself into a mood of wonder and gratitude.

Perhaps you, the reader, have upon occasion peered into this dark abyss. If so, you will find the following brief discussion, from *Some Problems of Philosophy* by William James (1842-1910), a stimulating (or nauseating) one.

WILLIAM JAMES

The Problem of Being

HOW COMES the world to be here at all instead of the nonentity which might be imagined in its place? Schopenhauer's remarks on this question may be considered classical. 'Apart from man,' he says, 'no being wonders at its own existence. When man first becomes conscious, he takes himself for granted, as something needing no explanation. But not for long; for, with the rise of the first reflection, that wonder begins which is the mother of metaphysics, and which made Aristotle say that men now and always seek to philosophize because of wonder— The lower a man stands in intellectual respects the less of a riddle does existence seem to him . . . but, the clearer his consciousness becomes the more the problem grasps him in its greatness. In fact the unrest which keeps the never stopping clock of metaphysics going is the thought that the non-existence of this world is just as possible as its existence. Nay more, we soon conceive the world as something the non-existence of which not only is conceivable but would indeed be preferable to its existence; so that our wonder passes easily into a brooding over that fatality which nevertheless could call such a world into being, and mislead the immense force that could produce and preserve it into an activity so hostile to its own interests. The philosophic wonder thus becomes a sad astonishment, and like the overture to Don Giovanni, philosophy begins with a minor chord.'

One need only shut oneself in a closet and begin to think

of the fact of one's being there, of one's queer bodily shape in the darkness (a thing to make children scream at, as Stevenson says), of one's fantastic character and all, to have the wonder steal over the detail as much as over the general fact of being, and to see that it is only familiarity that blunts it. Not only that *anything* should be, but that *this* very thing should be, is mysterious! Philosophy stares, but brings no reasoned solution, for from nothing to being there is no logical bridge.

Attempts are sometimes made to banish the question rather than to give it an answer. Those who ask it, we are told, extend illegitimately to the whole of being the contrast to a supposed alternative non-being which only particular beings possess. These, indeed, were not, and now are. But being in general, or in some shape, always was, and you cannot rightly bring the whole of it into relation with a primordial nonentity. Whether as God or as material atoms, it is itself primal and eternal. But if you call any being whatever eternal, some philosophers have always been ready to taunt you with the paradox inherent in the assumption. Is past eternity completed? they ask: If so, they go on, it must have had a beginning; for whether your imagination traverses it forwards or backwards, it offers an identical content or stuff to be measured; and if the amount comes to an end in one way, it ought to come to an end in the other. In other words, since we now witness its end, some past moment must have witnessed its beginning. If, however, it had a beginning, when was that, and why?

You are up against the previous nothing, and do not see how it ever passed into being. This dilemma, of having to choose between a regress which, although called infinite, has nevertheless come to a termination, and an absolute first, has played a great part in philosophy's history.

Other attempts still are made at exorcising the question. Non-being is not, said Parmenides and Zeno; only being is. Hence what is, is necessarily being—being, in short, is necessary. Others, calling the idea of nonentity no real idea, have said that on the absence of an idea can no genuine problem be founded. More curtly still, the whole ontological wonder has been called diseased, a case of *Grübelsucht* like asking, 'Why am I myself?' or 'Why is a triangle a triangle?'

Rationalistic minds here and there have sought to reduce the mystery. Some forms of being have been deemed more natural, so to say, or more inevitable and necessary than others. Empiricists of the evolutionary type—Herbert Spencer seems a good example—have assumed that whatever had the least of reality, was weakest, faintest, most imperceptible, most nascent, might come easiest first, and be the earliest successor to nonentity. Little by little the fuller grades of being might have added themselves in the same gradual way until the whole universe grew up.

To others not the minimum, but the maximum of being has seemed the earliest First for the intellect to accept. 'The perfection of a thing does not keep it from existing,' Spinoza said, 'on the contrary, it founds its existence.' It is mere prejudice to assume that it is harder for the great than for the little to be, and that easiest of all it is to be nothing. What makes things difficult in any line is the alien obstructions that are met with, and the smaller and weaker the thing the more powerful over it these become. Some things are so great and inclusive that to be is implied in their very nature. The anselmian or ontological proof of God's existence, sometimes called the cartesian proof, criticised by Saint Thomas, rejected by Kant, re-defended by Hegel, follows this line of thought. What is conceived as imperfect may lack being among its other lacks, but if God, who is expressly defined as *Ens perfectissimum*, lacked anything whatever, he would contradict his own definition. He cannot lack being therefore: He is *Ens necessarium, Ens realissimum*, as well as *Ens perfectissimum*.

Hegel in his lordly way says: 'It would be strange if God were not rich enough to embrace so poor a category as Being, the poorest and most abstract of all.' This is somewhat in line with Kant's saying that a real dollar does not contain one cent more than an imaginary dollar. At the beginning of his logic Hegel seeks in another way to mediate nonentity with being. Since 'being' in the abstract, mere being, means nothing in particular, it is indistinguishable from 'nothing'; and he seems dimly to think that this constitutes an identity between the two notions, of which some use may be made in getting from one to the other. Other still queerer attempts show well the rationalist temper. Mathematically you can deduce 1 from 0 by the following process: $\frac{0}{0}=\frac{1-1}{1-1}=1$. Or physically if all being has (as it seems to have) a 'polar' construction, so that every positive part of it has its negative, we get the simple equation: $+1-1=0$, *plus* and *minus* being the signs of polarity in physics.

It is not probable that the reader will be satisfied with any of these solutions, and contemporary philosophers, even rationalistically minded ones, have on the whole agreed that no one has intelligibly banished the mystery of *fact*. Whether the original nothing burst into God and vanished, as night vanishes in day, while God thereupon became the creative principle of all lesser beings; or whether all things have foisted or shaped themselves imperceptibly into existence, the same amount of existence has in the end to be assumed and begged by the philosopher. To comminute the difficulty is not to quench it. If you are a rationalist you beg a kilogram of being at once, we will say; if you are an empiricist you beg a thousand successive grams; but you beg the same amount in each case, and you are the same beggar whatever you may pretend. You leave the logical riddle untouched, of how the coming of whatever is, came it all at once, or came it piecemeal, can be intellectually understood.[1]

If being gradually *grew*, its quantity was of course not always the same, and may not be the same hereafter. To most philosophers this view has seemed absurd, neither God, nor primordial matter, nor energy being supposed to admit of increase or decrease. The orthodox opinion is that

[1] In more technical language, one may say that fact or being is 'contingent,' or matter of 'chance,' so far as our intellect is concerned. The conditions of its appearance are uncertain, unforeseeable, when future, and when past, elusive.

the quantity of reality must at all costs be conserved, and the waxing and waning of our phenomenal experiences must be treated as surface appearances which leave the deeps untouched.

Nevertheless, within experience, phenomena come and go. There are novelties; there are losses. The world seems, on the concrete and proximate level at least, really to grow. So the question recurs: How do our finite experiences come into being from moment to moment? By inertia? By perpetual creation? Do the new ones come at the call of the old ones? Why do not they all go out like a candle?

Who can tell off-hand? The question of being is the darkest in all philosophy. All of us are beggars here, and no school can speak disdainfully of another or give itself superior airs. For all of us alike, Fact forms a datum, gift, or *Vorgefundenes*, which we cannot burrow under, explain or get behind. It makes itself somehow, and our business is far more with its What than with its Whence or Why.

HAVELOCK ELLIS

TODAY, when empirical sex studies are bestsellers and few dispute the right of scientists to inquire into sexual matters, it is easy to forget how recently this right has been won. When the first volume of Henry Havelock Ellis' (1859-1939) path-breaking *Studies in the Psychology of Sex* appeared in England it was banned by the courts as "wicked, bawdy, and scandalous." Fortunately, Ellis found a courageous American publisher, and it was in this country that the seven volumes of the *Studies* finally appeared. They were the first notable effort to survey the entire field of sexuality, normal and abnormal, from a scientific point of view.

The following selection, a slightly abridged chapter from one of the *Studies*, discusses five factors that Ellis considers basic in a woman's sexual appeal. (In his own case, it is interesting to note, he married and passionately loved a woman of masculine appearance and strong lesbian impulses.) Ellis once wrote that he tried to present his findings "in that cold and dry light through which alone the goal of knowledge may truly be seen." But Ellis was wrong. No scientific writing was ever less cold or dry. In addition to his sound medical training and a thorough knowledge of the relevant sciences of his day, his writings also reflect an incredibly broad acquaintance with world literature. Add to this the intimate revelations of his case histories and the charm of an elegant style, and it is not difficult to understand why the *Studies* remain a delight to read long after they have been superseded by more accurate (and colder and dryer) investigations.

HAVELOCK ELLIS

What Makes a Woman Beautiful?

IN THE CONSTITUTION of our ideals of masculine and feminine beauty it was inevitable that the sexual characters should from a very early period in the history of man form an important element. From a primitive point of view a sexually desirable and attractive person is one whose sexual characters are either naturally prominent or artificially rendered so. The beautiful woman is one endowed, as Chaucer expresses it,

"With buttokes brode and brestës rounde and hye";

that is to say, she is the woman obviously best fitted to bear children and to suckle them. These two physical characters, indeed, since they represent aptitude for the two essential acts of motherhood, must necessarily tend to be regarded as beautiful among all peoples and in all stages of culture, even in high stages of civilization when more refined and perverse ideals tend to find favor, and at Pompeii as a decoration on the east side of the Purgatorium of the Temple of Isis we find a representation of Perseus rescuing Andromeda, who is shown as a woman with a very small head, small hands and feet, but with a fully developed body, large breasts, and large projecting nates.

To a certain extent—and, as we shall see, to a certain extent only—the primary sexual characters are objects of admiration among primitive peoples. In the primitive dances of many peoples, often of sexual significance, the display of the sexual organs on the part of both men and women is frequently a prominent feature. Even down to mediæval times in Europe the garments of men sometimes permitted the sexual organs to be visible. In some parts of the world, also, the artificial enlargement of the female sexual organs is practiced, and thus enlarged they are considered an important and attractive feature of beauty.

This insistence on the naked sexual organs as objects of attraction is, however, comparatively rare, and confined to peoples in a low state of culture. Very much more widespread is the attempt to beautify and call attention to the sexual organs by tattooing, by adornment and by striking peculiarities of clothing. The tendency for beauty of clothing to be accepted as a substitute for beauty of body appears early in the history of mankind, and, as we know, tends to be absolutely accepted in civilization. "We exclaim," as Goethe remarks, " 'What a beautiful little foot!' when we have merely seen a pretty shoe; we admire the lovely waist when nothing has met our eyes but an elegant girdle." Our realities and our traditional ideals are hopelessly at variance; the Greeks represented their statues without pubic hair because in real life they had adopted the oriental custom of removing the hairs; we compel our sculptors and painters to make similar representations, though they no longer correspond either to realities or to our own ideas of what is beautiful and fitting in real life. Our artists are themselves equally ignorant and confused, and, as Stratz has repeatedly shown, they constantly reproduce in all innocence the deformations and pathological characters of defective models. If we were honest, we should say—like the little boy before a picture of the Judgment of Paris, in answer to his mother's question as to which of the three goddesses he thought most beautiful—"I can't tell, because they haven't their clothes on."

The concealment actually attained was not, however, it would appear, originally sought. Various authors have brought together evidence to show that the main primitive purpose of adornment and clothing among savages is not to conceal the body, but to draw attention to it and to render it more attractive. Westermarck, especially, brings forward numerous examples of savage adornments which serve to attract attention to the sexual regions of man and woman.[1] He further argues that the primitive object of various savage peoples in practicing circumcision, as other similar mutilations, is really to secure sexual attractiveness, whatever religious significance they may sometimes have developed subsequently. A more recent view represents the magical influence of both adornment and mutilation as primary, as a method of guarding and insulating dangerous bodily functions. Frazer, in *The Golden Bough,* is the most able and brilliant champion of this view, which undoubtedly embodies a large element of truth, although it must not be accepted to the absolute exclusion of the influence of sexual attractiveness. The two are largely woven in together.

There is, indeed, a general tendency for the sexual functions to take on a religious character and for the sexual organs to become sacred at a very early period in culture. Generation, the reproductive force in man, animals, and plants, was realized by primitive man to be a fact of the first magnitude, and he symbolized it in the sexual organs of man and woman, which thus attained to a solemnity which was entirely independent of purposes of sexual allurement. Phallus worship may almost be said to be a universal phenomenon; it is found even among races of high culture, among the Romans of the Empire and the Japanese to-day; it has, indeed, been thought by some that one of the origins of the cross is to be found in the phallus.

Apart from the religious and magical properties so widely

---

[1] *History of Human Marriage,* Chapter IX, especially p. 201. We have a striking and comparatively modern European example of an article of clothing designed to draw attention to the sexual sphere in the codpiece (the French *braguette*), familiar to us through fifteenth and sixteenth century pictures and numerous allusions in Rabelais and in Elizabethan literature. This was originally a metal box for the protection of the sexual organs in war, but subsequently gave place to a leather case only worn by the lower classes, and became finally an elegant article of fashionable apparel, often made of silk and adorned with ribbons, even with gold and jewels.

accorded to the primary sexual characters, there are other reasons why they should not often have gained or long retained any great importance as objects of sexual allurement. They are unnecessary and inconvenient for this purpose. The erect attitude of man gives them here, indeed, an advantage possessed by very few animals, among whom it happens with extreme rarity that the primary sexual characters are rendered attractive to the eye of the opposite sex, though they often are to the sense of smell. The sexual regions constitute a peculiarly vulnerable spot, and remain so even in man, and the need for their protection which thus exists conflicts with the prominent display required for a sexual allurement. This end is far more effectively attained, with greater advantage and less disadvantage, by concentrating the chief ensigns of sexual attractiveness on the upper and more conspicuous parts of the body. This method is well-nigh universal among animals as well as in man.

There is another reason why the sexual organs should be discarded as objects of sexual allurement, a reason which always proves finally decisive as a people advances in culture. They are not æsthetically beautiful. It is fundamentally necessary that the intromittent organ of the male and the receptive canal of the female should retain their primitive characteristics; they cannot, therefore, be greatly modified by sexual or natural selection, and the exceedingly primitive character they are thus compelled to retain, however sexually desirable and attractive they may become to the opposite sex under the influence of emotion, can rarely be regarded as beautiful from the point of view of æsthetic contemplation. Under the influence of art there is a tendency for the sexual organs to be diminished in size, and in no civilized country has the artist ever chosen to give an erect organ to his representations of ideal masculine beauty. It is mainly because the unæsthetic character of a woman's sexual region is almost imperceptible in any ordinary and normal position of the nude body that the feminine form is a more æsthetically beautiful object of contemplation than the masculine. Apart from this character we are probably bound, from a strictly æsthetic point of view, to regard the male form as more æsthetically beautiful. The female form, moreover, usually overpasses very swiftly the period of the climax of its beauty, often only retaining it during a few weeks.

The primary sexual characters of man and woman have thus never at any time played a very large part in sexual allurement. With the growth of culture, indeed, the very methods which had been adopted to call attention to the sexual organs were by a further development retained for the purpose of concealing them. From the first the secondary sexual characters have been a far more widespread method of sexual allurement than the primary sexual characters, and in the most civilized countries to-day they still constitute the most attractive of such methods to the majority of the population.

Thus we find, among most of the peoples of Europe, Asia, and Africa, the chief continents of the world, that the large hips and buttocks of women are commonly regarded as an important feature of beauty. This secondary sexual character represents the most decided structural deviation of the feminine type from the masculine, a deviation demanded by the reproductive function of women, and in the admiration it arouses sexual selection is thus working in a line with natural selection. It cannot be said that, except in a very moderate degree, it has always been regarded as at the same time in a line with claims of purely æsthetic beauty. The European artist frequently seeks to attenuate rather than accentuate the protuberant lines of the feminine hips, and it is noteworthy that the Japanese also regard small hips as beautiful. Nearly everywhere else large hips and buttocks are regarded as a mark of beauty, and the average man is of this opinion even in the most æsthetic countries. The contrast of this exuberance with the more closely knit male form, the force of association, and the unquestionable fact that such development is the condition needed for healthy motherhood, have served as a basis for an ideal of sexual attractiveness which appeals to nearly all people more strongly than a more narrowly æsthetic ideal, which must inevitably be somewhat hermaphroditic in character. . . .

The special characteristics of the feminine hips and buttocks become conspicuous in walking and may be further emphasized by the special method of walking or carriage. The women of some southern countries are famous for the beauty of their way of walk; "the goddess is revealed by her walk," as Virgil said. In Spain, especially, among European countries, the walk very notably gives expression to the hips and buttocks. The spine is in Spain very curved, producing what is termed *ensellure,* or saddle-back—a characteristic which gives great flexibility to the back and prominence to the gluteal regions, sometimes slightly simulating steatopygia. The vibratory movement naturally produced by walking and sometimes artificially heightened thus becomes a trait of sexual beauty. Outside of Europe such vibration of the flanks and buttocks is more frankly displayed and cultivated as a sexual allurement. The Papuans are said to admire this vibratory movement of the buttocks in their women. Young girls are practiced in it by their mothers for hours at a time as soon as they have reached the age of 7 or 8, and the Papuan maiden walks thus whenever she is in the presence of men, subsiding into a simpler gait when no men are present. In some parts of tropical Africa the women walk in this fashion. It is also known to the Egyptians, and by the Arabs is called *ghung.* As Mantegazza remarks, the essentially feminine character of this gait makes it a method of sexual allurement. It should be observed that it rests on feminine anatomical characteristics, and that the natural walk of a femininely developed woman is inevitably different from that of a man.

An occasional development of the idea of sexual beauty as associated with developed hips is found in the tendency to regard the pregnant woman as the most beautiful type. Stratz observes that a woman artist once remarked to him that since motherhood is the final aim of woman, and a woman reaches her full flowering period in pregnancy, she ought to be most beautiful when pregnant. This is so, Stratz replied, if the period of her full physical bloom chances to correspond with the early months of pregnancy, for with the onset of pregnancy metabolism is heightened, the tissues become active, the tone of the skin softer and brighter, the breasts firmer, so that the charm of fullest bloom is increased until the moment when the expansion of the womb begins to destroy the harmony of the form. At one period of European culture, however,—at a moment and among a people not very sensitive to the most exquisite æsthetic sensations,—the ideal of beauty has even involved the character of advanced pregnancy. In northern Europe during the centuries immediately preceding the Renaissance the ideal of beauty, as we may see by the pictures of the time, was a pregnant woman, with protuberant abdomen and body more or less extended backward. This is notably apparent in the work of the Van Eycks: in the Eve in the Brussels Gallery; in the wife of Arnolfini in the highly finished portrait group in the National Gallery; even the virgins in the great masterpiece of the Van Eycks in the Cathedral at Ghent assume the type of the pregnant woman.

With the Renaissance this ideal of beauty disappeared from art. But in real life we still seem to trace its survival in the fashion for that class of garments which involved an immense amount of expansion below the waist and secured such expansion by the use of whalebone hoops and similar devices. The Elizabethan farthingale was such a garment. This was originally a Spanish invention, as indicated by the name (from *verdugardo,* provided with hoops), and reached England through France. We find the fashion at its most extreme point in the fashionable dress of Spain in the seventeenth century, such as it has been immortalized by Velasquez. In England hoops died out during the reign of George III but were revived for a time, half a century later, in the Victorian crinoline.

Only second to the pelvis and its integuments as a secondary sexual character in woman we must place the breasts. Among barbarous and civilized peoples the beauty of the breast is usually highly esteemed. Among Europeans, indeed, the importance of this region is so highly esteemed that the general rule against the exposure of the body is in its favor abrogated, and the breasts are the only portion of the body, in the narrow sense, which a European lady in full dress is allowed more or less to uncover. Moreover, at various periods and notably in the eighteenth century, women naturally deficient in this respect have sometimes worn artificial busts made of wax. Savages, also, sometimes show admiration for this part of the body, and in the Papuan folk-tales, for instance, the sole distinguishing mark of a beautiful woman is breasts that stand up. On the other hand, various savage peoples even appear to regard the development of the breasts as ugly and adopt devices for flattening this part of the body. The feeling that prompts this practice is not unknown in modern Europe, for the Bulgarians are said to regard developed breasts as ugly; in mediæval Europe, indeed, the general ideal of feminine slenderness was opposed to developed breasts, and the garments tended to compress them. But in a very high degree of civilization this feeling is unknown, as, indeed, it is unknown to most barbarians, and the beauty of a woman's breasts, and of any natural or artificial object which suggests the gracious curves of the bosom, is a universal source of pleasure.

The general admiration accorded to developed breasts and a developed pelvis is evidenced by a practice which, as embodied in the corset, is all but universal in many European countries, as well as the extra-European countries inhabited by the white race, and in one form or another is by no means unknown to peoples of other than the white race.

The tightening of the waist girth was little known to the Greeks of the best period, but it was practiced by the Greeks of the decadence and by them transmitted to the Romans; there are many references in Latin literature to this practice, and the ancient physician wrote against it in the same sense as modern doctors. So far as Christian Europe is concerned it would appear that the corset arose to gratify an ideal of asceticism rather than of sexual allurement. The bodice in early mediæval days bound and compressed the breasts and thus tended to efface the specifically feminine character of a woman's body. Gradually, however, the bodice was displaced downward, and its effect, ultimately, was to render the breasts more prominent instead of effacing them. Not only does the corset render the breasts more prominent; it has the further effect of displacing the breathing activity of the lungs in an upward direction, the advantage from the point of sexual allurement thus gained being that additional attention is drawn to the bosom from the respiratory movement thus imparted to it. So marked and so constant is this artificial respiratory effect, under the influence of the waist compression habitual among civilized women, that until recent years it was commonly supposed that there is a real and fundamental difference in breathing between men and women, that women's breathing is thoracic and men's abdominal. It is now known that under natural and healthy conditions there is no such difference, but that men and women breathe in a precisely identical manner. The corset may thus be regarded as the chief instrument of sexual allurement which the armory of costume supplies to a woman, for it furnishes her with a method of heightening at once her two chief sexual secondary characters, the bosom above, the hips and buttocks below. We cannot be surprised that all the scientific evidence in the world of the evil of the corset is powerless not merely to cause its abolition, but even to secure the general adoption of its comparatively harmless modifications.

The breasts and the developed hips are characteristics of women and are indications of functional effectiveness as well as sexual allurement. Another prominent sexual character which belongs to man, and is not obviously an index of function, is furnished by the hair on the face. The beard may be regarded as purely a sexual adornment, and thus comparable to the somewhat similar growth on the heads of many male animals. From this point of view its history is interesting, for it illustrates the tendency with increase of civilization not merely to dispense with sexual allurement in the primary sexual organs, but even to disregard those growths which would appear to have been developed solely to act as sexual allurements. The cultivation of the beard belongs peculiarly to barbarous races. Among these races it is frequently regarded as the most sacred and beautiful part of the person, as an object to swear by, an object to which the slightest insult must be treated as deadly. Holding such a position, it must doubtless act as a sexual allurement. "Allah has specially created an angel in Heaven," it is said in the *Arabian Nights*, "who has no other occupation than to sing the praises of the Creator for giving a beard to men and long hair to women." The sexual character of the beard and the other hirsute appendage is significantly indicated by the fact that the ascetic spirit in Christianity has always sought to minimize or to hide the hair. Altogether apart, however, from this religious influence, civilization tends to be opposed to the growth of hair on the masculine face and especially to the beard. It is part of the well-marked tendency with civilization to the abolition of sexual differences. We find this general tendency among the Greeks and Romans, and, on the whole, with certain variations and fluctuations of fashion, in modern Europe also. Schopenhauer frequently referred to this disappearance of the beard as a mark of civilization, "a barometer of culture." The absence of facial hair heightens æsthetic beauty of form, and is not felt to remove any substantial sexual attraction.

We have seen that there is good reason for assuming a certain fundamental tendency whereby the most various peoples of the world, at all events in the person of their most intelligent members, recognize and accept a common ideal of feminine beauty, so that to a certain extent beauty may be said to have an objectively æsthetic basis. We have further found that this æsthetic human ideal is modified, and very variously modified in different countries and even in the same country at different periods, by a tendency, prompted by a sexual impulse which is not necessarily in harmony with æsthetic canons, to emphasize, or even to repress, one or other of the prominent secondary sexual characters of the body. We now come to another tendency which is apt to an even greater extent to limit the cultivation of the purely æsthetic ideal of beauty: the influences of national or racial type.

To the average man of every race the woman who most completely embodies the type of his race is usually the most beautiful, and even mutilations and deformities often have their origin, as Humboldt long since pointed out, in the effort to accentuate the racial type. Eastern women possess by nature large and conspicuous eyes, and this characteristic they seek still further to heighten by art. The Ainu are the hairiest of races, and there is nothing which they consider so beautiful as hair. It is difficult to be sexually attracted to persons who are fundamentally unlike ourselves in racial constitution. . . .

An interesting question, which in part finds its explanation here and is of considerable significance from the point of view of sexual selection, concerns the relative admiration bestowed on blondes and brunettes. The question is not, indeed, one which is entirely settled by racial characteristics. There is something to be said on the matter from the objective standpoint of æsthetic considerations. Stratz, in a chapter on beauty of coloring in woman, points out that fair hair is more beautiful because it harmonizes better with the soft outlines of woman, and, one may add, it is more brilliantly conspicuous; a golden object looks larger than a black object. The hair of the armpit, also, Stratz considers should be light. On the other hand, the pubic hair should be dark in order to emphasize the breadth of the pelvis and the obtusity of the angle between the mons veneris and the thighs. The eyebrows and eyelashes should also be dark in order to increase the apparent size of the orbits. Stratz adds that among many thousand women he has only seen one who, together with an otherwise perfect form, has also possessed these excellencies in the highest measure. With an equable and matt complexion she had blonde, very long, smooth hair, with sparse, blonde, and curly axillary hair; but, although her eyes were blue, the eyebrows and eyelashes were black, as also was the not overdeveloped pubic hair. . . .

The main cause, however, in determining the relative

amount of admiration accorded in Europe to blondes and to brunettes is the fact that the population of Europe must be regarded as predominantly fair, and that our conception of beauty in feminine coloring is influenced by an instinctive desire to seek this type in its finest forms. In the north of Europe there can, of course, be no question concerning the predominant fairness of the population, but in portions of the centre and especially in the south it may be considered a question. It must, however, be remembered that the white population occupying all the shores of the Mediterranean have the black peoples of Africa immediately to the south of them. They have been liable to come in contact with the black peoples and in contrast with them they have tended not only to be more impressed with their own whiteness, but to appraise still more highly its blondest manifestations as representing a type the farthest removed from the Negro. It must be added that the northerner who comes into the south is apt to overestimate the darkness of the southerner because of the extreme fairness of his own people. The differences are, however, less extreme than we are apt to suppose; there are more dark people in the north than we commonly assume, and more fair people in the south. Thus, if we take Italy, we find in its fairest part, Venetia, according to Raseri, that there are 8 per cent. communes in which fair hair predominates, 81 per cent. in which brown predominates, and only 11 per cent. in which black predominates; as we go farther south black hair becomes more prevalent, but there are in most provinces a few communes in which fair hair is not only frequent, but even predominant. It is somewhat the same with light eyes, which are also most abundant in Venetia and decrease to a slighter extent as we go south. It is possible that in former days the blondes prevailed to a greater degree than to-day in the south of Europe. Among the Berbers of the Atlas Mountains, who are probably allied to the South Europeans, there appears to be a fairly considerable proportion of blondes, while on the other hand there is some reason to believe that blondes die out under the influence of civilization as well as of a hot climate.

However this may be, the European admiration for blondes dates back to early classic times. Gods and men in Homer would appear to be frequently described as fair. Venus is nearly always blonde, as was Milton's Eve. Lucian refers to women who dye their hair. The Greek sculptors gilded the hair of their statues, and the figurines in many cases show very fair hair. The Roman custom of dyeing the hair light, as Renier has shown, was not due to the desire to be like the fair Germans, and when Rome fell it would appear that the custom of dyeing the hair persisted, and never died out; it is mentioned by Anselm, who died at the beginning of the twelfth century.

In the poetry of the people in Italy brunettes, as we should expect, receive much commendation, though even here the blondes are preferred. When we turn to the painters and poets of Italy, and the æsthetic writers on beauty from the Renaissance onward, the admiration for fair hair is unqualified, though there is no correspondingly unanimous admiration for blue eyes. Angelico and most of the pre-Raphaelite artists usually painted their women with flaxen and light-golden hair, which often became brown with the artists of the Renaissance period. Firenzuola, in his admirable dialogue on feminine beauty, says that a woman's hair should be like gold or honey or the rays of the sun. Luigini also, in his *Libro della bella Donna,* says that hair must be golden. So also thought Petrarch and Ariosto. There is, however, no corresponding predilection among these writers for blue eyes. Firenzuola said that the eyes must be dark, though not black. Luigini said that they must be bright and black. Niphus had previously said that the eyes should be "black like those of Venus" and the skin ivory, even a little brown. He mentions that Avicenna had praised the mixed, or gray eye.

In France and other northern countries the admiration for very fair hair is just as marked as in Italy, and dates back to the earliest ages of which we have a record. "Even before the thirteenth century," remarks Houdoy, in his very interesting study of feminine beauty in northern France during mediæval times, "and for men as well as for women, fair hair was an essential condition of beauty; gold is the term of comparison almost exclusively used." He mentions that in the *Acta Sanctorum* it is stated that Saint Godelive of Bruges, though otherwise beautiful, had black hair and eyebrows and was hence contemptuously called a crow. In the *Chanson de Roland* and all the French mediæval poems the eyes are invariably *vairs*. This epithet is somewhat vague. It comes from *varius*, and signifies mixed, which Houdoy regards as showing various irradiations, the same quality which later gave rise to the term *iris* to describe the pupillary membrane. *Vair* would thus describe not so much the color of the eye as its brilliant and sparkling quality. While Houdoy may have been correct, it still seems probable that the eye described as *vair* was usually assumed to be "various" in color also, of the kind we commonly call gray, which is usually applied to blue eyes encircled with a ring of faintly sprinkled brown pigment. Such eyes are fairly typical of northern France and frequently beautiful. That this was the case seems to be clearly indicated by the fact that, as Houdoy himself points out, a few centuries later the *vair* eye was regarded as *vert*, and green eyes were celebrated as the most beautiful. The etymology was false, but a false etymology will hardly suffice to change an ideal. At the Renaissance Jehan Lemaire, when describing Venus as the type of beauty, speaks of her green eyes, and Ronsard, a little later, sang:

> "Noir je veux l'œil et brun le teint,
> Bien que l'œil verd toute la France adore."

Early in the sixteenth century Brantôme quotes some lines current in France, Spain, and Italy according to which a woman should have a white skin, but black eyes and eyebrows, and adds that personally he agrees with the Spaniard that "a brunette is sometimes equal to a blonde," but there is also a marked admiration for green eyes in Spanish literature; not only in the typical description of a Spanish beauty in the *Celestina* (Act. I) are the eyes green, but Cervantes, for example, when referring to the beautiful eyes of a woman, frequently speaks of them as green.

It would thus appear that in Continental Europe generally, from south to north, there is a fair uniformity of opinion as regards the pigmentary type of feminine beauty. Such variation as exists seemingly involves a somewhat greater degree of darkness for the southern beauty in harmony with the greater racial darkness of the southerner, but the variations fluctuate within a narrow range; the extremely dark type is always excluded, and so it would seem probable is the extremely fair type, for blue eyes have not, on the whole, been considered to form part of the admired type.

If we turn to England no serious modification of this conclusion is called for. Beauty is still fair. Indeed, the very word "fair" in England itself means beautiful. That in the seventeenth century it was generally held essential that beauty should be blonde is indicated by a passage in the *Anatomy of Melancholy,* where Burton argues that "golden hair was ever in great account," and quotes many examples from classic and more modern literature. That this remains the case is sufficiently evidenced by the fact that the ballet and chorus on the English stage wear yellow wigs, and the heroine of the stage is blonde, while the female villain of melodrama is a brunette.

While, however, this admiration of fairness as a mark of beauty unquestionably prevails in England, I do not think it can be said—as it probably can be said of the neighboring and closely allied country of France—that the most beautiful women belong to the fairest group of the community. In most parts of Europe the coarse and unbeautiful plebeian type tends to be very dark; in England it tends to be very fair. England is, however, somewhat fairer generally than most parts of Europe; so that, while it may be said that a very beautiful woman in France or in Spain may belong to the blondest section of the community, a very beautiful woman in England, even though of the same degree of blondness as her Continental sister, will not belong to the extremely blonde section of the English community. It thus comes about that when we are in northern France we find that gray eyes, a very fair but yet unfreckled complexion, brown hair, finely molded features, and highly sensitive facial expression combine to constitute a type which is more beautiful than any other we meet in France, and it belongs to the fairest section of the French population. When we cross over to England, however, unless we go to a so-called "Celtic" district, it is hopeless to seek among the blondest section of the community for any such beautiful and refined type. The English beautiful woman, though she may still be fair, is by no means very fair, and from the English standpoint she may even sometimes appear somewhat dark. In determining what I call the index of pigmentation—or degree of darkness of the eyes and hair—of different groups in the National Portrait Gallery I found that the "famous beauties" (my own personal criterion of beauty not being taken into account) were somewhat nearer to the dark than to the light end of the scale. If we consider, at random, individual instances of famous English beauties they are not extremely fair. Lady Venetia Stanley, in the early seventeenth century, who became the wife of Sir Kenelm Digby, was somewhat dark, with brown hair and eyebrows. Mrs. Overall, a little later in the same century, a Lancashire woman, the wife of the Dean of St. Paul's, was, says Aubrey, "the greatest beauty in her time in England," though very wanton, with "the loveliest eyes that were ever seen"; if we may trust a ballad given by Aubrey she was dark with black hair. The Gunnings, the famous beauties of the eighteenth century, were not extremely fair, and Lady Hamilton, the most characteristic type of English beauty, had blue, brown-flecked eyes and dark chestnut hair. Coloration is only one of the elements of beauty, though an important one. Other things being equal, the most blonde is most beautiful; but it so happens that among the races of Great Britain the other things are very frequently not equal, and that, notwithstanding a conviction ingrained in the language, with us the fairest of women is not always the "fairest." So magical, however, is the effect of brilliant coloring that it serves to keep alive in popular opinion an unqualified belief in the universal European creed of the beauty of blondness.

We have seen that underlying the conception of beauty, more especially as it manifests itself in woman to man, are to be found at least three fundamental elements: First there is the general beauty of the species . . . then there is the beauty due to the full development or even exaggeration of the sexual and more especially the secondary sexual characters; and last there is the beauty due to the complete embodiment of the particular racial or national type. To make the analysis fairly complete must be added at least one other factor: the influence of individual taste. Every individual, at all events in civilization, within certain narrow limits, builds up a feminine ideal of his own, in part on the basis of his own special organization and its demands, in part on the actual accidental attractions he has experienced. It is unnecessary to emphasize the existence of this factor, which has always to be taken into account in every consideration of sexual selection in civilized man. But its variations are numerous and in impassioned lovers it may even lead to the idealization of features which are in reality the reverse of beautiful. It may be said of many a man, as d'Annunzio says of the hero of his *Trionfo della Morte* in relation to the woman he loved, that "he felt himself bound to her by the real qualities of her body, and not only by those which were most beautiful, but specially by *those which were least beautiful*" (the novelist italicizes these words), so that his attention was fixed upon her defects, and emphasized them, thus arousing within himself an impetuous state of desire. Without invoking defects, however, there are endless personal variations which may all be said to come within the limits of possible beauty or charm. "There are no two women," as Stratz remarks, "who in exactly the same way stroke back a rebellious lock from their brows, no two who hold the hand in greeting in exactly the same way, no two who gather up their skirts as they walk with exactly the same movement." Among the multitude of minute differences—which yet can be seen and felt—the beholder is variously attracted or repelled according to his own individual idiosyncrasy, and the operations of sexual selection are effected accordingly.

Another factor in the constitution of the ideal of beauty, but one perhaps exclusively found under civilized conditions, is the love of the unusual, the remote, the exotic. It is commonly stated that rarity is admired in beauty. This is not strictly true, except as regards combinations and characters which vary only in a very slight degree from the generally admired type. "*Jucundum nihil est quod non reficit varietas,*" according to the saying of Publilius Syrus. The greater nervous restlessness and sensibility of civilization heightens this tendency, which is not infrequently found also among men of artistic genius. One may refer, for instance, to Baudelaire's profound admiration for the mulatto type of beauty. In every great centre of civilization the national ideal of beauty tends to be somewhat modified in exotic directions, and foreign ideals, as well as foreign fashions, become preferred to those that are native. It is significant of this tendency that when, a few years since, an enterprising Parisian journal hung in its *salle* the portraits of one hundred and thirty-one actresses, etc., and invited the votes of the public by ballot as to the most beautiful of them, not one of the three women who came out at the head of the poll was French. A dancer of Belgian origin (Cléo de Merode) was by far at the head with over 3000 votes, followed by an American from San Francisco (Sybil Sanderson), and then a Polish woman.

JEAN HENRI FABRE

Before Fabre wrote the incomparable ten volumes of his *Souvenirs Entomologiques,* only a few entomologists had found it worthwhile to publish a popular book on insects. When they did, the result was largely a dreary catalog of anatomical features. In contrast, Fabre had little interest in the dead specimens of museums. He wanted to spy on these tiny invertebrates as they went about their short but busy lives. And spy on them he did—carefully, lovingly, for more than half a century, recording the grotesque details in essays of such wit and charm that they have become as much a part of the literature of France as of the history of entomology.

Jean Henri Casimir Fabre (1823-1915) was nearly fifty when he gave up teaching physics to devote the rest of his long life to insect studies. Most of his observations were made at his home at Sérignan, on a barren patch of land that is now the site of a national shrine. With awe-inspiring perseverance he watched and waited, waited and watched. Now and then he tried a simple experiment. Occasionally he drew unwarranted conclusions, but the accuracy of his observations has stood magnificently the test of later research.

The selection here is the opening essay of his classic *Souvenirs.* Did Fabre choose to start his series with the sacred scarab because of the high symbolic comedy of its rituals? Chesterton once described the hornbill, a small bird attached to a huge beak, as one of God's "mysterious jokes." Perhaps in some similar spirit of reverent ribaldry Fabre found in the slapstick antics of these glossy little scavengers a hint of that laughter that is too loud for us to hear.

JEAN HENRI FABRE

The Sacred Beetle

It happened like this. There were five or six of us: myself, the oldest, officially their master but even more their friend and comrade; they, lads with warm hearts and joyous imaginations, overflowing with that youthful vitality which makes us so enthusiastic and so eager for knowledge. We started off one morning down a path fringed with dwarf elder and hawthorn, whose clustering blossoms were already a paradise for the Rose-chafer ecstatically drinking in their bitter perfumes. We talked as we went. We were going to see whether the Sacred Beetle had yet made his appearance on the sandy plateau of Les Angles, whether he was rolling that pellet of dung in which ancient Egypt beheld an image of the world; we were going to find out whether the stream at the foot of the hill was not hiding under its mantle of duckweed young Newts with gills like tiny branches of coral; whether that pretty little fish of our rivulets, the Stickleback, had donned his wedding scarf of purple and blue; whether the newly arrived Swallow was skimming the meadows on pointed wing, chasing the Crane-flies, who scatter their eggs as they dance through the air; if the Eyed Lizard was sunning his blue-speckled body on the threshold of a burrow dug in the sandstone; if the Laughing Gull, travelling from the sea in the wake of the legions of fish that ascend the Rhone to milt in its waters, was hovering in his hundreds over the river, ever and anon uttering his cry so like a maniac's laughter; if . . . but that will do. To be brief, let us say that, like good simple folk who find pleasure in all living things, we were off to spend a morning at the most wonderful of festivals, life's springtime awakening.

Our expectations were fulfilled. The Stickleback was dressed in his best: his scales would have paled the lustre of silver; his throat was flashing with the brightest vermilion. On the approach of the great black Horse-leech, the spines on his back and sides started up, as though worked by a spring. In the face of this resolute attitude, the bandit turns tail and slips ignominiously down among the water-weeds. The placid mollusc tribe—Planorbes, Limnæi and other Water-snails—were sucking in the air on the surface of the water. The Hydrophilus and her hideous larva, those pirates of the ponds, darted amongst them, wringing a neck or two as they passed. The stupid crowd did not seem even to notice it. But let us leave the plain and its waters and clamber up the bluff to the plateau above us. Up there, Sheep are grazing and Horses being exercised for the approaching races, while all are distributing manna to the enraptured Dung-beetles.

Here are the scavengers at work, the Beetles whose proud mission it is to purge the soil of its filth. One would never weary of admiring the variety of tools wherewith they are supplied, whether for shifting, cutting up and shaping the stercoral matter or for excavating deep burrows in which they will seclude themselves with their booty. This equipment resembles a technical museum where every digging-implement is represented. It includes things that seem copied from those appertaining to human industry and others of so original a type that they might well serve us as models for new inventions.

The Spanish Copris carries on his forehead a powerful pointed horn, curved backwards, like the long blade of a mattock. In addition to a similar horn, the Lunary Copris has two strong spikes, curved like a ploughshare, springing from the thorax and also, between the two, a jagged protuberance which does duty as a broad rake. *Bubas bubalis* and *B. bison*, both exclusively Mediterranean species, have their forehead armed with two stout diverging horns, between which juts a horizontal dagger, supplied by the corselet. *Minotaurus typhœus* carries on the front of his thorax three ploughshares, which stick straight out, parallel to one another, the side ones longer than the middle one. The Bull Onthophagus has as his tool two long curved pieces that remind us of the horns of a Bull; the Cow Onthophagus, on the other hand, has a two-pronged fork standing erect on his flat head. Even the poorest have, either on their head or on their corselet, hard knobs that make implements which the patient insect can turn to good use, notwithstanding their bluntness. All are supplied with a shovel, that is to say, they have a broad, flat head with a sharp edge; all use a rake, that is to say, they collect materials with their toothed fore-legs.

As some sort of compensation for their unsavoury task, several of them give out a powerful scent of musk, while their bellies shine like polished metal. The Mimic Geotrupes has gleams of copper and gold beneath; the Stercoraceous Geotrupes has a belly of amethystine violet. But generally their colouring is black. The Dung-beetles in gorgeous raiment, those veritable living gems, belong to the tropics. Upper Egypt can show us under its Camel-dung a Beetle rivalling the emerald's brilliant green; Guiana, Brazil and Senegambia boast of Copres that are a metallic red, rich as copper and ruby-bright. The Dung-beetles of our climes cannot flaunt such jewellery, but they are no less remarkable for their habits.

What excitement over a single patch of Cow-dung! Never did adventurers hurrying from the four corners of the earth display such eagerness in working a Californian claim. Before the sun becomes too hot, they are there in their hundreds, large and small, of every sort, shape and size, hastening to carve themselves a slice of the common cake. There are some that labour in the open air and scrape the surface; there are others that dig themselves galleries in the thick of the heap, in search of choice veins; some work the lower stratum and bury their spoil without delay in the ground just below; others again, the smallest, keep on one side and crumble a morsel that has slipped their way during the mighty excavations of their more powerful fellows. Some, newcomers and doubtless the hungriest, consume their meal on the spot; but the greater number dream of accumulating stocks that will allow them to spend long days in affluence, down in some safe retreat. A nice, fresh patch of dung is not found just when you want it, in the barren plains overgrown with thyme; a windfall of this sort is as manna from the sky; only fortune's favourites receive so fair a portion. Wherefore the riches of to-day are prudently hoarded for the morrow. The stercoraceous scent has carried the glad tidings half a mile around; and all have hastened up to get a store of provisions. A few laggards are still arriving, on the wing or on foot.

Who is this that comes trotting towards the heap, fearing lest he reach it too late? His long legs move with awkward jerks, as though driven by some mechanism within his belly; his little red antennæ unfurl their fan, a sign of anxious greed. He is coming, he has come, not without sending a few banqueters sprawling. It is the Sacred Beetle, clad all in black, the biggest and most famous of our Dung-beetles. Behold him at table, beside his fellow-guests, each of whom is giving the last touches to his ball with the flat of his broad fore-legs or else enriching it with yet one more layer before retiring to enjoy the fruit of his labours in peace. Let us follow the construction of the famous ball in all its phases.

The clypeus, or shield, that is the edge of the broad, flat head, is notched with six angular teeth arranged in a semicircle. This constitutes the tool for digging and cutting up, the rake that lifts and casts aside the unnutritious vegetable fibres, goes for something better, scrapes and collects it. A choice is thus made, for these connoisseurs differentiate between one thing and another, making a rough selection when the Beetle is occupied with his own provender, but an extremely scrupulous one when it is a matter of constructing the maternal ball, which has a central cavity in which the egg will hatch. Then every scrap of fibre is conscientiously rejected and only the stercoral quintessence is gathered as the material for building the inner layer of the cell. The young larva, on issuing from the egg, thus finds in the very walls of its lodging a food of special delicacy which strengthens its digestion and enables it afterwards to attack the coarse outer layers.

Where his own needs are concerned, the Beetle is less particular and contents himself with a very general sorting. The notched shield then does its scooping and digging, its casting aside and scraping together more or less at random. The fore-legs play a mighty part in the work. They are flat, bow-shaped, supplied with powerful nervures and armed on the outside with five strong teeth. If a vigorous effort be needed to remove an obstacle or to force a way through the thickest part of the heap, the Dung-beetle makes use of his elbows, that is to say, he flings his toothed legs to right and left and clears a semicircular space with an energetic sweep. Room once made, a different kind of work is found for these same limbs: they collect armfuls of the stuff raked together by the shield and push it under the insect's belly, between the four hinder legs. These are formed for the turner's trade. They are long and slender, especially the last pair, slightly bowed and finished with a very sharp claw. They are at once recognized as compasses, capable of embracing a globular body in their curved branches and of verifying and correcting its shape. Their function is, in fact, to fashion the ball.

Armful by armful, the material is heaped up under the belly, between the four legs, which, by a slight pressure, impart their own curve to it and give it a preliminary outline. Then, every now and again, the rough-hewn pill is set spinning between the four branches of the double pair of spherical compasses; it turns under the Dung-beetle's belly until it is rolled into a perfect ball. Should the surface layer lack plasticity and threaten to peel off, should some too-stringy part refuse to yield to the action of the lathe, the fore-legs touch up the faulty places; their broad paddles pat the ball to give consistency to the new layer and to work the recalcitrant bits into the mass.

Under a hot sun, when time presses, one stands amazed at the turner's feverish activity. And so the work proceeds apace: what a moment ago was a tiny pellet is now a ball the size of a walnut; soon it will be the size of an apple. I have seen some gluttons manufacture a ball the size of a man's fist. This indeed means food in the larder for days to come!

The Beetle has his provisions. The next thing is to withdraw from the fray and transport the victuals to a suitable place. Here the Scarab's most striking characteristics begin to show themselves. Straightway he begins his journey; he clasps his sphere with his two long hind-legs, whose terminal claws, planted in the mass, serve as pivots; he obtains a purchase with the middle pair of legs; and, with his toothed fore-arms, pressing in turn upon the ground, to do duty as levers, he proceeds with his load, he himself moving backwards, body bent, head down and hind-quarters in the air. The rear legs, the principal factor in the mechanism, are in continual movement backwards and forwards, shifting the claws to change the axis of rotation, to keep the load balanced and to push it along by alternate thrusts to right and left. In this way the ball finds itself touching the ground by turns with every point of its surface, a process which perfects its shape and gives an even consistency to its outer layer by means of pressure uniformly distributed.

And now to work with a will! The thing moves, it begins to roll; we shall get there, though not without difficulty. Here is a first awkward place: the Beetle is wending his way athwart a slope and the heavy mass tends to follow the incline; the insect, however, for reasons best known to itself, prefers to cut across this natural road, a bold project which may be brought to naught by a false step or by a grain of sand that disturbs the balance of the load. The false step is made: down goes the ball to the bottom of the valley; and the insect, toppled over by the shock, is lying on its back, kicking. It is soon up again and hastens to harness itself once more to its load. The machine works better than ever. But look out, you dunderhead! Follow the dip of the valley: that will save labour and mishaps; the road is good and level; your ball will roll quite easily. Not a bit of it! The Beetle prepares once again to mount the slope that has already been his undoing. Perhaps it suits him to return to the heights.

Against that I have nothing to say: the Scarab's judgment is better than mine as to the advisability of keeping to lofty regions; he can see farther than I can in these matters. But at least take this path, which will lead you up by a gentle incline! Certainly not! Let him find himself near some very steep slope, impossible to climb, and that is the very path which the obstinate fellow will choose. Now begins a Sisyphean labour. The ball, that enormous burden, is painfully hoisted, step by step, with infinite precautions, to a certain height, always backwards. We wonder by what miracle of statics a mass of this size can be kept upon the slope. Oh! An ill-advised movement frustrates all this toil: the ball rolls down, dragging the Beetle with it. Once more the heights are scaled and another fall is the sequel. The attempt is renewed, with greater skill this time at the difficult points; a wretched grass-root, the cause of the previous falls, is carefully got over. We are almost there; but steady now, steady! It is a dangerous ascent and the merest trifle may yet ruin everything. For see, a leg slips on a smooth bit of gravel! Down come ball and Beetle, all mixed up together. And the insect begins over again, with indefatigable obstinacy. Ten times, twenty times, he will attempt the hopeless ascent, until his persistence vanquishes all obstacles, or until, wisely recognizing the futility of his efforts, he adopts the level road.

The Scarab does not always push his precious ball alone: sometimes he takes a partner; or, to be accurate, the partner takes him. This is the way in which things usually happen: once his ball is ready, a Dung-beetle issues from the crowd and leaves the workyard, pushing his prize backwards. A neighbour, a newcomer, whose own task is hardly begun, abruptly drops his work and runs to the moving ball, to lend a hand to the lucky owner, who seems to accept the proffered aid kindly. Henceforth the two work in partnership. Each does his best to push the pellet to a place of safety. Was a compact really concluded in the workyard, a tacit agreement to share the cake between them? While one was kneading and moulding the ball, was the other tapping rich veins whence to extract choice materials and add them to the common store? I have never observed any such collaboration; I have always seen each Dung-beetle occupied solely with his own affairs in the works. The last-comer, therefore, has no acquired rights.

Can it then be a partnership between the two sexes, a couple intending to set up house? I thought so for a time. The two Beetles, one before, one behind, pushing the heavy ball with equal fervour, reminded me of a song which the hurdy-gurdies used to grind out some years ago:

Pour monter notre ménage, hélas! comment ferons-nous?
Toi devant et moi derrière, nous pousserons le tonneau.[1]

The evidence of the scalpel compelled me to abandon my belief in this domestic idyll. There is no outward difference between the two sexes in the Scarabæi. I therefore dissected the pair of Dung-beetles engaged in trundling one and the same ball; and they very often proved to be of the same sex.

Neither community of family nor community of labour! Then what is the motive for this apparent partnership? It is purely and simply an attempt at robbery. The zealous fellow-worker, on the false plea of lending a helping hand, cherishes a plan to purloin the ball at the first opportunity. To make one's own ball at the heap means hard work and patience; to steal one ready-made, or at least to foist one's self as a guest, is a much easier matter. Should the owner's vigilance slacken, you can run away with his property; should you be too closely watched, you can sit down to table uninvited, pleading services rendered. It is 'Heads I win, tails you lose' in these tactics, so that pillage is practiced as one of the most lucrative of trades. Some go to work craftily, in the way which I have described: they come to the aid of a comrade who has not the least need of them and hide the most barefaced greed under the cloak of charitable assistance. Others, bolder perhaps, more confident in their strength, go straight to their goal and commit robbery with violence.

[1] 'When you and I start housekeeping, alas, what shall we do?
You in front and I behind, we'll shove the tub along!'

Scenes are constantly happening such as this: a Scarab goes off, peacefully, by himself, rolling his ball, his lawful property, acquired by conscientious work. Another comes flying up, I know not whence, drops down heavily, folds his dingy wings under their cases and, with the back of his toothed fore-arms, knocks over the owner, who is powerless to ward off the attack in his awkward position, harnessed as he is to his property. While the victim struggles to his feet, the other perches himself atop the ball, the best position from which to repel an assailant. With his fore-arms crossed over his breast, ready to hit back, he awaits events. The dispossessed one moves round the ball, seeking a favourable spot at which to make the assault; the usurper spins round on the roof of the citadel, facing his opponent all the time. If the latter raise himself in order to scale the wall, the robber gives him a blow that stretches him on his back. Safe at the top of his fortress, the besieged Beetle could foil his adversary's attempts indefinitely if the latter did not change his tactics. He turns sapper so as to reduce the citadel with the garrison. The ball, shaken from below, totters and begins rolling, carrying with it the thieving Dung-beetle, who makes violent efforts to maintain his position on the top. This he succeeds in doing —though not invariably—thanks to hurried gymnastic feats which land him higher on the ball and make up for the ground which he loses by its rotation. Should a false movement bring him to earth, the chances become equal and the struggle turns into a wrestling-match. Robber and robbed grapple with each other, breast to breast. Their legs lock and unlock, their joints intertwine, their horny armour clashes and grates with the rasping sound of metal under the file. Then the one who succeeds in throwing his opponent and releasing himself scrambles to the top of the ball and there takes up his position. The siege is renewed, now by the robber, now by the robbed, as the chances of the hand-to-hand conflict may decree. The former, a brawny desperado, no novice at the game, often has the best of the fight. Then, after two or three unsuccessful attempts, the defeated Beetle wearies and returns philosophically to the heap, to make himself a new pellet. As for the other, with all fear of a surprise attack at an end, he harnesses himself to the conquered ball and pushes it whither he pleases. I have sometimes seen a third thief appear upon the scene and rob the robber. Nor can I honestly say that I was sorry.

I ask myself in vain what Proudhon introduced into Scarabæan morality the daring paradox that 'property means plunder,' or what diplomatist taught the Dung-beetle the savage maxim that 'might is right.' I have no data that would enable me to trace the origin of these spoliations, which have become a custom, of this abuse of strength to capture a lump of ordure. All that I can say is that theft is a general practice among the Scarabs. These dung-rollers rob one another with a calm effrontery which, to my knowledge, is without a parallel. I leave it to future observers to elucidate this curious problem in animal psychology and I go back to the two partners rolling their ball in concert.

But first let me dispel a current error in the text-books. I find in M. Émile Blanchard's magnificent work, *Métamorphoses, mœurs et instincts des insectes,* the following passage:

'Sometimes our insect is stopped by an insurmountable obstacle; the ball has fallen into a hole. At such moments the Ateuchus gives evidence of a really astonishing grasp of the situation as well as of a system of ready communication between individuals of the same species which is even more remarkable. Recognizing the impossibility of coaxing the ball out of the hole, the Ateuchus seems to abandon it and flies away. If you are sufficiently endowed with that great and noble virtue called patience, stay by the forsaken ball: after a while, the Ateuchus will return to the same spot and will not return alone; he will be accompanied by two, three, four or five companions, who will all alight at the place indicated and will combine their efforts to raise the load. The Ateuchus has been to fetch reinforcements; and this explains why it is such a common sight, in the dry fields, to see several Ateuchi joining in the removal of a single ball.'

Lastly, I read in Illiger's *Entomological Magazine:*

'A *Gymnopleurus pilularius,* while constructing the ball of dung destined to contain her eggs, let it roll into a hole, whence she strove for a long time to extract it unaided. Finding that she was wasting her time in vain efforts, she ran to a neighbouring heap of manure to fetch three individuals of her own species, who, uniting their strength to hers, succeeded in withdrawing the ball from the cavity into which it had fallen and then returned to their manure to continue their work.'

I crave a thousand pardons of my illustrious master, M. Blanchard, but things certainly do not happen as he says. To begin with, the two accounts are so much alike that they must have had a common origin. Illiger, on the strength of observations not continuous enough to deserve blind confidence, put forward the case of his Gymnopleurus; and the same story was repeated about the Scarabæi because it is, in fact, quite usual to see two of these insects occupied together either in rolling a ball or in getting it out of a troublesome place. But this cooperation in no way proves that the Dung-beetle who found himself in difficulties went to requisition the aid of his mates. I have had no small measure of the patience recommended by M. Blanchard; I have lived laborious days in close intimacy, if I may say so, with the Sacred Beetle; I have done everything that I could think of in order to enter as thoroughly as possible into his ways and habits and to study them from life; and I have never seen anything that suggested either nearly or remotely the idea of companions summoned to lend assistance. As I shall presently relate, I have subjected the Dung-beetle to far more serious trials than that of getting his ball into a hole; I have confronted him with much graver difficulties than that of mounting a slope, which is sheer sport to the obstinate Sisyphus, who seems to delight in the rough gymnastics involved in climbing steep places, as if the ball thereby grew firmer and accordingly increased in value; I have created artificial situations in which the insect had the uttermost need of help; and never did my eyes detect any evidence of friendly services rendered by comrade to comrade. I have seen Beetles robbed and Beetles robbing and nothing more. If a number of them were gathered around the same pill, it meant that a battle was taking place. My humble opinion, therefore, is that the incident of a number of Scarabæi collected around the same ball with thieving intentions has given rise to these stories of comrades called in to lend a hand. Imperfect observations are responsible for this transformation of the bold highwayman into a helpful companion who has left his work to do another a friendly turn.

It is no light matter to attribute to an insect a really astonishing grasp of a situation, combined with an even more amazing power of communication between individuals of the same species. Such an admission involves more than one imagines. That is why I insist on my point. What! Are we to believe that a Beetle in distress will conceive the idea of going in quest of help? We are to imagine him flying off and scouring the country to find fellow-workers on some patch of dung; when he has found them, we are to suppose that he addresses them, in a sort of pantomime, by gestures with his antennæ more particularly, in some such words as these:

'I say, you fellows, my load's upset in a hole over there; come and help me get it out. I'll do as much for you one day!'

And we are to believe that his comrades understand! And, more incredible still, that they straightway leave their work, the pellet which they have just begun, the beloved pill exposed to the cupidity of others and certain to be filched in their absence, and go to the help of the suppliant! I am profoundly incredulous of such unselfishness; and my incredulity is confirmed by what I have witnessed for years and years, not in glass-cases but in the very places where the Scarab works. Apart from its maternal solicitude, in which respect it is nearly always admirable, the insect cares for nothing but itself, unless it lives in societies, like the Hive-bees, the Ants and the rest.

But let me end this digression, which is excused by the importance of the subject. I was saying that a Sacred

Beetle, in possession of a ball which he is pushing backwards, is often joined by another, who comes hurrying up to lend an assistance which is anything but disinterested, his intention being to rob his companion if the opportunity present itself. Let us call the two workers partners, though that is not the proper name for them, seeing that the one forces himself upon the other, who probably accepts outside help only for fear of a worse evil. The meeting, by the way, is absolutely peaceful. The owner of the ball does not cease work for an instant on the arrival of the newcomer; and his uninvited assistant seems animated by the best intentions and sets to work on the spot. The way in which the two partners harness themselves differs. The proprietor occupies the chief position, the place of honour: he pushes at the rear, with his hind-legs in the air and his head down. His subordinate is in front, in the reverse posture, head up, toothed arms on the ball, long hind-legs on the ground. Between the two, the ball rolls along, one driving it before him, the other pulling it towards him.

The efforts of the couple are not always very harmonious, the more so as the assistant has his back to the road to be traversed, while the owner's view is impeded by the load. The result is that they are constantly having accidents, absurd tumbles, taken cheerfully and in good part: each picks himself up quickly and resumes the same position as before. On level ground this system of traction does not correspond with the dynamic force expended, through lack of precision in the combined movements: the Scarab at the back would do as well and better if left to himself. And so the helper, having given a proof of his good-will at the risk of throwing the machinery out of gear, now decides to keep still, without letting go of the precious ball, of course. He already looks upon that as his: a ball touched is a ball gained. He won't be so silly as not to stick to it: the other might give him the slip!

So he gathers his legs flat under his belly, encrusting himself, so to speak, on the ball and becoming one with it. Henceforth, the whole concern—the ball and the Beetle clinging to its surface—is rolled along by the efforts of the lawful owner.

The intruder sits tight and lies low, heedless whether the load pass over his body, whether he be at the top, bottom or side of the rolling ball. A queer sort of assistant, who gets a free ride so as to make sure of his share of the victuals!

But a steep ascent heaves in sight and gives him a fine part to play. He takes the lead now, holding up the heavy mass with his toothed arms, while his mate seeks a purchase in order to hoist the load a little higher. In this way, by a combination of well-directed efforts, the Beetle above gripping, the one below pushing, I have seen a couple mount hills which would have been too much for a single carter, however persevering. But in times of difficulty not all show the same zeal: there are some who, on awkward slopes where their assistance is most needed, seem blissfully unaware of the trouble. While the unhappy Sisyphus exhausts himself in attempts to get over the bad part, the other quietly leaves him to it: imbedded in the ball, he rolls down with it if it comes to grief and is hoisted up with it when they start afresh.

I have often tried the following experiment on the two partners in order to judge their inventive faculties when placed in a serious predicament. Suppose them to be on level ground, number two seated motionless on the ball, number one busy pushing. Without disturbing the latter, I nail the ball to the ground with a long, strong pin. It stops suddenly. The Beetle, unaware of my perfidy, doubtless believes that some natural obstacle, a rut, a tuft of couch-grass, a pebble, bars the way. He redoubles his efforts, struggles his hardest; nothing happens.

'What can the matter be? Let's go and see.'

The Beetle walks two or three times round his pellet. Discovering nothing to account for its immobility, he returns to the rear and starts pushing again. The ball remains stationary.

'Let's look up above.'

The Beetle goes up, to find nothing but his motionless colleague, for I had taken care to drive in the pin so deep that the head disappeared in the ball. He explores the whole upper surface and comes down again. Fresh thrusts are vigorously applied in front and at the sides, with the same absence of success. There is not a doubt about it: never before was Dung-beetle confronted with such a problem in inertia.

Now is the time, the very time, to claim assistance, which is all the easier as his mate is there, close at hand, squatting on the summit of the ball. Will the Scarab rouse him? Will he talk to him like this:

'What are you doing there, lazybones? Come and look at the thing: it's broken down!'

Nothing proves that he does anything of the kind, for I see him steadily shaking the unshakable, inspecting his stationary machine on every side, while all this time his companion sits resting. At long last, however, the latter becomes aware that something unusual is happening; he is apprised of it by his mate's restless tramping and by the immobility of the ball. He comes down, therefore, and in his turn examines the machine. Double harness does no better than single harness. This is beginning to look serious. The little fans of the Beetles' antennæ open and shut, open again, betraying by their agitation acute anxiety. Then a stroke of genius ends the perplexity:

'Who knows what's underneath?'

They now start exploring below the ball; and a little digging soon reveals the presence of the pin. They recognize at once that the trouble is there.

If I had had a voice in their deliberations, I should have said:

'We must make a hole in the ball and pull out that skewer which is holding it down.'

This most elementary of all proceedings and one so easy to such expert diggers was not adopted, was not even tried. The Dung-beetle was shrewder than man. The two colleagues, one on this side, one on that, slip under the ball, which begins to slide up the pin, getting higher and higher in proportion as the living wedges make their way underneath. The clever operation is made possible by the softness of the material, which gives easily and makes a channel under the head of the immovable stake. Soon the pellet is suspended at a height equal to the thickness of the Scarabs' bodies. The rest is not such plain sailing. The Dung-beetles, who at first were lying flat, rise gradually to their feet, still pushing with their backs. The work becomes harder and harder as the legs, in straightening out, lose their strength; but none the less they do it. Then comes a time when they can no longer push with their backs, the limit of their height having been reached. A last resource remains, but one much less favourable to the development of motive power. This is for the insect to adopt one or other of its postures when harnessed to the ball, head down or up, and to push with its hind- or fore-legs, as the case may be. Finally the ball drops to the ground, unless we have used too long a pin. The gash made by our stake is repaired, more or less, and the carting of the precious pellet is at once resumed.

But, should the pin really be too long, then the ball, which remains firmly fixed, ends by being suspended at a height above that of the insect's full stature. In that case, after vain evolutions around the unconquerable greased pole, the Dung-beetles throw up the sponge, unless we are sufficiently kind-hearted to finish the work ourselves and restore their treasure to them. Or again we can help them by raising the floor with a small flat stone, a pedestal from the top of which it is possible for the Beetle to continue his labours. Its use does not appear to be immediately understood, for neither of the two is in any hurry to take advantage of it. Nevertheless, by accident or design, one or other at last finds himself on the stone. Oh, joy! As he passed, he felt the ball touch his back. At that contact, courage returns; and his efforts begin once more. Standing on his helpful platform, the Scarab stretches his joints, rounds his shoulders, as one might say, and shoves the pellet upwards. When his shoulders no longer avail, he works with his legs, now upright, now head downwards. There is a fresh pause, accompanied by fresh signs of uneasiness, when the limit of extension is reached. Thereupon, without disturbing the creature, we place a second little stone on the top of the first. With the aid of this new step, which provides a fulcrum for its levers, the insect pursues its task. Thus adding story upon story as required, I have seen the Scarab, hoisted to the summit of a tottering pile three or four fingers'-breadth in height, persevere in his work until the ball was completely detached.

Had he some vague consciousness of the service performed by the gradual raising of the pedestal? I venture to doubt it, though he cleverly took advantage of my platform of little stones. As a matter of fact, if the very elementary idea of using a higher support in order to reach something placed above one's grasp were not beyond the Beetle's comprehension, how is it that, when there are two of them, neither thinks of lending the other his back so as to raise him by that much and make it possible for him to go on working? If one helped the other in this way, they could reach twice as high. They are very far, however, from any such cooperation. Each pushes the ball, with all his might, I admit, but he pushes as if he were alone and seems to have no notion of the happy result that would follow a combined effort. In this instance, when the ball is nailed to the ground by a pin, they do exactly what they do in corresponding circumstances, as, for example, when the load is brought to a standstill by some obstacle, caught in a loop of couch-grass or transfixed by some spiky bit of stalk that has run into the soft, rolling mass. I produced artificially a stoppage which is not really very different from those occurring naturally when the ball is being rolled amid the thousand and one irregularities of the ground; and the Beetle behaves, in my experimental tests, as he would have behaved in any other circumstances in which I had no part. He uses his back as a wedge and a lever and pushes with his feet, without introducing anything new into his methods, even when he has a companion and can avail himself of his assistance.

When he is all alone in face of the difficulty, when he has no assistant, his dynamic operations remain absolutely the same; and his efforts to move his transfixed ball end in success, provided that we give him the indispensable support of a platform, built up little by little. If we deny him this succour, then, no longer encouraged by the contact of his beloved ball, he loses heart and sooner or later flies away, doubtless with many regrets, and disappears. Where to? I do not know. What I do know is that he does not return with a gang of fellow-labourers whom he has begged to help him. What would he do with them, he who cannot make use of even one comrade?

But perhaps my experiment, which leaves the ball suspended at an inaccessible height and the insect with its means of action exhausted, is a little too far removed from ordinary conditions. Let us try instead a miniature pit, deep enough and steep enough to prevent the Dung-beetle, when placed at the bottom, from rolling his load up the side. These are exactly the conditions stated by Messrs. Blanchard and Illiger. Well, what happens? When dogged but utterly fruitless efforts have convinced him of his helplessness, the Beetle takes wing and disappears. Relying upon what these learned writers said, I have waited long hours for the insect to return reinforced by a few friends. I have always waited in vain. Many a time also I have found the pellet several days later just where I left it, stuck at the top of a pin or in a hole, proving that nothing fresh had happened in my absence. A ball abandoned from necessity is a ball abandoned for good, with no attempt at salvage with the aid of others. A dexterous use of wedge and lever to set the ball rolling again is therefore, when all is said, the greatest intellectual effort which I have observed in the Sacred Beetle. To make up for what the experiment refutes, namely, an appeal for help among fellow-workers, I gladly chronicle this feat of mechanical prowess for the Dung-beetles' greater glory.

Directing their steps at random, over sandy plains thick with thyme, over cart-ruts and steep places, the two Beetle brethren roll the ball along for some time, thus giving its substance a certain consistency which may be to their liking. While still on the road, they select a favourable spot. The rightful owner, the Beetle who throughout has kept the place of honour, behind the ball, the one in short who has done almost all the carting by himself, sets to work to dig the dining-room. Beside him is the ball, with number two clinging to it, shamming dead. Number one attacks the sand with his sharp-edged forehead and his toothed legs; he flings armfuls of it behind him; and the work of excavating proceeds apace. Soon the Beetle has disappeared from view in the half-dug cavern. Whenever he returns to the upper air with a load, he invariably glances at his ball to see if all is well. From time to time he brings it nearer the threshold of the burrow; he feels it and seems to acquire new vigour from the contact. The other, lying demure and motionless on the ball, continues to inspire confidence. Meanwhile the underground hall grows larger and deeper; and the digger's field of operations is now too vast for any but very occasional appearances. Now is the time. The crafty sleeper awakens and hurriedly decamps with the ball, which he pushes behind him with the speed of a pickpocket anxious not to be caught in the act. This breach of trust rouses my indignation, but the historian triumphs for the moment over the moralist and I leave him alone: I shall have time enough to intervene on the side of law and order if things threaten to turn out badly.

The thief is already some yards away. His victim comes out of the burrow, looks around and finds nothing. Doubtless an old hand himself, he knows what this means. Scent and sight soon put him on the track. He makes haste and catches up the robber; but the artful dodger, when he feels his pursuer close on his heels, promptly changes his posture, gets on his hind-legs and clasps the ball with his toothed arms, as he does when acting as an assistant.

You rogue, you! I see through your tricks: you mean to plead as an excuse that the pellet rolled down the slope and that you are only trying to stop it and bring it back home. I, however, an impartial witness, declare that the ball was quite steady at the entrance to the burrow and did not roll of its own accord. Besides, the ground is level. I declare that I saw you set the thing in motion and make off with unmistakable intentions. It was an attempt at larceny, or I've never seen one!

My evidence is not admitted. The owner cheerfully accepts

the other's excuses; and the two bring the ball back to the burrow as though nothing had happened.

If the thief, however, has time to get far enough away, or if he manages to cover his trail by adroitly doubling back, the injury is irreparable. To collect provisions under a blazing sun, to cart them a long distance, to dig a comfortable banqueting-hall in the sand, and then—just when everything is ready and your appetite, whetted by exercise, lends an added charm to the approaching feast—suddenly to find yourself cheated by a crafty partner is, it must be admitted, a reverse of fortune that would dishearten most of us. The Dung-beetle does not allow himself to be cast down by this piece of ill-luck: he rubs his cheeks, spreads his antennæ, sniffs the air and flies to the nearest heap to begin all over again. I admire and envy this cast of character.

Suppose the Scarab fortunate enough to have found a loyal partner; or, better still, suppose that he has met no self-incited companion. The burrow is ready. It is a shallow cavity, about the size of one's fist, dug in soft earth, usually in sand, and communicating with the outside by a short passage just wide enough to admit the ball. As soon as the provisions are safely stored away, the Scarab shuts himself in by stopping up the entrance to his dwelling with rubbish kept in a corner for the purpose. Once the door is closed, nothing outside betrays the existence of the banqueting-chamber. And, now, hail mirth and jollity! All is for the best in the best of all possible worlds! The table is sumptuously spread; the ceiling tempers the heat of the sun and allows only a moist and gentle warmth to penetrate; the undisturbed quiet, the darkness, the Crickets' concert overhead are all pleasant aids to digestion. So complete has been the illusion that I have caught myself listening at the door, expecting to hear the revellers burst into the famous snatch in *Galatée:*

> *Ah! qu'il est doux de ne rien faire,*
> *Quand tout s'agite autour de nous!*[1]

[1] 'Ah, how sweet is *far niente*,
When round us throbs the busy world!'

Who would dare disturb the bliss of such a banquet? But the desire for knowledge is capable of all things; and I had the necessary daring. I will set down here the result of my violation of the home.

The ball by itself fills almost the whole room; the rich repast rises from floor to ceiling. A narrow passage runs between it and the walls. Here sit the banqueters, two at most, very often only one, belly to table, back to the wall. Once the seat is chosen, no one stirs; all the vital forces are absorbed by the digestive faculties. There is no fidgeting, which might mean the loss of a mouthful; no dainty toying with the food, which might cause some to be wasted. Everything has to pass through, properly and in order. To see them seated so solemnly around a ball of dung, one would think that they were conscious of their function as cleansers of the earth and that they were deliberately devoting themselves to that marvellous chemistry which out of filth brings forth the flower that delights our eyes and the Beetles' wing-case that jewels our lawns in spring. For this supreme work which turns into living matter the refuse which neither the Horse nor the Mule can utilize, despite the perfection of their digestive organs, the Dung-beetle must needs be specially equipped. And indeed anatomy compels us to admire the prodigious length of his coiled intestine, which slowly elaborates the materials in its manifold windings and exhausts them to the very last serviceable atom. Matter from which the ruminant's stomach could extract nothing, yields to this powerful alembic riches that, at a mere touch, are transmuted into ebon mail in the Sacred Scarab and a breastplate of gold and rubies in other Dung-beetles.

Now this wonderful metamorphosis of ordure has to be accomplished in the shortest possible time: the public health demands it. And so the Scarab is endowed with matchless digestive powers. Once housed in the company of food, he goes on eating and digesting, day and night, until the provisions are exhausted. There is no difficulty in proving this. Open the cell to which the Dung-beetle has retired from the world. At any hour of the day, we shall find the insect seated at table and, behind it, still hanging to it, a continuous cord, roughly coiled like a pile of cables. One can easily guess, without embarrassing explanations, what this cord represents. The great ball of dung passes mouthful by mouthful through the Beetle's digestive canals, yielding up its nutritive essences, and reappears at the opposite end spun into a cord. Well, this unbroken cord, which is always found hanging from the aperture of the drawplate, is ample proof, without further evidence, that the digestive processes go on without ceasing. When the provisions are coming to an end, the cable unrolled is of an astounding length: it can be measured in feet. Where shall we find the like of this stomach which, to avoid any loss when life's balance-sheet is made out, feasts for a week or a fortnight, without stopping, on such distasteful fare?

When the whole ball has passed through the machine, the hermit comes back to the daylight, tries his luck afresh, finds another patch of dung, fashions a new ball and starts eating again. This life of pleasure lasts for a month or two, from May to June; then, with the coming of the fierce heat beloved of the Cicadæ, the Sacred Beetles take up their summer quarters and bury themselves in the cool earth. They reappear with the first autumn rains, less numerous and less active than in spring, but now seemingly absorbed in the most important work of all, the future of the species.

GILBERT KEITH CHESTERTON

IT MAY COME as a shock to many readers to find a selection by Gilbert Keith Chesterton (1874-1936) included here. The rotund British writer was not noted for his knowledge of things scientific. He never, for example, could quite bring himself to accept the theory of man's descent from lower animals. Yet there are times, as in the following selection, when he startles you with unexpected scientific insight.

Chesterton's topic is nothing less than the fundamental contrast between deductive logic, true of all possible worlds, and inductive logic, capable only of telling us how we may reasonably expect *this* world to behave. Let us hasten to add that Chesterton's analysis is in full agreement with the views of modern logicians. Perhaps his "test of the imagination" is not strictly accurate—who can "imagine" the four-dimensional constructions of relativity?—but in essence his position is unassailable. Logical and mathematical statements are true by definition. They are "empty tautologies," to use a current phrase, like the impressive maxim that there are always six eggs in half a dozen. Nature, on the other hand, is under no similar constraints. Fortunately, her "weird repetitions," as GK calls them, often conform to surprisingly low-order equations. But as Hume and others before Hume made clear, there is no *logical* reason why she should behave so politely.

The following selection is taken from, of all places, *Orthodoxy*, Chesterton's most famous work of Christian (it was published fourteen years before he became a Catholic) apologetics. The style is that for which the author is justly famous—brilliant, witty, alliterative, dazzling in its metaphors and verbal swordplay, and a joy to read even when you disagree with him.

GILBERT KEITH CHESTERTON

The Logic of Elfland

My first and last philosophy, that which I believe in with unbroken certainty, I learnt in the nursery. I generally learnt it from a nurse; that is, from the solemn and star-appointed priestess at once of democracy and tradition. The things I believed most then, the things I believe most now, are the things called fairy tales. They seem to me to be the entirely reasonable things. They are not fantasies: compared with them other things are fantastic. Compared with them religion and rationalism are both abnormal, though religion is abnormally right and rationalism abnormally wrong. Fairyland is nothing but the sunny country of common sense. It is not earth that judges heaven, but heaven that judges earth; so for me at least it was not earth that criticised elfland, but elfland that criticised the earth. I knew the magic beanstalk before I had tasted beans; I was sure of the Man in the Moon before I was certain of the moon. This was at one with all popular tradition. Modern minor poets are naturalists, and talk about the bush or the brook; but the singers of the old epics and fables were supernaturalists, and talked about the gods of brook and bush. That is what the moderns mean when they say that the ancients did not "appreciate Nature," because they said that Nature was divine. Old nurses do not tell children about the grass, but about the fairies that dance on the grass; and the old Greeks could not see the trees for the dryads.

But I deal here with what ethic and philosophy come from

being fed on fairy tales. If I were describing them in detail I could note many noble and healthy principles that arise from them. There is the chivalrous lesson of "Jack the Giant Killer"; that giants should be killed because they are gigantic. It is a manly mutiny against pride as such. For the rebel is older than all the kingdoms, and the Jacobin has more tradition than the Jacobite. There is the lesson of "Cinderella," which is the same as that of the Magnificat—*exaltavit humiles*. There is the great lesson of "Beauty and the Beast"; that a thing must be loved *before* it is lovable. There is the terrible allegory of the "Sleeping Beauty," which tells how the human creature was blessed with all birthday gifts, yet cursed with death; and how death also may perhaps be softened to a sleep. But I am not concerned with any of the separate statutes of elfland, but with the whole spirit of its law, which I learnt before I could speak, and shall retain when I cannot write. I am concerned with a certain way of looking at life, which was created in me by the fairy tales, but has since been meekly ratified by the mere facts.

It might be stated this way. There are certain sequences or developments (cases of one thing following another), which are, in the true sense of the word, reasonable. They are, in the true sense of the word, necessary. Such are mathematical and merely logical sequences. We in fairyland (who are the most reasonable of all creatures) admit that reason and that necessity. For instance, if the Ugly Sisters are older than Cinderella, it is (in an iron and awful sense) *necessary* that Cinderella is younger than the Ugly Sisters. There is no getting out of it. Haeckel may talk as much fatalism about that fact as he pleases: it really must be. If Jack is the son of a miller, a miller is the father of Jack. Cold reason decrees it from her awful throne: and we in fairyland submit. If the three brothers all ride horses, there are six animals and eighteen legs involved: that is true rationalism, and fairyland is full of it. But as I put my head over the hedge of the elves and began to take notice of the natural world, I observed an extraordinary thing. I observed that learned men in spectacles were talking of the actual things that happened—dawn and death and so on—as if *they* were rational and inevitable. They talked as if the fact that trees bear fruit were just as *necessary* as the fact that two and one trees make three. But it is not. There is an enormous difference by the test of fairyland; which is the test of the imagination. You cannot *imagine* two and one not making three. But you can easily imagine trees not growing fruit; you can imagine them growing golden candlesticks or tigers hanging on by the tail. These men in spectacles spoke much of a man named Newton, who was hit by an apple, and who discovered a law. But they could not be got to see the distinction between a true law, a law of reason, and the mere fact of apples falling. If the apple hit Newton's nose, Newton's nose hit the apple. That is a true necessity: because we cannot conceive the one occurring without the other. But we can quite well conceive the apple not falling on his nose; we can fancy it flying ardently through the air to hit some other nose, of which it had a more definite dislike. We have always in our fairy tales kept this sharp distinction between the science of mental relations, in which there really are laws, and the science of physical facts, in which there are no laws, but only weird repetitions. We believe in bodily miracles, but not in mental impossibilities. We believe that a Bean-stalk climbed up to Heaven; but that does not at all confuse our convictions on the philosophical question of how many beans make five.

Here is the peculiar perfection of tone and truth in the nursery tales. The man of science says, "Cut the stalk, and the apple will fall"; but he says it calmly, as if the one idea really led up to the other. The witch in the fairy tale says, "Blow the horn, and the ogre's castle will fall"; but she does not say it as if it were something in which the effect obviously arose out of the cause. Doubtless she has given the advice to many champions, and has seen many castles fall, but she does not lose either her wonder or her reason. She does not muddle her head until it imagines a necessary mental connection between a horn and a falling tower. But the scientific men do muddle their heads, until they imagine a necessary mental connection between an apple leaving the tree and an apple reaching the ground. They do really talk as if they had found not only a set of marvellous facts, but a truth connecting those facts. They do talk as if the connection of two strange things physically connected them philosophically. They feel that because one incomprehensible thing constantly follows another incomprehensible thing the two together somehow make up a comprehensible thing. Two black riddles make a white answer.

In fairyland we avoid the word "law"; but in the land of science they are singularly fond of it. Thus they will call some interesting conjecture about how forgotten folks pronounced the alphabet, Grimm's Law. But Grimm's Law is far less intellectual than Grimm's Fairy Tales. The tales are, at any rate, certainly tales; while the law is not a law. A law implies that we know the nature of the generalisation and enactment; not merely that we have noticed some of the effects. If there is a law that pick-pockets shall go to prison, it implies that there is an imaginable mental connection between the idea of prison and the idea of picking pockets. And we know what the idea is. We can say why we take liberty from a man who takes liberties. But we cannot say why an egg can turn into a chicken any more than we can say why a bear could turn into a fairy prince. As *ideas,* the egg and the chicken are further off each other than the bear and the prince; for no egg in itself suggests a chicken, whereas some princes do suggest bears. Granted, then, that certain transformations do happen, it is essential that we should regard them in the philosophic manner of fairy tales, not in the unphilosophic manner of science and the "Laws of Nature." When we are asked why eggs turn to birds or fruits fall in autumn, we must answer exactly as the fairy godmother would answer if Cinderella asked her why mice turned to horses or her clothes fell from her at twelve o'clock. We must answer that it is *magic.* It is not a "law," for we do not understand its general formula. It is not a necessity, for though we can count on it happening practically, we have no right to say that it must always happen. It is no argument for unalterable law (as Huxley fancied) that we count on the ordinary course of things. We do not count on it; we bet on it. We risk the remote possibility of a miracle as we do that of a poisoned pancake or a world-destroying comet. We leave it out of account, not because it is a miracle, and therefore an impossibility, but because it is a miracle, and therefore an exception. All the terms used in the science books, "law," "necessity," "order," "tendency," and so on, are really unintellectual, because they assume an inner synthesis which we do not possess. The only words that ever satisfied me as describing Nature are the terms used in the fairy books, "charm," "spell," "enchantment." They express the arbitrariness of the fact and its mystery. A tree grows fruit because it is a *magic* tree. Water runs downhill because it is bewitched. The sun shines because it is bewitched.

I deny altogether that this is fantastic or even mystical. We may have some mysticism later on; but this fairy-tale language about things is simply rational and agnostic. It is the only way I can express in words my clear and definite perception that one thing is quite distinct from another; that there is no logical connection between flying and laying eggs. It is the man who talks about "a law" that he has never seen who is the mystic. Nay, the ordinary scientific man is strictly a sentimentalist. He is a sentimentalist in this essential sense, that he is soaked and swept away by mere associations. He has so often seen birds fly and lay eggs that he feels as if there must be some dreamy, tender connection between the two ideas, whereas there is none. A forlorn lover might be unable to dissociate the moon from lost love; so the materialist is unable to dissociate the moon from the tide. In both cases there is no connection, except that one has seen them together. A sentimentalist might shed tears at the smell of apple-blossom, because, by a dark association of his own, it reminded him of his boyhood. So the materialist professor (though he conceals his tears) is yet a sentimentalist, because, by a dark association of his own, apple-blossoms remind him of apples. But the cool rationalist from fairyland does not see why, in the abstract, the apple tree should not grow crimson tulips; it sometimes does in his country.

This elementary wonder, however, is not a mere fancy derived from the fairy tales; on the contrary, all the fire of the fairy tales is derived from this. Just as we all like love tales because there is an instinct of sex, we all like astonishing tales because they touch the nerve of the ancient instinct of astonishment. This is proved by the fact that when we are very young children we do not need fairy tales: we only need tales. Mere life is interesting enough. A child of seven is excited by being told that Tommy opened a door and saw a dragon. But a child of three is excited by being told that Tommy opened a door. Boys like romantic tales; but babies like realistic tales—because they find them romantic. In fact, a baby is about the only person, I should think, to whom a modern realistic novel could be read without boring him. This proves that even nursery tales only echo an almost pre-natal leap of interest and amazement. These tales say that apples were golden only to refresh the forgotten moment when we found that they were green. They make rivers run with wine only to make us remember, for one wild moment, that they run with water. I have said that this is wholly reasonable and even agnostic. And, indeed, on this point I am all for the higher agnosticism; its better name is Ignorance. We have all read in scientific books, and, indeed, in all romances, the story of the man who has forgotten his name. This man walks about the streets and can see and appreciate everything; only he cannot remember who he is. Well, every man is that man in the story. Every man has forgotten who he is. One may understand the cosmos, but never the ego; the self is more distant than any star. Thou shalt love the Lord thy God; but thou shalt not know thyself. We are all under the same mental calamity; we have all forgotten our names. We have all forgotten what we really are. All that we call common sense and rationality and practicality and positivism only means that for certain dead levels of our life we forget that we have forgotten. All that we call spirit and art and ecstasy only means that for one awful instant we remember that we forget.

CARL SAGAN

FOLLOWING THE literary paths broken by British astronomers like Sir Robert Ball, Sir Arthur Stanley Eddington, and Sir James Jeans—renowned scientists with that rare ability to write eloquently—Carl Sagan has become one of the world's great science popularizers. Who can say how many millions have gained their first insights into the adventure of science by reading his lyrical books, hearing him on the Johnny Carson show, or watching his flamboyant television productions?

One of Sagan's major passions, as everyone knows, is the search for intelligent life on planets beyond the solar system. He is much too knowledgeable to take seriously the public's idiotic mania for close encounters with UFOs, but listening for messages from higher minds "out there" is something else. As cosmologist Philip Morrison has said, we will never know if they are there unless we listen. And of course we will never find out if we destroy ourselves before we hear anything. Another Sagan passion is making citizens aware of this possibility—one that grows more probable every year as the awesome weapons accumulate.

Thomas Huxley's best-known essay, "On a Piece of Chalk," took its departure from the great beds of white limestone that lie beneath most of southern England. In Sagan's essay the starting point is a grain of salt. It starts him wondering about some of the deepest questions in the philosophy of science. How is it that nature displays such orderly structure that its laws become knowable by our crude animal minds? Do we *really* know, or is science merely a constantly shifting collection of myths, never getting closer to ultimate truth? How much of the cosmos is knowable? Will science ever understand everything, or will its search be endless?

Whatever the answers, few can quarrel with Sagan's conclusion to another essay in *Broca's Brain,* from which our selection is taken: "We have entered, almost without noticing, an age of exploration and discovery unparalleled since the Renaissance."

CARL SAGAN

Can We Know the Universe? Reflections on a Grain of Salt

Nothing is rich but the inexhaustible wealth of nature. She shows us only surfaces, but she is a million fathoms deep.

RALPH WALDO EMERSON

SCIENCE IS a way of thinking much more than it is a body of knowledge. Its goal is to find out how the world works, to seek what regularities there may be, to penetrate to the connections of things—from subnuclear particles, which may be the constituents of all matter, to living organisms, the human social community, and thence to the cosmos as a whole. Our intuition is by no means an infallible guide. Our perceptions may be distorted by training and prejudice or merely because of the limitations of our sense organs, which, of course, perceive directly but a small fraction of the phenomena of the world. Even so straightforward a question as whether in the absence of friction a pound of lead falls faster than a gram of fluff was answered incorrectly by Aristotle and almost everyone else before the time of Galileo. Science is based on experiment, on a willingness to challenge old dogma, on an openness to see the universe as it really is. Accordingly, science sometimes requires courage—at the very least the courage to question the conventional wisdom.

Beyond this the main trick of science is to *really* think of something: the shape of clouds and their occasional sharp bottom edges at the same altitude everywhere in the sky; the formation of a dewdrop on a leaf; the origin of a name or a word—Shakespeare, say, or "philanthropic"; the reason for human social customs—the incest taboo, for example; how it is that a lens in sunlight can make paper burn; how a "walking stick" got to look so much like a twig; why the Moon seems to follow us as we walk; what prevents us from digging a hole down to the center of the Earth; what the definition is of "down" on a spherical Earth; how it is possible for the body to convert yesterday's lunch into today's muscle and sinew; or how far is up—does the universe go on forever, or if it does not, is there any meaning to the question of what lies on the other side? Some of these questions are pretty easy. Others, especially the last, are mysteries to which no one even today knows the answer. They are natural questions to ask. Every culture has posed such questions in one way or another. Almost always the proposed answers are in the nature of "Just So Stories," attempted explanations divorced from experiment, or even from careful comparative observations.

But the scientific cast of mind examines the world critically as if many alternative worlds might exist, as if other things might be here which are not. Then we are forced to ask why what we see is present and not something else. Why are the Sun and the Moon and the planets spheres? Why not pyramids, or cubes, or dodecahedra? Why not irregular, jumbly shapes? Why so symmetrical, worlds? If you spend any time spinning hypotheses, checking to see whether they make sense, whether they conform to what else we know, thinking of tests you can pose to substantiate or deflate your hypotheses, you will find yourself doing science. And as you come to practice this habit of thought more and more you will get better and better at it. To penetrate into the heart of the thing—even a little thing, a blade of grass, as Walt Whitman said—is to experience a kind of exhilaration that, it may be, only human beings of all the beings on this planet can feel. We are an intelligent species and the use of our intelligence quite properly gives us pleasure. In this respect the brain is like a muscle. When we think well, we feel good. Understanding is a kind of ecstasy.

But to what extent can we *really* know the universe around us? Sometimes this question is posed by people who hope the answer will be in the negative, who are fearful of a universe in which everything might one day be known. And sometimes we hear pronouncements from scientists who confidently state that everything worth knowing will soon be known—or even is already known—and who paint pictures of a Dionysian or Polynesian age in which the zest for intellectual discovery has withered, to be replaced by a kind of subdued languor, the lotus eaters drinking fermented coconut milk or some other mild hallucinogen. In addition to maligning both the Polynesians, who were intrepid explorers (and whose brief respite in paradise is now sadly ending), as well as the inducements to intellectual discovery provided by some hallucinogens, this contention turns out to be trivially mistaken.

Let us approach a much more modest question: not whether we can know the universe or the Milky Way Galaxy or a star or a world. Can we know, ultimately and in detail, a grain of salt? Consider one microgram of table salt, a speck just barely large enough for someone with keen eyesight to make out without a microscope. In that grain of salt there are about 10^{16} sodium and chlorine atoms. This is a 1 followed by 16 zeros, 10 million billion atoms. If we wish to know a grain of salt, we must know at least the three-dimensional positions of each of these atoms. (In fact, there is much more to be known—for example, the nature of the forces between the atoms—but we are making only a modest calculation.) Now, is this number more or less than the number of things which the brain can know?

How much *can* the brain know? There are perhaps 10^{11} neurons in the brain, the circuit elements and switches that are responsible in their electrical and chemical activity for the functioning of our minds. A typical brain neuron has perhaps a thousand little wires, called dendrites, which connect it with its fellows. If, as seems likely, every bit of information in the brain corresponds to one of these connections, the total number of things knowable by the brain is no more than 10^{14}, one hundred trillion. But this number is only one percent of the number of atoms in our speck of salt.

So in this sense the universe is intractable, astonishingly immune to any human attempt at full knowledge. We cannot on this level understand a grain of salt, much less the universe.

But let us look a little more deeply at our microgram of salt. Salt happens to be a crystal in which, except for defects in the structure of the crystal lattice, the position of every sodium and chlorine atom is predetermined. If we could shrink ourselves into this crystalline world, we would see rank upon rank of atoms in an ordered array, a regularly alternating structure—sodium, chlorine, sodium, chlorine, specifying the sheet of atoms we are standing on and all the sheets above us and below us. An absolutely pure crystal of salt could have the position of every atom specified by something like 10 bits of information.* This would not strain the information-carrying capacity of the brain.

If the universe had natural laws that governed its behavior to the same degree of regularity that determines a crystal of salt, then, of course, the universe would be knowable. Even if there were many such laws, each of considerable complexity, human beings might have the capability to understand them all. Even if such knowledge exceeded the information-carrying capacity of the brain, we might store the additional information outside our bodies—in books, for example, or in computer memories—and still, in some sense, know the universe.

Human beings are, understandably, highly motivated to find regularities, natural laws. The search for rules, the only possible way to understand such a vast and complex universe, is called science. The universe forces those who live in it to understand it. Those creatures who find everyday experience a muddled jumble of events with no predictability, no regularity, are in grave peril. The universe belongs to those who, at least to some degree, have figured it out.

It is an astonishing fact that there *are* laws of nature, rules

*Chlorine is a deadly poison gas employed on European battlefields in World War I. Sodium is a corrosive metal which burns upon contact with water. Together they make a placid and unpoisonous material, table salt. Why each of these substances has the properties it does is a subject called chemistry, which requires more than 10 bits of information to understand.

that summarize conveniently—not just qualitatively but quantitatively—how the world works. We might imagine a universe in which there are no such laws, in which the 10^{80} elementary particles that make up a universe like our own behave with utter and uncompromising abandon. To understand such a universe we would need a brain at least as massive as the universe. It seems unlikely that such a universe could have life and intelligence, because beings and brains require some degree of internal stability and order. But even if in a much more random universe there were such beings with an intelligence much greater than our own, there could not be much knowledge, passion or joy.

Fortunately for us, we live in a universe that has at least important parts that are knowable. Our commonsense experience and our evolutionary history have prepared us to understand something of the workaday world. When we go into other realms, however, common sense and ordinary intuition turn out to be highly unreliable guides. It is stunning that as we go close to the speed of light our mass increases indefinitely, we shrink toward zero thickness in the direction of motion, and time for us comes as near to stopping as we would like. Many people think that this is silly, and every week or two I get a letter from someone who complains to me about it. But it is a virtually certain consequence not just of experiment but also of Albert Einstein's brilliant analysis of space and time called the Special Theory of Relativity. It does not matter that these effects seem unreasonable to us. We are not in the habit of traveling close to the speed of light. The testimony of our common sense is suspect at high velocities.

Or consider an isolated molecule composed of two atoms shaped something like a dumbbell—a molecule of salt, it might be. Such a molecule rotates about an axis through the line connecting the two atoms. But in the world of quantum mechanics, the realm of the very small, not all orientations of our dumbbell molecule are possible. It might be that the molecule could be oriented in a horizontal position, say, or in a vertical position, but not at many angles in between. Some rotational positions are forbidden. Forbidden by what? By the laws of nature. The universe is built in such a way as to limit, or quantize, rotation. We do not experience this directly in everyday life; we would find it startling as well as awkward in sitting-up exercises, to find arms outstretched from the sides or pointed up to the skies permitted but many intermediate positions forbidden. We do not live in the world of the small, on the scale of 10^{-13} centimeters, in the realm where there are twelve zeros between the decimal place and the one. Our common-sense intuitions do not count. What does count is experiment—in this case observations from the far infrared spectra of molecules. They show molecular rotation to be quantized.

The idea that the world places restrictions on what humans might do is frustrating. Why *shouldn't* we be able to have intermediate rotational positions? Why *can't* we travel faster than the speed of light? But so far as we can tell, this is the way the universe is constructed. Such prohibitions not only press us toward a little humility; they also make the world more knowable. Every restriction corresponds to a law of nature, a regularization of the universe. The more restrictions there are on what matter and energy can do, the more knowledge human beings can attain. Whether in some sense the universe is ultimately knowable depends not only on how many natural laws there are that encompass widely divergent phenomena, but also on whether we have the openness and the intellectual capacity to understand such laws. Our formulations of the regularities of nature are surely dependent on how the brain is built, but also, and to a significant degree, on how the universe is built.

For myself, I like a universe that includes much that is unknown and, at the same time, much that is knowable. A universe in which everything is known would be static and dull, as boring as the heaven of some weak-minded theologians. A universe that is unknowable is no fit place for a thinking being. The ideal universe for us is one very much like the universe we inhabit. And I would guess that this is not really much of a coincidence.

JOSEPH WOOD KRUTCH

WILLIAM JAMES, in a previous essay in this volume, attempted to convey something of the wonder of the fact that anything should exist at all. In the essay to follow, Joseph Krutch (1893-1970) is concerned with the wonder of the fact that existence is so strangely bifurcated into living and non-living. A disciple of John Dewey or Alfred North Whitehead, since each was impressed in his own way by continuity, might quickly remind us that there is no sharp boundary between life and non-life. Are viruses living? They are or are not depending on your definition of "life," and no matter how precise this definition is made, science is almost certain to come up with something about which the application of the definition will be debatable.

To this we must all agree. Nevertheless we should also remember that the fact of twilight does not make meaningless the phrase "as different as night and day." New qualities have a queer habit of emerging along continuums, otherwise we could not talk at all. There is no way of distinguishing one thing from another in a night in which all terms are gray. Nature, then, is indeed bifurcated, and for Krutch, the non-living seems even more startling and absurd than the living. It has a kind of frozen, ultimate horror in its senseless patterns. To the scientist in a scientist these are meaningless reflections. But to the poet in a scientist—one is tempted to say, to the human being in a scientist—they may strike deep inner chords of feeling.

Mr. Krutch began his career as an English professor, dramatic critic, and author of a psychoanalytic study of Edgar Allan Poe. In 1929 his book *The Modern Temper* voiced such a gloomy outlook on life that Bertrand Russell, in *The Con-*

quest of Happiness, felt obliged to devote an entire chapter to refuting it. "Ours is a lost cause," Krutch concluded then, "and there is no place for us in the natural universe. . . ." Russell denied that this pessimism of *Ecclesiastes* was the inevitable accompaniment of a scientific outlook. He advised Krutch to stop writing, flee from the literary coteries of New York, and become a pirate.

Mr. Krutch continued, however, to write; but after publishing several excellent books of criticism and biography, he did move from Manhattan to the country, and more recently, for reasons of health, to the deserts of the Southwest. Simultaneously, his attention shifted from books and plays to birds and animals. A sympathetic study of Thoreau was followed by a series of delightful nature books, including *The Best of Two Worlds* from which the following essay is taken. Judging from this and other recent writing, one would guess that his temper also shifted—from heroic despair to a mood of modest hope.

JOSEPH WOOD KRUTCH

The Colloid and the Crystal

THE FIRST REAL SNOW was soon followed by a second. Over the radio the weatherman talked lengthily about cold masses and warm masses, about what was moving out to sea and what wasn't. Did Benjamin Franklin, I wondered, know what he was starting when it first occurred to him to trace by correspondence the course of storms? From my stationary position the most reasonable explanation seemed to be simply that winter had not quite liked the looks of the landscape as she first made it up. She was changing her sheets.

Another forty-eight hours brought one of those nights ideal for frosting the panes. When I came down to breakfast, two of the windows were almost opaque and the others were etched with graceful, fernlike sprays of ice which looked rather like the impressions left in rocks by some of the antediluvian plants, and they were almost as beautiful as anything which the living can achieve. Nothing else which has never lived looks so much as though it were actually informed with life.

I resisted, I am proud to say, the almost universal impulse to scratch my initials into one of the surfaces. The effect, I knew, would not be an improvement. But so, of course, do those less virtuous than I. That indeed is precisely why they scratch. The impulse to mar and to destroy is as ancient and almost as nearly universal as the impulse to create. The one is an easier way than the other of demonstrating power. Why else should anyone not hungry prefer a dead rabbit to a live one? Not even those horrible Dutch painters of bloody still—or shall we say stilled?—lifes can have really believed that their subjects were more beautiful dead.

Indoors it so happened that a Christmas cactus had chosen this moment to bloom. Its lush blossoms, fuchsia-shaped but pure red rather than magenta, hung at the drooping ends of strange, thick stems and outlined themselves in blood against the glistening background of the frosty pane—jungle flower against frostflower; the warm beauty that breathes and lives and dies competing with the cold beauty that burgeons, not because it wants to, but merely because it is obeying the laws of physics which require that crystals shall take the shape they have always taken since the world began. The effect of red flower against white tracery was almost too theatrical, not quite in good taste perhaps. My eye recoiled in shock and sought through a clear area of the glass the more normal out-of-doors.

On the snow-capped summit of my bird-feeder a chickadee pecked at the new-fallen snow and swallowed a few of the flakes which serve him in lieu of the water he sometimes sadly lacks when there is nothing except ice too solid to be picked at. A downy woodpecker was hammering at a lump of suet and at the coconut full of peanut butter. One nuthatch was dining while the mate waited his—or was it her?—turn. The woodpecker announces the fact that he is a male by the bright red spot on the back of his neck, but to me, at least, the sexes of the nuthatch are indistinguishable. I shall never know whether it is the male or the female who eats first. And that is a pity. If I knew, I could say, like the Ugly Duchess, "and the moral of that is . . ."

But I soon realized that at the moment the frosted windows were what interested me most—especially the fact that there is no other natural phenomenon in which the lifeless mocks so closely the living. One might almost think that the frostflower had got the idea from the leaf and the branch if one did not know how inconceivably more ancient the first is. No wonder that enthusiastic biologists in the nineteenth century, anxious to conclude that there was no qualitative difference between life and chemical processes, tried to believe that the crystal furnished the link, that its growth was actually the same as the growth of a living organism. But excusable though the fancy was, no one, I think, believes anything of the sort today. Protoplasm is a colloid and the colloids are fundamentally different from the crystalline substances. Instead of crystallizing they jell, and life in its simplest known form is a shapeless blob of rebellious jelly rather than a crystal eternally obeying the most ancient law.

No man ever saw a dinosaur. The last of these giant reptiles was dead eons before the most dubious halfman surveyed the world about him. Not even the dinosaurs ever cast their dim eyes upon many of the still earlier creatures which preceded them. Life changes so rapidly that its later phases know nothing of those which preceded them. But the frost-flower is older than the dinosaur, older than the protozoan, older no doubt than the enzyme or the ferment. Yet it is precisely what it has always been. Millions of years before there were any eyes to see it, millions of years before any life existed, it grew in its own special way, crystallized along its preordained lines of cleavage, stretched out its pseudo-branches and pseudo-leaves. It was beautiful before beauty itself existed.

We find it difficult to conceive a world except in terms of purpose, of will, or of intention. At the thought of the something without beginning and presumably without end, of something which is, nevertheless, regular though blind, and organized without any end in view, the mind reels. Constituted as we are it is easier to conceive how the slime floating upon the waters might become in time Homo sapiens than it is to imagine how so complex a thing as a crystal could have always been and can always remain just what it is—complicated and perfect but without any meaning, even for itself. How can the lifeless even obey a law?

To a mathematical physicist I once confessed somewhat

shamefacedly that I had never been able to understand how inanimate nature managed to follow so invariably and so promptly her own laws. If I flip a coin across a table, it will come to rest at a certain point. But before it stops at just that point, many factors must be taken into consideration. There is the question of the strength of the initial impulse, of the exact amount of resistance offered by the friction of that particular table top, and of the density of the air at the moment. It would take a physicist a long time to work out the problem and he could achieve only an approximation at that. Yet presumably the coin will stop exactly where it should. Some very rapid calculations have to be made before it can do so, and they are, presumably, always accurate.

And then, just as I was blushing at what I supposed he must regard as my folly, the mathematician came to my rescue by informing me that Laplace had been puzzled by exactly the same fact. "Nature laughs at the difficulties of integration," he remarked—and by "integration" he meant, of course, the mathematician's word for the process involved when a man solves one of the differential equations to which he has reduced the laws of motion.

When my Christmas cactus blooms so theatrically a few inches in front of the frost-covered pane, it also is obeying laws but obeying them much less rigidly and in a different way. It blooms at about Christmastime because it has got into the habit of doing so, because, one is tempted to say, it wants to. As a matter of fact it was, this year, not a Christmas cactus but a New Year's cactus, and because of this unpredictability I would like to call it "he," not "it." His flowers assume their accustomed shape and take on their accustomed color. But not as the frostflowers follow their predestined pattern. Like me, the cactus has a history which stretches back over a long past full of changes and developments. He has not always been merely obeying fixed laws. He has resisted and rebelled; he has attempted novelties, passed through many phases. Like all living things he has had a will of his own. He has made laws, not merely obeyed them.

"Life," so the platitudinarian is fond of saying, "is strange."

But from our standpoint it is not really so strange as those things which have no life and yet nevertheless move in their predestined orbits and "act" though they do not "behave." At the very least one ought to say that if life is strange there is nothing about it more strange than the fact that it has its being in a universe so astonishingly shared on the one hand by "things" and on the other by "creatures," that man himself is both a "thing" which obeys the laws of chemistry or physics and a "creature" who to some extent defies them. No other contrast, certainly not the contrast between the human being and the animal, or the animal and the plant, or even the spirit and the body, is so tremendous as this contrast between what lives and what does not.

To think of the lifeless as merely inert, to make the contrast merely in terms of a negative, is to miss the real strangeness. Not the shapeless stone which seems to be merely waiting to be acted upon but the snowflake or the frostflower is the true representative of the lifeless universe as opposed to ours. They represent plainly, as the stone does not, the fixed and perfect system of organization which includes the sun and its planets, includes therefore this earth itself, but against which life has set up its seemingly puny opposition. Order and obedience are the primary characteristics of that which is not alive. The snowflake eternally obeys its one and only law: "Be thou six pointed"; the planets their one and only: "Travel thou in an ellipse." The astronomer can tell where the North Star will be ten thousand years hence; the botanist cannot tell where the dandelion will bloom tomorrow.

Life is rebellious and anarchial, always testing the supposed immutability of the rules which the nonliving changelessly accepts. Because the snowflake goes on doing as it was told, its story up to the end of time was finished when it first assumed the form which it has kept ever since. But the story of every living thing is still in the telling. It may hope and it may try. Moreover, though it may succeed or fail, it will certainly change. No form of frostflower ever became extinct. Such, if you like, is its glory. But such also is the fact which makes it alien. It may melt but it cannot die.

If I wanted to contemplate what is to me the deepest of all mysteries, I should choose as my object lesson a snowflake under a lens and an amoeba under the microscope. To a detached observer—if one can possibly imagine any observer who *could* be detached when faced with such an ultimate choice—the snowflake would certainly seem the "higher" of the two. Against its intricate glistening perfection one would have to place a shapeless, slightly turbid glob, perpetually oozing out in this direction or that but not suggesting so strongly as the snowflake does, intelligence and plan. Crystal and colloid, the chemist would call them, but what an inconceivable contrast those neutral terms imply! Like the star, the snowflake seems to declare the glory of God, while the promise of the amoeba, given only perhaps to itself, seems only contemptible. But its jelly holds, nevertheless, not only its promise but ours also, while the snowflake represents some achievement which we cannot possibly share. After the passage of billions of years, one can see and be aware of the other, but the relationship can never be reciprocal. Even after these billions of years no aggregate of colloids can be as beautiful as the crystal always was, but it can know, as the crystal cannot, what beauty is.

Even to admire too much or too exclusively the alien kind of beauty is dangerous. Much as I love and am moved by the grand, inanimate forms of nature, I am always shocked and a little frightened by those of her professed lovers to whom landscape is the most important thing, and to whom landscape is merely a matter of forms and colors. If they see or are moved by an animal or flower, it is to them merely a matter of a picturesque completion and their fellow creatures are no more than decorative details. But without some continuous awareness of the two great realms of the inanimate and the animate there can be no love of nature as I understand it, and what is worse, there must be a sort of disloyalty to our cause, to us who are colloid, not crystal. The pantheist who feels the oneness of all living things, I can understand; perhaps indeed he and I are in essential agreement. But the ultimate All is not one thing, but two. And because the alien half is in its way as proud and confident and successful as our half, its fundamental difference may not be disregarded with impunity. Of us and all we stand for, the enemy is not so much death as the not-living, or rather that great system which succeeds without ever having had the need to be alive. The frostflower is not merely a wonder; it is also a threat and a warning. How admirable, it seems to say, not living can be! What triumphs mere immutable law can achieve!

Some of Charles Peirce's strange speculations about the possibility that "natural law" is not law at all but merely a set of habits fixed more firmly than any habits we know anything about in ourselves or in the animals suggest the possibility that the snowflake was not, after all, always inanimate, that it merely surrendered at some time impossibly remote the life which once achieved its perfect organization. Yet even if we can imagine such a thing to be true, it serves only to warn us all the more strongly against the possibility that what we call the living might in the end succumb also to the seduction of the immutably fixed.

No student of the anthill has ever failed to be astonished either into admiration or horror by what is sometimes called the perfection of its society. Though even the anthill can change its ways, though even ant individuals—ridiculous as the conjunction of the two words may seem—can sometimes make choices, the perfection of the techniques, the regularity of the habits almost suggest the possibility that the insect is on its way back to inanition, that, vast as the difference still is, an anthill crystallizes somewhat as a snowflake does. But not even the anthill, nothing else indeed in the whole known universe is so perfectly planned as one of these same snowflakes. Would, then, the ultimately planned society be, like the anthill, one in which no one makes plans, any more than a snowflake does? From the cradle in which it is not really born to the grave where it is only a little deader than it always was, the ant-citizen follows a plan to the making of which he no longer contributes anything.

Perhaps we men represent the ultimate to which the rebellion, begun so long ago in some amoeba-like jelly, can go. And perhaps the inanimate is beginning the slow process of subduing us again. Certainly the psychologist and the philosopher are tending more and more to think of us as creatures who obey laws rather than as creatures of will and responsibility. We are, they say, "conditioned" by this or by that. Even the greatest heroes are studied on the assumption that they can be "accounted for" by something outside themselves. They are, it is explained, "the product of forces." All the emphasis is placed, not upon that power to resist and rebel which we were once supposed to have, but upon the "influences" which "formed us." Men are made by society, not society by men. History as well as character "obeys laws." In their view, we crystallize in obedience to some dictate from without instead of moving in conformity with something within.

And so my eye goes questioningly back to the frosted pane. While I slept the graceful pseudo-fronds crept across the glass, assuming, as life itself does, an intricate organization. "Why live," they seem to say, "when we can be beautiful, complicated, and orderly without the uncertainty and effort required of a living thing? Once we were all that was. Perhaps some day we shall be all that is. Why not join us?"

Last summer no clod or no stone would have been heard if it had asked such a question. The hundreds of things which walked and sang, the millions which crawled and twined were all having their day. What was dead seemed to exist only in order that the living might live upon it. The plants were busy turning the inorganic into green life and the animals were busy turning that green into red. When we moved, we walked mostly upon grass. Our pre-eminence was unchallenged.

On this winter day nothing seems so successful as the frostflower. It thrives on the very thing which has driven some of us indoors or underground and which has been fatal to many. It is having now its hour of triumph, as we before had ours. Like the cactus flower itself, I am a hothouse plant.

Even my cats gaze dreamily out of the window at a universe which is no longer theirs.

How are we to resist, if resist we can? This house into which I have withdrawn is merely an expedient and it serves only my mere physical existence. What mental or spiritual convictions, what will to maintain to my own kind of existence can I assert? For me it is not enough merely to say, as I do say, that I shall resist the invitation to submerge myself into a crystalline society and to stop planning in order that I may be planned for. Neither is it enough to go further, as I do go, and to insist that the most important thing about a man is not that part of him which is "the product of forces" but that part, however small it may be, which enables him to become something other than what the most accomplished sociologist, working in conjunction with the most accomplished psychologist, could predict that he would be.

I need, so I am told, a faith, something outside myself to which I can be loyal. And with that I agree, in my own way. I am on what I call "our side," and I know, though vaguely, what I think that is. Wordsworth's God had his dwelling in the light of setting suns. But the God who dwells there seems to me most probably the God of the atom, the star, and the crystal. Mine, if I have one, reveals Himself in another class of phenomena. He makes the grass green and the blood red.

JOSÉ ORTEGA Y GASSET

THE TITLE of Ortega y Gasset's best known work, *The Revolt of the Masses,* suggests that it might be a Marxian exhortation to the proletariat to shake off their chains. It is nothing of the sort. The book is a searing indictment of the increasing power of the common man in twentieth century industrial society. True democracy, Ortega maintains, flourishes only when citizens of widely differing views are willing to delegate responsibilities of government to a superior minority. Today we see it everywhere degenerating into a "hyper-democracy" in which the average man himself insists upon holding the reins. Since this "mass man," whether rich or poor, hates everyone unlike himself, he tries to stamp his mediocrity and vulgarity upon everyone. He may do it quietly, through a variety of pressure groups, or violently by a Communist or Fascist revolution. In either case the result is the same: a homogenized society of identical, other-directed, middle class blanks.

This critique of western culture is of course far from new. Some of its presuppositions go back to Plato, and in recent times many American writers, including H. L. Mencken and Walter Lippmann, have played variants on the theme. But in Ortega's book, first published in 1930, it found a crackling, jabbing expression that made the book a profoundly disturbing one.

At the time of his death, José Ortega y Gasset (1883-1955) was Spain's most distinguished philosopher and man of letters. When civil war broke out in 1936, Ortega, then professor of philosophy at the University of Madrid and one of the intellectual bulwarks of the Republican government, left Spain as a voluntary exile and did not return until 1945.

His last decade was what he sadly called a kind of "non-existence." He wrote little, took part in nothing. In philosophy he was a vitalist, holding views similar to those of Henri Bergson and William James.

The following chapter, from *The Revolt of the Masses,* has the distinction of being the most uncomplimentary piece of writing ever directed against the modern scientist. Ortega saw him as a "learned ignoramus," arrogant in his illusion that because he knows one small thing well he is therefore qualified to pronounce upon all things. It is a much stronger attack than, say, the preposterous spoofing of Charles Fort or those recent books that contrast "scientism" unfavorably with humane letters. Even a working scientist can skim through Fort with amusement and the recent books with only mild annoyance. But there are few scientists who will read this selection without acute discomfort and the dark suspicion that much of what Ortega says is true.

JOSÉ ORTEGA Y GASSET

The Barbarism of "Specialization"

MY THESIS was that Nineteenth-Century civilisation has automatically produced the mass-man. It will be well not to close the general exposition without analysing, in a particular case, the mechanism of that production. In this way, by taking concrete form, the thesis gains in persuasive force.

This civilisation of the Nineteenth Century, I said, may be summed up in the two great dimensions: liberal democracy and technicism. Let us take for the moment only the latter. Modern technicism springs from the union between capitalism and experimental science. Not all technicism is scientific. That which made the stone axe in the Chelian period was lacking in science, and yet a technique was created. China reached a high degree of technique without in the least suspecting the existence of physics. It is only modern European technique that has a scientific basis, from which it derives its specific character, its possibility of limitless progress. All other techniques—Mesopotamian, Egyptian, Greek, Roman, Oriental—reach up to a point of development beyond which they cannot proceed, and hardly do they reach it when they commence to display a lamentable retrogression.

This marvellous Western technique has made possible the proliferation of the European species. Recall the fact from which this essay took its departure and which, as I said, contains in germ all these present considerations. From the Sixth Century to 1800, Europe never succeeds in reaching a population greater than 180 millions. From

1800 to 1914 it rises to more than 460 millions. The jump is unparalleled in our history. There can be no doubt that it is technicism—in combination with liberal democracy—which has engendered mass-man in the quantitative sense of the expression. But these pages have attempted to show that it is also responsible for the existence of mass-man in the qualitative and pejorative sense of the term.

By mass—as I pointed out at the start—is not to be specially understood the workers; it does not indicate a social class, but a kind of man to be found to-day in all social classes, who consequently represents our age, in which he is the predominant, ruling power. We are now about to find abundant evidence for this.

Who is it that exercises social power to-day? Who imposes the forms of his own mind on the period? Without a doubt, the man of the middle class. Which group, within that middle class, is considered the superior, the aristocracy of the present? Without a doubt, the technician: engineer, doctor, financier, teacher, and so on. Who, inside the group of technicians, represents it at its best and purest? Again, without a doubt, the man of science. If an astral personage were to visit Europe to-day and, for the purpose of forming judgment on it, inquire as to the type of man by which it would prefer to be judged, there is no doubt that Europe, pleasantly assured of a favourable judgment, would point to her men of science. Of course, our astral personage would not inquire for exceptional individuals, but would seek the generic type of "man of science," the high-point of European humanity.

And now it turns out that the actual scientific man is the prototype of the mass-man. Not by chance, not through the individual failings of each particular man of science, but because science itself—the root of our civilisation—automatically converts him into mass-man, makes of him a primitive, a modern barbarian. The fact is well known; it has made itself clear over and over again; but only when fitted into its place in the organism of this thesis does it take on its full meaning and its evident seriousness.

Experimental science is initiated towards the end of the Sixteenth Century (Galileo), it is definitely constituted at the close of the Seventeenth (Newton), and it begins to develop in the middle of the Eighteenth. The development of anything is not the same as its constitution; it is subject to different considerations. Thus, the constitution of physics, the collective name of the experimental sciences, rendered necessary an effort towards unification. Such was the work of Newton and other men of his time. But the development of physics introduced a task opposite in character to unification. In order to progress, science demanded specialisation, not in herself, but in men of science. Science is not specialist. If it were, it would *ipso facto* cease to be true. Not even empirical science, taken in its integrity, can be true if separated from mathematics, from logic, from philosophy. But scientific work does, necessarily, require to be specialised.

It would be of great interest, and of greater utility than at first sight appears, to draw up the history of physical and biological sciences, indicating the process of increasing specialisation in the work of investigators. It would then be seen how, generation after generation, the scientist has been gradually restricted and confined into narrower fields of mental occupation. But this is not the important point that such a history would show, but rather the reverse side of the matter: how in each generation the scientist, through having to reduce the sphere of his labour, was progressively losing contact with other branches of science, with that integral interpretation of the universe which is the only thing deserving the names of science, culture, European civilisation.

Specialisation commences precisely at a period which gives to civilised man the title "encyclopaedic." The Nineteenth Century starts on its course under the direction of beings who lived "encyclopaedically," though their production has already some tinge of specialism. In the following generation, the balance is upset and specialism begins to dislodge culture from the individual scientist. When by 1890 a third generation assumes intellectual command in Europe we meet with a type of scientist unparalleled in history. He is one who, out of all that has to be known in order to be a man of judgment, is only acquainted with one science, and even of that one only knows the small corner in which he is an active investigator. He even proclaims it as a virtue that he take no cognizance of what lies outside the narrow territory specially cultivated by himself, and gives the name of "dilettantism" to any curiosity for the general scheme of knowledge.

What happens is that, enclosed within the narrow limits of his visual field, he does actually succeed in discovering new facts, and advancing the progress of the science which he hardly knows, and incidentally the encyclopedia of thought of which he is conscientiously ignorant. How has such a thing been possible, how is it still possible? For it is necessary to insist upon this extraordinary but undeniable fact: experimental science has progressed thanks in great part to the work of men astoundingly mediocre, and even less than mediocre. That is to say, modern science, the root and symbol of our actual civilisation, finds a place for the intellectually commonplace man and allows him to work therein with success. The reason of this lies in what is at the same time the great advantage and the gravest peril of the new science, and of the civilisation directed and represented by it, namely, mechanisation. A fair amount of the things that have to be done in physics or in biology is mechanical work of the mind which can be done by anyone, or almost anyone. For the purpose of innumerable investigations it is possible to divide science into small sections, to enclose oneself in one of these, and to leave out of consideration all the rest. The solidity and exactitude of the methods allow of this temporary but quite real disarticulation of knowledge. The work is done under one of these methods as with a machine, and in order to obtain quite abundant results it is not even necessary to have rigorous notions of their meaning and foundations. In this way the majority of scientists help the general advance of science while shut up in the narrow cell of their laboratory, like the bee in the cell of its hive, or the turnspit in its wheel.

But this creates an extraordinarily strange type of man. The investigator who has discovered a new fact of Nature must necessarily experience a feeling of power and self-assurance. With a certain apparent justice he will look upon himself as "a man who knows." And in fact there is in him a portion of something which, added to many other portions not existing in him, does really constitute knowledge. This is the true inner nature of the specialist, who in the first years of this century has reached the wildest stage of exaggeration. The specialist "knows" very well his own, tiny corner of the universe; he is radically ignorant of all the rest.

Here we have a precise example of this strange new man, whom I have attempted to define, from both of his two opposite aspects. I have said that he was a human product unparalleled in history. The specialist serves as a striking concrete example of the species, making clear to us the radical nature of the novelty. For, previously, men could be divided simply into the learned and the ignorant, those more or less the one, and those more or less the other. But your specialist cannot be brought in under either of these two categories. He is not learned, for he is formally ignorant of all that does not enter into his specialty; but neither is he ignorant, because he is "a scientist," and "knows" very well his own tiny portion of the universe. We shall have to say that he is a learned ignoramus, which is a very serious matter, as it implies that he is a person who is ignorant, not in the fashion of the ignorant man, but with all the petulance of one who is learned in his own special line.

And such in fact is the behaviour of the specialist. In politics, in art, in social usages, in the other sciences, he will adopt the attitudes of primitive, ignorant man; but he will adopt them forcefully and with self-sufficiency, and will not admit of—this is the paradox—specialists in those matters. By specialising him, civilisation has made him hermetic and self-satisfied within his limitations; but this very inner feeling of dominance and worth will induce him to wish to predominate outside his specialty. The result is that even in this case, representing a maximum of qualification in man—specialisation—and therefore the thing most opposed to the mass-man, the result is that he will behave in almost all spheres of life as does the unqualified, the mass-man.

This is no mere wild statement. Anyone who wishes can observe the stupidity of thought, judgment, and action shown to-day in politics, art, religion, and the general problems of life and the world by the "men of science," and of course, behind them, the doctors, engineers, financiers, teachers, and so on. That state of "not listening," of not submitting to higher courts of appeal which I have repeatedly put forward as characteristic of the mass-man, reaches its height precisely in these partially qualified men. They symbolise, and to a great extent constitute, the actual dominion of the masses, and their barbarism is the most immediate cause of European demoralisation. Furthermore, they afford the clearest, most striking example of how the civilisation of the last century, *abandoned to its own devices*, has brought about this rebirth of primitivism and barbarism.

The most immediate result of this *unbalanced* specialisation has been that to-day, when there are more "scientists" than ever, there are much less "cultured" men than, for example, about 1750. And the worst is that with these turnspits of science not even the real progress of science itself is assured. For science needs from time to time, as a necessary regulator of its own advance, a labour of reconstitution, and, as I have said, this demands an effort towards unification, which grows more and more difficult, involving, as it does, ever-vaster regions of the world of knowledge. Newton was able to found his system of physics without knowing much philosophy, but Einstein needed to saturate himself with Kant and Mach before he could reach his own keen synthesis. Kant and Mach—the names are mere symbols of the enormous mass of philosophic and psychological thought which has influenced Einstein—have served to *liberate* the mind of the latter and leave the way open for his innovation. But Einstein is not sufficient. Physics is entering on the gravest crisis of its history, and can only be saved by a new "Encyclopaedia" more systematic than the first.

The specialisation, then, that has made possible the progress of experimental science during a century, is approaching a stage where it can no longer continue its advance unless a new generation undertakes to provide it with a more powerful form of turnspit.

But if the specialist is ignorant of the inner philosophy of the science he cultivates, he is much more radically ignorant of the historical conditions requisite for its continuation; that is to say: how society and the heart of man are to be organised in order that there may continue to be investigators. The decrease in scientific vocations noted in recent years, to which I have alluded, is an anxious symptom for anyone who has a clear idea of what civilisation is, an idea generally lacking to the typical "scientist," the highpoint of our present civilisation. He also believes that civilisation *is there* in just the same way as the earth's crust and the forest primeval.

THOMAS HENRY HUXLEY

THE MOST FAMOUS of debates over a theory of modern science took place in 1860 when Bishop Wilberforce shared a platform with Thomas Henry Huxley (1825-1895). The Bishop concluded his windy attack on evolution by asking Huxley whether his descent from the ape was on his father's or his mother's side.

> The lean tall figure of Huxley quietly rose.
> He looked, for a moment, thoughtfully, at the crowd;
> Saw rows of hostile faces; caught the grin
> Of ignorant curiosity; here and there,
> A hopeful gleam of friendship; and, far back,
> The young, swift-footed, waiting for the fire.
> He fixed his eyes on these—then, in low tones,
> Clear, cool, incisive, "*I have come here,*" he said,
> "*In the cause of Science only.*"
>
> (Alfred Noyes, *The Book of Earth*)

Everyone knows the substance of Huxley's crushing reply. His own account, from a recently discovered letter, is as follows: "If then, said I, the question is put to me would I rather have a miserable ape for a grandfather or a man highly endowed by nature and possessing great means and influence and yet who employs those faculties and that influence for the mere purpose of introducing ridicule into a grave scientific discussion—I unhesitatingly affirm my preference for the ape. Whereupon there was unextinguishable laughter among the people, and they listened to the rest of my argument with the greatest attention."

It would be difficult to find two men who exemplified more

thoroughly than Wilberforce and Huxley the products of contrasting kinds of education. The bishop had received a classical training at Oxford, his knowledge of Greek and Latin surpassed only by his sublime ignorance of science. Huxley was without formal schooling, but through his own efforts had acquired a broad background in both science and letters. Looking back now on the two men, it is Huxley who seems to us the paragon of culture; Wilberforce who takes on the coloration of the Philistine.

In Huxley's day, it must be remembered, science was only beginning to be recognized as indispensable in general education. To this crusade Huxley devoted himself with tireless spirit. Although his contributions to biology and paleontology were immense, he liked to think of himself less as a man of research than as a popularizer. He was in fact the great "science writer" of his time, writing with a clarity and eloquence that exerted an enormous influence on his generation.

The following selection is an address delivered by Huxley in 1880 at the opening of Sir Josiah Mason's Science College, in Birmingham. It was his most controversial attack on the neglect of science, especially social science, in the education of his contemporaries. The Arnold to whom he refers is of course Matthew Arnold, whose educational views Huxley found so contrary to his own.

THOMAS HENRY HUXLEY

Science and Culture

SIX YEARS AGO, as some of my present hearers may remember, I had the privilege of addressing a large assemblage of the inhabitants of this city, who had gathered together to do honour to the memory of their famous townsman, Joseph Priestley; and, if any satisfaction attaches to posthumous glory, we may hope that the manes of the burnt-out philosopher were then finally appeased.

No man, however, who is endowed with a fair share of common sense, and not more than a fair share of vanity, will identify either contemporary or posthumous fame with the highest good; and Priestley's life leaves no doubt that he, at any rate, set a much higher value upon the advancement of knowledge, and the promotion of that freedom of thought which is at once the cause and the consequence of intellectual progress.

Hence I am disposed to think that, if Priestley could be amongst us to-day, the occasion of our meeting would afford him even greater pleasure than the proceedings which celebrated the centenary of his chief discovery. The kindly heart would be moved, the high sense of social duty would be satisfied, by the spectacle of well-earned wealth, neither squandered in tawdry luxury and vainglorious show, nor scattered with the careless charity which blesses neither him that gives nor him that takes, but expended in the execution of a well-considered plan for the aid of present and future generations of those who are willing to help themselves.

We shall all be of one mind thus far. But it is needful to share Priestley's keen interest in physical science; and to have learned, as he had learned, the value of scientific training in fields of inquiry apparently far remote from physical science; in order to appreciate, as he would have appreciated, the value of the noble gift which Sir Josiah Mason has bestowed upon the inhabitants of the Midland district.

For us children of the nineteenth century, however, the establishment of a college under the conditions of Sir Josiah Mason's Trust, has a significance apart from any which it could have possessed a hundred years ago. It appears to be an indication that we are reaching the crisis of the battle, or rather of the long series of battles, which have been fought over education in a campaign which began long before Priestley's time, and will probably not be finished just yet.

In the last century, the combatants were the champions of ancient literature on the one side, and those of modern literature on the other; but, some thirty years ago, the contest became complicated by the appearance of a third army, ranged round the banner of Physical Science.

I am not aware that any one has authority to speak in the name of this new host. For it must be admitted to be somewhat of a guerilla force, composed largely of irregulars, each of whom fights pretty much for his own hand. But the impressions of a full private, who has seen a good deal of service in the ranks, respecting the present position of affairs and the conditions of a permanent peace, may not be devoid of interest; and I do not know that I could make a better use of the present opportunity than by laying them before you.

From the time that the first suggestion to introduce physical science into ordinary education was timidly whispered, until now, the advocates of scientific education have met with opposition of two kinds. On the one hand, they have been pooh-poohed by the men of business who pride themselves on being the representatives of practicality; while, on the other hand, they have been excommunicated by the classical scholars, in their capacity of Levites in charge of the ark of culture and monopolists of liberal education.

The practical men believed that the idol whom they worship—rule of thumb—has been the source of the past prosperity, and will suffice for the future welfare of the arts and manufactures. They were of opinion that science is speculative rubbish; that theory and practice have nothing to do with one another; and that the scientific habit of mind is an impediment, rather than an aid, in the conduct of ordinary affairs.

I have used the past tense in speaking of the practical men—for although they were very formidable thirty years ago, I am not sure that the pure species has not been extirpated. In fact, so far as mere argument goes, they have been subjected to such a *feu d'enfer* that it is a miracle if any have escaped. But I have remarked that your typical practical man has an unexpected resemblance to one of Milton's angels. His spiritual wounds, such as are inflicted by logical weapons, may be as deep as a well and as wide as a church door, but beyond shedding a few drops of ichor, celestial or otherwise, he is no whit the worse. So, if any of these opponents be left, I will not waste time in vain repetition of the demonstrative evidence of the practical value of science; but knowing that a parable will sometimes penetrate where syllogisms fail to effect an entrance, I will offer a story for their consideration.

Once upon a time, a boy, with nothing to depend upon but his own vigorous nature, was thrown into the thick of the struggle for existence in the midst of a great manufacturing population. He seems to have had a hard fight, inasmuch as, by the time he was thirty years of age, his total disposable funds amounted to twenty pounds. Nevertheless, middle life found him giving proof of his comprehension of the practical problems he had been roughly called upon to solve, by a career of remarkable prosperity.

Finally, having reached old age with its well-earned surroundings of "honour, troops of friends," the hero of my story bethought himself of those who were making a like start in life, and how he could stretch out a helping hand to them.

After long and anxious reflection this successful practical man of business could devise nothing better than to provide them with the means of obtaining "sound, extensive, and practical scientific knowledge." And he devoted a large part of his wealth and five years of incessant work to this end.

I need not point the moral of a tale which, as the solid and spacious fabric of the Scientific College assures us, is no fable, nor can anything which I could say intensify the force of this practical answer to practical objections.

We may take it for granted then, that, in the opinion of those best qualified to judge, the diffusion of thorough scientific education is an absolutely essential condition of industrial progress; and that the College which has been opened to-day will confer an inestimable boon upon those whose livelihood is to be gained by the practise of the arts and manufactures of the district.

The only question worth discussion is, whether the conditions, under which the work of the College is to be carried out, are such as to give it the best possible chance of achieving permanent success.

Sir Josiah Mason, without doubt most wisely, has left very large freedom of action to the trustees, to whom he proposes ultimately to commit the administration of the College, so that they may be able to adjust its arrangements in accordance with the changing conditions of the future. But, with respect to three points, he has laid most explicit injunctions upon both administrators and teachers.

Party politics are forbidden to enter into the minds of either, so far as the work of the College is concerned; theology is as sternly banished from its precincts; and finally, it is especially declared that the College shall make no provision for "mere literary instruction and education."

It does not concern me at present to dwell upon the first two injunctions any longer than may be needful to express my full conviction of their wisdom. But the third prohibition brings us face to face with those other opponents of scientific education, who are by no means in the moribund condition of the practical man, but alive, alert, and formidable.

It is not impossible that we shall hear this express exclusion of "literary instruction and education" from a College which, nevertheless, professes to give a high and efficient education, sharply criticised. Certainly the time was that the Levites of culture would have sounded their trumpets against its walls as against an educational Jericho.

How often have we not been told that the study of physical science is incompetent to confer culture; that it touches none of the higher problems of life; and, what is worse, that the continual devotion to scientific studies tends to generate a narrow and bigoted belief in the applicability of scientific methods to the search after truth of all kinds? How frequently one has reason to observe that no reply to a troublesome argument tells so well as calling its author a "mere scientific specialist." And, as I am afraid it is not permissible to speak of this form of opposition to scientific education in the past tense; may we not expect to be told that this, not only omission, but prohibition, of "mere literary instruction and education" is a patent example of scientific narrow-mindedness?

I am not acquainted with Sir Josiah Mason's reasons for the action which he has taken; but if, as I apprehend is the case, he refers to the ordinary classical course of our schools and universities by the name of "mere literary instruction and education," I venture to offer sundry reasons of my own in support of that action.

For I hold very strongly by two convictions— The first is, that neither the discipline nor the subject-matter of classical education is of such direct value to the student of physical science as to justify the expenditure of valuable time upon either; and the second is, that for the purpose of attaining real culture, an exclusively scientific education is at least as effectual as an exclusively literary education.

I need hardly point out to you that these opinions, especially the latter, are diametrically opposed to those of the great majority of educated Englishmen, influenced as they are by school and university traditions. In their belief, culture is obtainable only by a liberal education; and a liberal education is synonymous, not merely with education and instruction in literature, but in one particular form of literature, namely, that of Greek and Roman antiquity. They hold that the man who has learned Latin and Greek, however little, is educated; while he who is versed in other branches of knowledge, however deeply, is a more or less respectable specialist, not admissible into the cultured caste. The stamp of the educated man, the University degree, is not for him.

I am too well acquainted with the generous catholicity of spirit, the true sympathy with scientific thought, which pervades the writings of our chief apostle of culture to identify him with these opinions; and yet one may cull from one and another of those epistles to the Philistines, which so much delight all who do not answer to that name, sentences which lend them some support.

Mr. Arnold tells us that the meaning of culture is "to know the best that has been thought and said in the world." It is the criticism of life contained in literature. That criticism regards "Europe as being, for intellectual and spiritual purposes, one great confederation, bound to a joint action and working to a common result; and whose members have, for their common outfit, a knowledge of Greek, Roman, and Eastern antiquity, and of one another. Special, local, and temporary advantages being put out of account, that modern nation will in the intellectual and spiritual sphere make most progress, which most thoroughly carries out this programme. And what is that but saying that we too, all of us, as individuals, the more thoroughly we carry it out, shall make the more progress?"

We have here to deal with two distinct propositions. The first, that a criticism of life is the essence of culture; the second, that literature contains the materials which suffice for the construction of such a criticism.

I think that we must all assent to the first proposition. For culture certainly means something quite different from learning or technical skill. It implies the possession of an ideal, and the habit of critically estimating the value of things by comparison with a theoretic standard. Perfect culture should supply a complete theory of life, based upon a clear knowledge alike of its possibilities and of its limitations.

But we may agree to all this, and yet strongly dissent from the assumption that literature alone is competent to supply this knowledge. After having learnt all that Greek, Roman, and Eastern antiquity have thought and said, and all that modern literatures have to tell us, it is not self-evident that we have laid a sufficiently broad and deep foundation for that criticism of life, which constitutes culture.

Indeed, to any one acquainted with the scope of physical science, it is not at all evident. Considering progress only in the "intellectual and spiritual sphere," I find myself wholly unable to admit that either nations or individuals will really advance, if their common outfit draws nothing from the stores of physical science. I should say that an army, without weapons of precision and with no particular base of operations, might more hopefully enter upon a campaign on the Rhine, than a man, devoid of a knowledge of what physical science has done in the last century, upon a criticism of life.

When a biologist meets with an anomaly, he instinctively turns to the study of development to clear it up. The rationale of contradictory opinions may with equal confidence be sought in history.

It is, happily, no new thing that Englishmen should employ their wealth in building and endowing institutions for educational purposes. But, five or six hundred years ago, deeds of foundation expressed or implied conditions as nearly as possible contrary to those which have been thought expedient by Sir Josiah Mason. That is to say, physical science was practically ignored, while a certain literary training was enjoined as a means to the acquirement of knowledge which was essentially theological.

The reason of this singular contradiction between the actions of men alike animated by a strong and disinterested desire to promote the welfare of their fellows, is easily discovered.

At that time, in fact, if any one desired knowledge beyond such as could be obtained by his own observation, or by common conversation, his first necessity was to learn the Latin language, inasmuch as all the higher knowledge of the western world was contained in works written in that language. Hence, Latin grammar, with logic and rhetoric, studied through Latin, were the fundamentals of education. With respect to the substance of the knowledge imparted through this channel, the Jewish and Christian Scriptures, as interpreted and supplemented by the Romish Church, were held to contain a complete and infallibly true body of information.

Theological dicta were, to the thinkers of those days, that which the axioms and definitions of Euclid are to the geometers of these. The business of the philosophers of the middle ages was to deduce from the data furnished by the theologians, conclusions in accordance with ecclesiastical decrees. They were allowed the high privilege of showing, by logical process, how and why that which the Church said was true, must be true. And if their demonstrations fell short of or exceeded this limit, the Church was maternally ready to check their aberrations; if need were by the help of the secular arm.

Between the two, our ancestors were furnished with a compact and complete criticism of life. They were told how the world began and how it would end; they learned that all material existence was but a base and insignificant blot upon the fair face of the spiritual world, and that nature was, to all intents and purposes, the play-ground of the devil; they learned that the earth is the centre of the visible universe, and that man is the cynosure of things terrestrial; and more especially was it inculcated that the course of nature had no fixed order, but that it could be, and constantly was, altered by the agency of innumerable spiritual beings, good and bad, according as they were moved by the deeds and prayers of men. The sum and substance of the whole doctrine was to produce the conviction that the only thing really worth knowing in this world was how to secure that place in a better which, under certain conditions, the Church promised.

Our ancestors had a living belief in this theory of life, and acted upon it in their dealings with education, as in all other matters. Culture meant saintliness—after the fashion of the saints of those days; the education that led to it was, of necessity, theological; and the way to theology lay through Latin.

That the study of nature—further than was requisite for the satisfaction of everyday wants—should have any bearing on human life was far from the thoughts of men thus trained. Indeed, as nature had been cursed for man's sake, it was an obvious conclusion that those who meddled with nature were likely to come into pretty close contact with Satan. And, if any born scientific investigator followed his instincts, he might safely reckon upon earning the reputation, and probably upon suffering the fate, of a sorcerer.

Had the western world been left to itself in Chinese isolation, there is no saying how long this state of things might have endured. But, happily, it was not left to itself. Even earlier than the thirteenth century, the development of Moorish civilisation in Spain and the great movement of the Crusades had introduced the leaven which, from that day to this, has never ceased to work. At first, through the intermediation of Arabic translations, afterwards by the study of the originals, the western nations of Europe became acquainted with the writings of the ancient philosophers and poets, and, in time, with the whole of the vast literature of antiquity.

Whatever there was of high intellectual aspiration or dominant capacity in Italy, France, Germany, and England, spent itself for centuries in taking possession of the rich inheritance left by the dead civilisations of Greece and Rome. Marvellously aided by the invention of printing, classical learning spread and flourished. Those who possessed it prided themselves on having attained the highest culture then within the reach of mankind.

And justly. For, saving Dante on his solitary pinnacle, there was no figure in modern literature at the time of the Renascence to compare with the men of antiquity; there was no art to compete with their sculpture; there was no physical science but that which Greece had created. Above all, there was no other example of perfect intellectual freedom—of the unhesitating acceptance of reason as the sole guide to truth and the supreme arbiter of conduct.

The new learning necessarily soon exerted a profound influence upon education. The language of the monks and schoolmen seemed little better than gibberish to scholars fresh from Virgil and Cicero, and the study of Latin was placed upon a new foundation. Moreover, Latin itself ceased to afford the sole key to knowledge. The student who sought the highest thought of antiquity found only a second-hand reflection of it in Roman literature, and turned his face to the full light of the Greeks. And after a battle, not altogether dissimilar to that which is at present being fought over the teaching of physical science, the study of Greek was recognised as an essential element of all higher education.

Thus the Humanists, as they were called, won the day; and the great reform which they effected was of incalculable service to mankind. But the Nemesis of all reformers is finality; and the reformers of education, like those of religion, fell into the profound, however common, error of mistaking the beginning for the end of the work of reformation.

The representatives of the Humanists, in the nineteenth century, take their stand upon classical education as the sole avenue to culture, as firmly as if we were still in the age of Renascence. Yet, surely, the present intellectual relations of the modern and the ancient worlds are profoundly different from those which obtained three centuries ago. Leaving aside the existence of a great and characteristically modern literature, of modern painting, and, especially, of modern music, there is one feature of the present state of the civilised world which separates it more widely from the Renascence, than the Renascence was separated from the middle ages.

This distinctive character of our own times lies in the vast and constantly increasing part which is played by natural knowledge. Not only is our daily life shaped by it, not only does the prosperity of millions of men depend upon it, but our whole theory of life has long been influenced, consciously or unconsciously, by the general conceptions of the universe, which have been forced upon us by physical science.

In fact, the most elementary acquaintance with the results of scientific investigation shows us that they offer a broad and striking contradiction to the opinion so implicitly credited and taught in the middle ages.

The notions of the beginning and the end of the world entertained by our forefathers are no longer credible. It is very certain that the earth is not the chief body in the material universe, and that the world is not subordinated to man's use. It is even more certain that nature is the expression of a definite order with which nothing interferes, and that the chief business of mankind is to learn that order and govern themselves accordingly. Moreover this scientific "criticism of life" presents itself to us with different credentials from any other. It appeals not to authority, nor to what anybody may have thought or said, but to nature. It admits that all our interpretations of natural fact are more or less imperfect and symbolic, and bids the learner seek for truth not among words but among things. It warns us that the assertion which outstrips evidence is not only a blunder but a crime.

The purely classical education advocated by the representatives of the Humanists in our day, gives no inkling of all this. A man may be a better scholar than Erasmus, and know no more of the chief causes of the present intellectual fermentation than Erasmus did. Scholarly and pious persons, worthy of all respect, favour us with allocutions upon the sadness of the antagonism of science to their mediæval way of thinking, which betray an ignorance of the first principles of scientific investigation, an incapacity for understanding what a man of science means by veracity, and an unconsciousness of the weight of established scientific truths, which is almost comical.

There is no great force in the *tu quoque* argument, or else the advocates of scientific education might fairly enough retort upon the modern Humanists that they may be learned specialists, but that they possess no such sound foundation for a criticism of life as deserves the name of culture. And, indeed, if we were disposed to be cruel, we might urge that the Humanists have brought this reproach upon themselves, not because they are too full of the spirit of the ancient Greek, but because they lack it.

The period of the Renascence is commonly called that of the "Revival of Letters," as if the influences then brought to bear upon the mind of Western Europe had been wholly exhausted in the field of literature. I think it is very commonly forgotten that the revival of science, effected by the same agency, although less conspicuous, was not less momentous.

In fact, the few and scattered students of nature of that day picked up the clue to her secrets exactly as it fell from the hands of the Greeks a thousand years before. The foundations of mathematics were so well laid by them, that our children learn their geometry from a book written for the schools of Alexandria two thousand years ago. Modern astronomy is the natural continuation and development of the work of Hipparchus and of Ptolemy; modern physics of that of Democritus and of Archimedes; it was long before modern biological science outgrew the knowledge bequeathed to us by Aristotle, by Theophrastus, and by Galen.

We cannot know all the best thoughts and sayings of the Greeks unless we know what they thought about natural phænomena. We cannot fully apprehend their criticism of life unless we understand the extent to which that criticism was affected by scientific conceptions. We falsely pretend to be the inheritors of their culture, unless we are penetrated, as the best minds among them were, with an unhesitating faith that the free employment of reason, in accordance with scientific method, is the sole method of reaching truth.

Thus I venture to think that the pretensions of our modern Humanists to the possession of the monopoly of culture and to the exclusive inheritance of the spirit of antiquity must be abated, if not abandoned. But I should be very sorry that anything I have said should be taken to imply a desire on my part to depreciate the value of classical education, as it might be and as it sometimes is. The native capacities of mankind vary no less than their opportunities; and while culture is one, the road by which one man may best reach it is widely different from that which is most advantageous to another. Again, while scientific education is yet inchoate and tentative, classical education is thoroughly well organised upon the practical experience of generations of teachers. So that, given ample time for learning and destination for ordinary life, or for a literary career, I do not think that a young Englishman in search of culture can do better than follow the course usually marked out for him, supplementing its deficiencies by his own efforts.

But for those who mean to make science their serious occupation; or who intend to follow the profession of medicine; or who have to enter early upon the business of life; for all these, in my opinion, classical education is a mistake; and it is for this reason that I am glad to see "mere literary education and instruction" shut out from the curriculum of Sir Josiah Mason's College, seeing that its inclusion would probably lead to the introduction of the ordinary smattering of Latin and Greek.

Nevertheless, I am the last person to question the importance of genuine literary education, or to suppose that intellectual culture can be complete without it. An exclusively scientific training will bring about a mental twist as surely as an exclusively literary training. The value of the cargo does not compensate for a ship's being out of trim; and I should be very sorry to think that the Scientific College would turn out none but lop-sided men.

There is no need, however, that such a catastrophe should happen. Instruction in English, French, and German is provided, and thus the three greatest literatures of the modern world are made accessible to the student.

French and German, and especially the latter language, are absolutely indispensable to those who desire full knowledge in any department of science. But even supposing that the knowledge of these languages acquired is not more than sufficient for purely scientific purposes, every Englishman has, in his native tongue, an almost perfect instrument of literary expression; and, in his own literature, models of every kind of literary excellence. If an Englishman cannot get literary culture out of his Bible, his Shakespeare, his Milton, neither, in my belief, will the profoundest study of Homer and Sophocles, Virgil and Horace, give it to him.

Thus, since the constitution of the College makes sufficient provision for literary as well as for scientific education, and since artistic instruction is also contemplated, it seems to me that a fairly complete culture is offered to all who are willing to take advantage of it.

But I am not sure that at this point the "practical" man, scotched but not slain, may ask what all this talk about culture has to do with an Institution, the object of which is defined to be "to promote the prosperity of the manufactures and the industry of the country." He may suggest that what is wanted for this end is not culture, nor even a purely scientific discipline, but simply a knowledge of applied science.

I often wish that this phrase, "applied science," had never been invented. For it suggests that there is a sort of scientific knowledge of direct practical use, which can be studied apart from another sort of scientific knowledge, which is of no practical utility, and which is termed "pure science." But there is no more complete fallacy than this. What people call applied science is nothing but the application of pure science to particular classes of problems. It consists of deductions from those general principles, established by reasoning and observation, which constitute pure science. No one can safely make these deductions until he has a firm grasp of the principles; and he can obtain that grasp only by personal experience of the operations of observation and of reasoning on which they are founded.

Almost all the processes employed in the arts and manufactures fall within the range either of physics or of chemistry. In order to improve them, one must thoroughly understand them; and no one has a chance of really understanding them, unless he has obtained that mastery of principles and that habit of dealing with facts, which is given by long-continued and well-directed purely scientific training in the physical and the chemical laboratory. So that there really is no question as to the necessity of purely scientific discipline, even if the work of the College were limited by the narrowest interpretation of its stated aims.

And, as to the desirableness of a wider culture than that yielded by science alone, it is to be recollected that the improvement of manufacturing processes is only one of the conditions which contribute to the prosperity of industry. Industry is a means and not an end; and mankind work only to get something which they want. What that something is depends partly on their innate, and partly on their acquired, desires.

If the wealth resulting from prosperous industry is to be spent upon the gratification of unworthy desires, if the increasing perfection of manufacturing processes is to be accompanied by an increasing debasement of those who carry them on, I do not see the good of industry and prosperity.

Now it is perfectly true that men's views of what is desirable depend upon their characters; and that the innate proclivities to which we give that name are not touched by any amount of instruction. But it does not follow that even mere intellectual education may not, to an indefinite extent, modify the practical manifestation of the characters of men in their actions, by supplying them with motives unknown to the ignorant. A pleasure-loving character will have pleasure of some sort; but, if you give him the choice, he may prefer pleasures which do not degrade him to those which do. And this choice is offered to every man, who possesses in literary or artistic culture a never-failing source of pleasures, which are neither withered by age, nor staled by custom, nor embittered in the recollection by the pangs of self-reproach.

If the Institution opened to-day fufils the intention of its founder, the picked intelligences among all classes of the population of this district will pass through it. No child born in Birmingham, henceforward, if he have the capacity to profit by the opportunities offered to him, first in the primary and other schools, and afterwards in the Scientific College, need fail to obtain, not merely the instruction, but the culture most appropriate to the conditions of his life.

Within these walls, the future employer and the future artisan may sojourn together for a while, and carry, through all their lives, the stamp of the influences then brought to bear upon them. Hence, it is not beside the mark to remind you, that the prosperity of industry depends not merely upon the improvement of manufacturing processes, not merely upon the ennobling of the individual character, but upon a third condition, namely, a clear understanding of the conditions of social life, on the part of both the capitalist and the operative, and their agreement upon common principles of social action. They must learn that social phænomena are as much the expression of natural laws as any others; that no social arrangements can be permanent unless they harmonise with the requirements of social statics and dynamics; and that, in the nature of things, there is an arbiter whose decisions execute themselves.

But this knowledge is only to be obtained by the application of the methods of investigation adopted in physical researches to the investigation of the phænomena of society. Hence, I confess, I should like to see one addition made to the excellent scheme of education propounded for the College, in the shape of provision for the teaching of Sociology. For though we are all agreed that party politics are to have no place in the instruction of the College; yet in this country, practically governed as it is now by universal suffrage, every man who does his duty must exercise political functions. And, if the evils which are inseparable from the good of political liberty are to be checked, if the perpetual oscillation of nations between anarchy and despotism is to be replaced by the steady march of self-restraining freedom; it will be because men will gradually bring themselves to deal with political, as they now deal with scientific questions; to be as ashamed of undue haste and partisan prejudice in the one case as in the other; and to believe that the machinery of society is at least as delicate as that of a spinning-jenny, and as little likely to be improved by the meddling of those who have not taken the trouble to master the principles of its action.

JOHN BURROUGHS

FROM MATTHEW ARNOLD'S classic essay "Literature and Science" to the angry books of Robert Hutchins there have been many high-sounding rebuttals to the point of view taken by Huxley in the previous essay. Too often these replies have come from men of exclusively literary background. The reader senses a concealed dislike of science or a vested metaphysical interest that makes it hard for him to lend a sympathetic ear. The essay to follow is happily free of these defects. It may lack the Greek quotations of an Arnold and the impudent humor of a Hutchins but it makes its points nonetheless, in the relaxed, gentle style so characteristic of the author.

John Burroughs (1837-1921) is little read today, yet as a nature writer he is unsurpassed in American letters. True, he made no original contributions to natural history. But he was a keen observer, more trustworthy for example than Thoreau who wrote in *Walden* that he once "stood in the very abutment of a rainbow's arch." "Why did he not dig for the pot of gold?" Burroughs asks, scolding Thoreau for this and other scientific howlers.

Like Thoreau, Burroughs loved the simple life, close to the elements, in which he could see the fire that warmed him, or better still, he once explained, a life in which he could cut the wood that fed the fire that warmed him. Unlike Thoreau he did not prefer the company of woodchucks to the company of men. Perhaps it was this reverence for all life, including human life, that explains Burroughs' passionate admiration for the poetry of his good friend Walt Whitman (his first book was an appreciation of Whitman) and the warmth of spirit that one feels in all his writing.

JOHN BURROUGHS

Science and Literature

INTERESTED as I am in all branches of natural science, and great as is my debt to these things, yet I suppose my interest in nature is not strictly a scientific one. I seldom, for instance, go into a natural history museum without feeling as if I were attending a funeral. There lie the birds and animals stark and stiff, or else, what is worse, stand up in ghastly mockery of life, and the people pass along and gaze at them through the glass with the same cold and unprofitable curiosity that they gaze upon the face of their dead neighbor in his coffin. The fish in the water, the bird in the tree, the animal in the fields or woods, what a different impression they make upon us!

To the great body of mankind, the view of nature presented through the natural sciences has a good deal of this lifeless funereal character of the specimens in the museum. It is dead dissected nature, a cabinet of curiosities carefully labeled and classified. "Every creature sundered from its natural surroundings," says Goethe, "and brought into strange company, makes an unpleasant impression on us, which disappears only by habit." Why is it that the hunter, the trapper, the traveler, the farmer, or even the schoolboy, can often tell us more of what we want to know about the bird, the flower, the animal, than the professor in all the pride of his nomenclature? Why, but that these give us a glimpse of the live creature as it stands related to other things, to the whole life of nature, and to the human heart, while the latter shows it to us as it stands related to some artificial system of human knowledge.

"The world is too much with us," said Wordsworth, and he intimated that our science and our civilization had put us "out of tune" with nature.

Great God! I'd rather be
A Pagan suckled in a creed outworn;
So might I, standing on this pleasant lea,
Have glimpses that would make me less forlorn,
Have sight of Proteus rising from the sea,
Or hear old Triton blow his wreathed horn.

To the scientific mind such language is simply nonsense, as are those other lines of the bard of Grasmere, in which he makes his poet—

Contented if he might enjoy
The things which others understand.

Enjoyment is less an end in science than it is in literature. A poem or other work of the imagination that failed to give us the joy of the spirit would be of little value, but from a work of science we expect only the satisfaction which comes with increased stores of exact knowledge.

Yet it may be questioned if the distrust with which science and literature seem to be more and more regarding each other in our day is well founded. That such distrust exists is very evident. Professor Huxley taunts the poets with "sensual caterwauling," and the poets taunt the professor and his ilk with gross materialism.

Science is said to be democratic, its aims and methods in keeping with the great modern movement; while literature is alleged to be aristocratic in its spirit and tendencies. Literature is for the few; science is for the many. Hence their opposition in this respect.

Science is founding schools and colleges from which the study of literature, as such, is to be excluded; and it is becoming clamorous for the positions occupied by the classics in the curriculum of the older institutions. As a reaction against the extreme partiality for classical studies, the study of names instead of things, which has so long been shown in our educational system, this new cry is wholesome and good; but so far as it implies that science is capable of taking the place of the great literatures as an instrument of high culture, it is mischievous and misleading.

About the intrinsic value of science, its value as a factor in our civilization, there can be but one opinion; but about its value to the scholar, the thinker, the man of letters, there is room for very divergent views. It is certainly true that the great ages of the world have not been ages of exact science; nor have the great literatures, in which so much of the power and vitality of the race have been stored, sprung from minds which held correct views of the physical universe. Indeed, if the growth and maturity of man's moral and intellectual stature were a question of material appliances or conveniences, or of accumulated stores of exact knowledge, the world of to-day ought to be able to show more eminent achievements in all fields of human activity than ever before. But this it cannot do. Shakespeare wrote his plays for people who believed in witches, and probably believed in them himself; Dante's immortal poem could never have been produced in a scientific age. Is it likely that the Hebrew Scriptures would have been any more precious to the race, or their influence any deeper, had they been inspired by correct views of physical science?

It is not my purpose to write a diatribe against the physical sciences. I would as soon think of abusing the dictionary. But as the dictionary can hardly be said to be an end in itself, so I would indicate that the final value of physical science is its capability to foster in us noble ideals, and to lead us to new and larger views of moral and spiritual truths. The extent to which it is able to do this measures its value to the spirit,—measures its value to the educator.

That the great sciences can do this, that they are capable of becoming instruments of pure culture, instruments to refine and spiritualize the whole moral and intellectual nature, is no doubt true; but that they can ever usurp the place of the humanities or general literature in this respect is one of those mistaken notions which seem to be gaining ground so fast in our time.

Can there be any doubt that contact with a great character, a great soul, through literature, immensely surpasses in educational value, in moral and spiritual stimulus, contact with any of the forms or laws of physical nature through science? Is there not something in the study of the great literatures of the world that opens the mind, inspires it with noble sentiments and ideals, cultivates and develops the intuitions, and reaches and stamps the character, to an extent that is hopelessly beyond the reach of science? They add something to the mind that is like leaf-mould to the soil, like the contribution from animal and vegetable life and from the rains and the dews. Until science is mixed with emotion, and appeals to the heart and imagination, it is like dead inorganic matter; and when it becomes so mixed and so transformed it is literature.

The college of the future will doubtless lay much less stress upon the study of the ancient languages; but the time thus gained will not be devoted to the study of the minutiæ of physical science, as contemplated by Mr. Herbert Spencer, but to the study of man himself, his deeds and his thoughts, as illustrated in history and embodied in the great literatures.

"Microscopes and telescopes, properly considered," says Goethe, "put our human eyes out of their natural, healthy, and profitable point of view." By which remark he probably meant that artificial knowledge obtained by the aid of instruments, and therefore by a kind of violence and inquisition, a kind of dissecting and dislocating process, is less innocent, is less sweet and wholesome, than natural knowledge, the fruits of our natural faculties and perceptions. And the reason is that physical science pursued in and for itself results more and more in barren analysis, becomes more and more separated from human and living currents and forces,—in fact, becomes more and more mechanical, and rests in a mechanical conception of the universe. And the universe, considered as a machine, however scientific it may be, has neither value to the spirit nor charm to the imagination.

The man of to-day is fortunate if he can attain as fresh and lively a conception of things as did Plutarch and Virgil. How alive the ancient observers made the world! They conceived of everything as living, being,—the primordial atoms, space, form, the earth, the sky. The stars and planets they thought of as requiring nutriment, and as breathing or exhaling. To them, fire did not consume things, but fed or preyed upon them, like an animal. It was not so much false science, as a livelier kind of science, which made them regard the peculiar quality of anything as a spirit. Thus there was a spirit in snow; when the snow melted the spirit escaped. This spirit, says Plutarch, "is nothing but the sharp point and finest scale of the congealed substance, endued with a virtue of cutting and dividing not only the flesh, but also silver and brazen vessels." "Therefore this piercing spirit, like a flame" (how much, in fact, frost is like flame!) "seizing upon those that travel in the snow, seems to burn their outsides, and like fire to enter and penetrate the flesh." There is a spirit of salt, too, and of heat, and of trees. The sharp, acrimonious quality of the fig-tree bespeaks of a fierce and strong spirit which it darts out into objects.

To the ancient philosophers, the eye was not a mere passive instrument, but sent forth a spirit, or fiery visual rays, that went to coöperate with the rays from outward objects. Hence the power of the eye, and its potency in love matters. "The mutual looks of nature's beauties, or that which comes from the eye, whether light or a stream of spirits, melt and dissolve the lovers with a pleasing pain, which they call the bitter-sweet of love." "There is such a communication, such a flame raised by one glance, that those must be altogether unacquainted with love that wonder at the Median naphtha that takes fire at a distance from the flame." "Water from the heavens," says Plutarch, "is light and aerial, and, being mixed with spirit, is the quicker passed and elevated into the plants by reason of its tenuity." Rain-water, he further says, "is bred in the air and wind, and falls pure and sincere." Science could hardly give an explanation as pleasing to the fancy as that. And it is true enough, too. Mixed with spirit, or the gases of the air, and falling pure and sincere, is undoubtedly the main secret of the matter. He said the ancients hesitated to put out a fire because of the relation it had to the sacred and eternal flame. "Nothing," he says, "bears such a resemblance to an animal as fire. It is moved and nourished by itself, and by its brightness, like the soul, discovers and makes everything apparent; but in its quenching it principally shows some power that seems to proceed from our vital principle, for it makes a noise and resists like an animal dying or violently slaughtered."

The feeling, too, with which the old philosophers looked upon the starry heavens is less antagonistic to science than it is welcome and suggestive to the human heart. Says Plutarch in his "Sentiments of Nature Philosophers delighted in:" "To men, the heavenly bodies that are so visible did give the knowledge of the Deity; when they contemplated that they are the causes of so great an harmony, that they regulate day and night, winter and summer, by their rising and setting, and likewise considered these things which by their influence in the earth do receive a being and do likewise fructify. It was manifest to men that the Heaven was the father of those things, and the Earth the mother: that the Heaven was the father is clear, since from the heavens there is the pouring down of waters, which have their spermatic faculty; the Earth the mother because she receives them and brings forth. Likewise men, considering that the stars are running in a perpetual motion, and that the sun and moon give us the power to view and contemplate, they call them all Gods."

The ancients had that kind of knowledge which the heart gathers; we have in superabundance that kind of knowledge which the head gathers. If much of theirs was made up of mere childish delusions, how much of ours is made up of hard, barren, and unprofitable details,—a mere desert of sand where no green thing grows or can grow! How much there is in books that one does not want to know, that it would be a mere weariness and burden to the spirit to know; how much of modern physical science is a mere rattling of dead bones, a mere threshing of empty straw! Probably we shall come round to as lively a conception of things by and by. Darwin has brought us a long way toward it. At any rate, the ignorance of the old writers is often more captivating than our exact but more barren knowledge.

The old books are full of this dew-scented knowledge,—knowledge gathered at first hand in the morning of the world. In our more exact scientific knowledge this pristine quality is generally missing; and hence it is that the results of science are far less available for literature than the results of experience.

Science is probably unfavorable to the growth of literature because it does not throw man back upon himself and concentrate him as the old belief did; it takes him away from himself, away from human relations and emotions, and leads him on and on. We wonder and marvel more, but we fear, dread, love, sympathize less. Unless, indeed, we finally come to see, as we probably shall, that after science has done its best the mystery is as great as ever, and the imagination and the emotions have just as free a field as before.

Science and literature in their aims and methods have but little in common. Demonstrable fact is the province of the one; sentiment is the province of the other. "The more a book brings sentiment into light," says M. Taine, "the more it is a work of literature;" and, we may add, the more it brings the facts and laws of natural things to light, the more it is a work of science. Or, as Emerson says in one of his early essays, "literature affords a platform whence we may command a view of our present life, a purchase by which we may move it." In like manner science affords a platform whence we may view our physical existence,—a purchase by which we may move the material world. The value of the one is in its ideality, that of the other in its exact demonstrations. The knowledge which literature most loves and treasures is knowledge of life; while science is intent upon a knowledge of things, not as they are in their relation to the mind and heart of man, but as they are in and of themselves, in their relations to each other and to the human body. Science is a capital or fund perpetually reinvested; it accumulates, rolls up, is carried forward by every new man. Every man of science has all the science before him to go upon, to set himself up in business with. What an enormous sum Darwin availed himself of and reinvested! Not so in literature; to every poet, to every artist, it is still the first day of creation, so far as the essentials of his task are concerned. Literature is not so much a fund to be reinvested as it is a crop to be ever new-grown. Wherein science furthers the eye, sharpens the ear, lengthens the arm, quickens the foot, or extends man farther into nature in the natural bent and direction of his faculties and powers, a service is undoubtedly rendered to literature. But so far as it engenders a habit of peeping and prying into nature, and blinds us to the festive splendor and meaning of the whole, our verdict must be against it.

It cannot be said that literature has kept pace with civilization, though science has; in fact, it may be said without exaggeration that science *is* civilization—the application of the powers of nature to the arts of life. The reason why literature has not kept pace is because so much more than mere knowledge, well-demonstrated facts, goes to the making of it, while little else goes to the making of pure science. Indeed, the kingdom of heaven, in literature as in religion, "cometh not with observation." This felicity is within you as much in the one case as in the other. It is the fruit of the spirit, and not of the diligence of the hands.

Because this is so, because modern achievements in letters are not on a par with our material and scientific triumphs, there are those who predict for literature a permanent decay, and think the field it now occupies is to be entirely usurped by science. But this can never be. Literature will have its period of decadence and of partial eclipse; but the chief interest of mankind in nature or in the universe can never be for any length of time a merely scientific interest,—an interest measured by our exact knowledge of these things; though it must undoubtedly be an interest consistent with the scientific view. Think of having one's interest in a flower, a bird, the landscape, the starry skies, dependent upon the stimulus afforded by the textbooks, or dependent upon our knowledge of the structure, habits, functions, relations of these objects!

This other and larger interest in natural objects, to which I refer, is an interest as old as the race itself, and which all men, learned and unlearned alike, feel in some degree,—an interest born of our relations to these things, of our associations with them. It is the human sentiments they awaken and foster in us, the emotion of love or admiration, or awe or fear, they call up; and is in fact the interest of literature as distinguished from that of science. The admiration one feels for a flower, for a person, for a fine view, for a noble deed, the pleasure one takes in a spring morning, in a stroll upon the beach, is the admiration and the pleasure literature feels and art feels; only in them the feeling is freely opened and expanded which in most minds is usually vague and germinal. Science has its own pleasure in these things; but it is not, as a rule, a pleasure in which the mass of mankind can share, because it is not directly related to the human affections and emotions. In fact, the scientific treatment of nature can no more do away with or supersede the literary treatment of it—the view of it as seen through our sympathies and emotions, and touched by the ideal, such as the poet gives us—than the compound of the laboratory can take the place of the organic compounds found in our food, drink, and air.

If Audubon had not felt other than a scientific interest in the birds,—namely, a human interest, an interest born of sentiment,—would he have ever written their biographies as he did?

It is too true that the ornithologists of our day for the most part look upon the birds only as so much legitimate game for expert dissection and classification, and hence have added no new lineaments to Audubon's and Wilson's portraits. Such a man as Darwin was full of what we may call the sentiment of science. Darwin was always pursuing an idea, always tracking a living, active principle. He is full of the ideal interpretation of fact, science fired with faith and enthusiasm, the fascination of the power and mystery of nature. All his works have a human and almost poetic side. They are undoubtedly the best feeders of literature we have yet had from the field of science. His book on the earthworm, or on the formation of vegetable mould, reads like a fable in which some high and beautiful philosophy is clothed. How alive he makes the plants and the trees!—shows all their movements, their sleeping and waking, and almost their very dreams—does, indeed, disclose and establish a kind of rudimentary soul or intelligence in the tip of the radicle of plants. No poet has ever made the trees so human. Mark, for instance, his discovery of the value of cross-fertilization in the vegetable kingdom, and the means Nature takes to bring it about. Cross-fertilization is just as important in the intellectual kingdom as in the vegetable. The thoughts of the recluse finally become pale and feeble. Without pollen from other minds, how can one have a race of vigorous seedlings of his own? Thus all Darwinian books have to me a literary or poetic substratum. The old fable of metamorphosis and transformation he illustrates afresh in his "Origin of Species," in the "Descent of Man." Darwin's interest in nature is strongly scientific, but our interest in him is largely literary; he is tracking a principle, the principle of organic life, following it through all its windings and turnings and doublings and redoublings upon itself, in the air, in the earth, in the water, in the vegetable, and in all the branches of the animal world; the footsteps of creative energy; not why, but how; and we follow him as we would follow a great explorer, or general, or voyager like Columbus, charmed by his candor, dilated by his mastery. He is said to have lost his taste for poetry, and to have cared little for what is called religion. His sympathies were so large and comprehensive; the mere science in him is so perpetually overarched by that which is not science, but faith, insight, imagination, prophecy, inspiration,—"substance of things hoped for, the evidence of things not seen;" his love of truth so deep and abiding; and his determination to see things, facts, in their relations, and as they issue in principle, so unsleeping,—that both his poetic and religious emotions, as well as his scientific proclivities, found full scope, and this demonstration becomes almost a song.

It is easy to see how such a mind as Goethe's would have followed him and supplemented him, not from its wealth of scientific lore, but from its poetic insight into the methods of nature.

Again, it is the fine humanism of such a man as Humboldt that gives his name and his teachings currency. Men who have not this humanism, who do not in any way relate their science to life or to the needs of the spirit, but pile up mere technical, desiccated knowledge, are for the most part a waste or a weariness. Humboldt's humanism makes him a stimulus and a support to all students of nature. The noble character, the poetic soul, shines out in all his works and gives them a value above and beyond their scientific worth, great as that undoubtedly is. To his desire for universal knowledge he added the love of beautiful forms, and his "Cosmos" is an attempt at an artistic creation, a harmonious representation of the universe that should satisfy the æsthetic sense as well as the understanding. It is a graphic description of nature, not a mechanical one. Men of pure science look askant at it, or at Humboldt, on this account. A sage of Berlin says he failed to reach the utmost height of science because of his want of "physico-mathematical knowledge;" he was not sufficiently content with the mere dead corpse of nature to weigh and measure it. Lucky for him and for the world that there was something that had a stronger attraction for him than the algebraic formulas. Humboldt was not content till he had escaped from the trammels of mechanical science into the larger and more vital air of literature, or the literary treatment of nature. What keeps his "Views of Nature" and his "Scientific Travels" alive is not so much the pure science which they hold as the good literature which they embody. The observations he records upon that wonderful tropical nature, that are the fruit of his own unaided perceptions, betraying only the wiser hunter, trapper, walker, farmer, etc., how welcome it all is! But the moment he goes behind the beautiful or natural reason and discourses as a geologist, mineralogist, physical geographer, etc., how one's interest flags! It is all of interest and value to specialists in those fields, but it has no human and therefore no literary interest or value. When he tells us that "monkeys are more melancholy in proportion as they have more resemblance to man;" that "their sprightliness diminishes as their intellectual faculties appear to increase,"—we read with more attention than when he discourses as a learned naturalist upon the different species of monkeys. It is a real addition to our knowledge of nature to learn that the extreme heat and dryness of the summer, within the equatorial zone in South America, produces effects analogous to those produced by the cold of our northern winters. The trees lose their leaves, the snakes and crocodiles and other reptiles bury themselves in the mud, and many phases of life, both animal and vegetable, are wrapped in a long sleep. This is not strictly scientific knowledge; it is knowledge that lies upon the surface, and that any eye and mind may gather. One feels inclined to skip the elaborate account of the physical features of the lake of Valencia and its surroundings, but the old Mestizo Indian who gave the travelers goat's milk, and who, with his beautiful daughter, lived on a little island in its midst, awakens lively curiosity. He guarded his daughter as a miser guards his treasure. When some hunters by chance passed a night on his island, he suspected some designs upon his girl, and he obliged her to climb up a very lofty acacia-tree, which grew in the plain at some distance from the hut, while he stretched himself at the foot of the tree, and would not permit her to descend till the young men had departed. Thus, throughout the work, when the scientific interest is paramount, the literary and human interest fail, and *vice versa.*

No man of letters was ever more hospitable to science than Goethe; indeed, some of the leading ideas of modern science were distinctly foreshadowed by him; yet they took the form and texture of literature, or of sentiment, rather than of exact science. They were the reachings forth of his spirit; his grasping for the ideal clews to nature, rather than logical steps of his understanding; and his whole interest in physics was a search for a truth above physics,—

to get nearer, if possible, to this mystery called nature. "The understanding will not reach her," he said to Eckermann; "man must be capable of elevating himself to the highest reason to come in contact with this divinity, which manifests itself in the primitive phenomena, which dwells behind them, and from which they proceed." Of like purport is his remark that the common observations which science makes upon nature and its procedure, "in whatever terms expressed, are really after all only *symptoms* which, if any real wisdom is to result from our studies, must be traced back to the physiological and pathological principles of which they are the exponents."

Literature, I say, does not keep pace with civilization. That the world is better housed, better clothed, better fed, better transported, better equipped for war, better armed for peace, more skilled in agriculture, in navigation, in engineering, in surgery, has steam, electricity, gunpowder, dynamite,—all of this, it seems, is of little moment to literature. Are men better? Are men greater? Is life sweeter? These are the test questions. Time has been saved, almost annihilated, by steam and electricity, yet where is the leisure? The more time we save the less we have. The hurry of the machine passes into the man. We can outrun the wind and the storm, but we cannot outrun the demon of Hurry. The farther we go the harder he spurs us. What we save in time we make up in space; we must cover more surface. What we gain in power and facility is more than added in the length of the task. The needlewoman has her sewing-machine, but she must take ten thousand stitches now where she took only ten before, and it is probably true that the second condition is worse than the first. In the shoe factory, knife factory, shirt factory, and all other factories, men and women work harder, look grimmer, suffer more in mind and body, than under the old conditions of industry. The iron of the machine enters the soul; man becomes a mere tool, a cog or spoke or belt or spindle. More work is done, but in what does it all issue? Certainly not in beauty, in power, in character, in good manners, in finer men and women; but mostly in giving wealth and leisure to people who use them to publish their own unfitness for leisure and wealth.

It may be said that science has added to the health and longevity of the race; that the progress in surgery, in physiology, in pathology, in therapeutics, has greatly mitigated human suffering and prolonged life. This is unquestionably true; but in this service science is but paying back with one hand what it robbed us of with the other. With its appliances, its machinery, its luxuries, its immunities, and its interference with the law of natural selection, it has made the race more delicate and tender, and, if it did not arm them better against disease also, we should all soon perish. An old physician said that if he bled and physicked now, as in his early practice, his patients would all die. Are we stronger, more hardy, more virile than our ancestors? We are more comfortable and better schooled than our fathers, but who shall say we are wiser or happier? "Knowledge comes, but wisdom lingers," just as it always has and always will. The essential conditions of human life are always the same; the non-essential change with every man and hour.

Literature is more interested in some branches of science than in others; more interested in meteorology than in mineralogy; more interested in the superior sciences, like astronomy and geology, than in the inferior experimental sciences; has a warmer interest in Humboldt the traveler than in Humboldt the mineralogist; in Audubon and Wilson than in the experts and feather-splitters who have finished their task; in Watts, Morse, Franklin, than in the masters of theories and formulas; and has a greater stake in virtue, heroism, character, beauty, than in all the knowledge in the world. There is no literature without a certain subtle and vital blending of the real and the ideal.

Unless knowledge in some way issues in life, in character, in impulse, in motive, in love, in virtue, in some live human quality or attribute, it does not belong to literature. Man, and man alone, is of perennial interest to man. In nature we glean only the human traits,—only those things that in some way appeal to, or are interpretative of, the meaning or ideal within us. Unless the account of your excursion to field and forest, or to the bowels of the earth, or to the bottom of the sea, has some human interest, and in some measure falls in with the festival of life, literature will have none of it.

All persons are interested in the live bird and in the live animal, because they dimly read themselves there, or see their own lives rendered in new characters on another plane. Flowers, trees, rivers, lakes, mountains, rocks, clouds, the rain, the sea, are far more interesting to literature, because they are more or less directly related to our natural lives, and serve as vehicles for the expression of our natural emotions. That which is more directly related to what may be called our artificial life, our need for shelter, clothing, food, transportation—such as the factory, the mill, the forge, the railway, and the whole catalogue of useful arts,—is of less interest, and literature is shyer of it. And it may be observed that the more completely the thing is taken out of nature and artificialized, the less interest we take in it. Thus the sailing vessel is more pleasing to contemplate than the steamer; the old grist-mill, with its dripping water-wheel, than the steam-mill; the open fire than the stove or register. Tools and implements are not so interesting as weapons; nor the trades as the pursuit of hunting, fishing, surveying, exploring. A jackknife is not so interesting as an arrow-head, a rifle as a war-club, a watch as an hour-glass, a threshing machine as the flying flail. Commerce is less interesting to literature than war, because it is more artificial; nature does not have such full swing in it. The blacksmith interests us more than the gunsmith; we see more of nature at his forge. The farmer is dearer to literature than the merchant; the gardener than the agricultural chemist; the drover, the herder, the fisherman, the lumberman, the miner, are more interesting to her than the man of more elegant and artificial pursuits.

The reason of all this is clear to see. We are embosomed in nature; we are an apple on the bough, a babe at the breast. In nature, in God, we live and move and have our being. Our life depends upon the purity, the closeness, the vitality of the connection. We want and must have nature at first hand; water from the spring, milk from the udder, bread from the wheat, air from the open. Vitiate our supplies, weaken our connection, and we fail. All our instincts, appetites, functions, must be kept whole and normal; in fact, our reliance is wholly upon nature, and this bears fruit in the mind. In art, in literature, in life, we are drawn by that which seems nearest to, and most in accord with, her. Natural or untaught knowledge,—how much closer it touches us than professional knowledge! Keep me close to nature, is the constant demand of literature; open the windows and let in the air, the sun, let in health and strength; my blood must have oxygen, my lungs must be momentarily filled with the fresh unhoused element. I cannot breathe the cosmic ether of the abstruse inquirer, nor thrive on the gases of the scientist in his laboratory; the air of hill and field alone suffices.

The life of the hut is of more interest to literature than the life of the palace, except so far as the same Nature has her way in both. Get rid of the artificial, the complex, and let in the primitive and the simple. Art and poetry never tire of the plow, the scythe, the axe, the hoe, the flail, the oar; but the pride and glory of the agricultural warehouse,—can that be sung? The machine that talks and walks and suffers and loves is still the best. Artifice, the more artifice there is thrust between us and Nature, the more appliances, conductors, fenders, the less freely her virtue passes. The direct rays of the open fire are better even for roasting a potato than conducted heat.

What we owe to science, as tending to foster a disinterested love of truth, as tending to clarify the mental vision, or sharpen curiosity, or cultivating the spirit of fearless inquiry, or stimulating the desire to see and know things as they really are, would not be easy to determine. A great deal, no doubt. But the value of the modern spirit, the modern emancipation, as a factor in the production of a great literature, remains to be seen.

Science will no doubt draw off, and has already drawn off, a vast deal of force and thought that has heretofore found an outlet in other pursuits, perhaps in law, criticism, or historical inquiries; but is it probable that it will nip in the bud any great poets, painters, romancers, musicians, orators? Certain branches of scientific inquiry drew Goethe strongly, but his aptitude in them was clearly less than in his own chosen field. Alexander Wilson left poetry for ornithology, and he made a wise choice. He became eminent in the one, and he was only mediocre in the other. Sir Charles Lyell also certainly chose wisely in abandoning verse-making for geology. In the latter field he ranks first, and in interpreting "Nature's infinite book of secrecy," as it lies folded in the geological strata, he found ample room for the exercise of all the imagination and power of interpretation he possessed. His conclusions have sky-room and perspective, and give us a sort of poetic satisfaction.

The true poet and the true scientist are not estranged. They go forth into nature like two friends. Behold them strolling through the summer fields and woods. The younger of the two is much the more active and inquiring; he is ever and anon stepping aside to examine some object more minutely, plucking a flower, treasuring a shell, pursuing a bird, watching a butterfly; now he turns over a stone, peers into the marshes, chips off a fragment of a rock, and everywhere seems intent on some special and particular knowledge of the things about him. The elder man has more an air of leisurely contemplation and enjoyment,—is less curious about special objects and features, and more desirous of putting himself in harmony with the spirit of the whole. But when his younger companion has any fresh and characteristic bit of information to impart to him, how attentively he listens, how sure and discriminating is his appreciation! The interests of the two in the universe are widely different, yet in no true sense are they hostile or mutually destructive.

ISAAC ASIMOV

AT LAST COUNT Isaac Asimov has written, apart from his science fiction, more than 200 books of nonfiction. Most of them are about science. We leave aside such tomes as his annotated editions of Shakespeare, the Bible, *Gulliver's Travels,* and Byron's *Don Juan,* his collections of original limericks, his two-volume autobiography, and other nonscience odds and ends.

How Asimov manages to write so profusely and so well is a mystery. I once visited him in his Manhattan study, and was surprised to observe that it had no windows. This, he told me, was by careful intent. If the room had windows he would be tempted now and then to get up from his typewriter (now a word processor) and look out. Would that not interrupt the steady flow of his thoughts?

The good doctor, as he is known in science-fiction circles, became hooked on the genre as a boy when he began to read the SF pulp magazines on the racks of his Russian immigrant father's candy store in New York City. After obtaining a doctorate in biochemistry, Asimov continued to pound out science fiction, having started to sell regularly to the pulps when he was in his late teens. In 1960 the commercial success of his *Intelligent Man's Guide to Science* (later given a nonsexist title) shifted his interest from science fiction to science fact.

Who writing today, anywhere on earth, is a more effective teacher of the physical and biological sciences? Who in history has more enthusiastically conveyed the range and excitement of science to so many laypersons? Who could be bored by anything the good doctor writes?

And why should I say more about a public figure so well

known and so universally admired? The selection I have chosen is from Asimov's 1983 book of essays, *The Roving Mind.* It is the best reply I know to that old canard, occasionally uttered by poets who know little about science, that somehow scientific knowledge destroys beauty by telling us things we really don't need to know.

ISAAC ASIMOV

Science and Beauty

ONE OF Walt Whitman's best-known poems is this one:

When I heard the learn'd astronomer,
When the proofs, the figures, were ranged in columns before me,
When I was shown the charts and diagrams, to add, divide and measure them,
When I sitting heard the astronomer where he lectured with much applause in the lecture-room,
How soon unaccountable I became tired and sick,
Till rising and gliding out I wander'd off by myself,
In the mystical moist night-air, and from time to time,
Look'd up in perfect silence at the stars.

I imagine that many people reading those lines tell themselves, exultantly, "How true! Science just sucks all the beauty out of everything, reducing it all to numbers and tables and measurements! Why bother learning all that junk when I can just go out and look at the stars?"

That is a very convenient point of view since it makes it not only unnecessary, but downright aesthetically wrong, to try to follow all that hard stuff in science. Instead, you can just take a look at the night sky, get a quick beauty fix, and go off to a nightclub.

The trouble is that Whitman is talking through his hat, but the poor soul didn't know any better.

I don't deny that the night sky is beautiful, and I have in my time spread out on a hillside for hours looking at the stars and being awed by their beauty (and receiving bug-bites whose marks took weeks to go away).

But what I see—those quiet, twinkling points of light—*is not all the beauty there is.* Should I stare lovingly at a single leaf and willingly remain ignorant of the forest? Should I be satisfied to watch the sun glinting off a single pebble and scorn any knowledge of a beach?

Those bright spots in the sky that we call planets are worlds. There are worlds with thick atmospheres of carbon dioxide and sulfuric acid; worlds of red-hot liquid with hurricanes that could gulp down the whole earth; dead worlds with quiet pock-marks of craters; worlds with volcanoes puffing plumes of dust into airlessness; worlds with pink and desolate deserts—each with a weird and unearthly beauty that boils down to a mere speck of light if we just gaze at the night sky.

Those other bright spots, which are stars rather than planets, are actually suns. Some of them are of incomparable grandeur, each glowing with the light of a thousand suns like ours; some of them are merely red-hot coals doling out their energy stingily. Some of them are compact bodies as massive as our sun, but with all that mass squeezed into a ball smaller than the earth. Some are more compact still, with the mass of the sun squeezed down into the volume of a small asteroid. And some are more compact still, with their mass shrinking down to a volume of zero, the site of which is marked by an intense gravitational field that swallows up everything and gives back nothing; with matter spiraling into that bottomless hole and giving out a wild death-scream of X-rays.

There are stars that pulsate endlessly in a great cosmic breathing; and others that, having consumed their fuel, expand and redden until they swallow up their planets, if they have any (and someday, billions of years from now, our sun will expand and the earth will crisp and sere and vaporize into a gas of iron and rock with no sign of the life it once bore). And some stars explode in a vast cataclysm whose ferocious blast of cosmic rays, hurrying outward at nearly the speed of light reaching across thousands of light-years to touch the earth and supply some of the driving force of evolution through mutations.

Those paltry few stars we see as we look up in perfect silence (some 2,500, and no more, on even the darkest and clearest night) are joined by a vast horde we don't see, up to as many as three hundred billion—300,000,000,000—to form an enormous pinwheel in space. This pinwheel, the Milky Way galaxy, stretches so widely that it takes light, moving at 186,282 miles each *second*, a hundred thousand *years* to cross it from end to end; and it rotates about its center in a vast and stately turn that takes two hundred million years to complete—and the sun and the earth and we ourselves all make that turn.

Beyond our Milky Way galaxy are others, a score or so of them bound to our own in a cluster of galaxies, most of them small, with no more than a few billion stars in each; but with one at least, the Andromeda galaxy, twice as large as our own.

Beyond our own cluster, other galaxies and other clusters exist; some clusters made up of thousands of galaxies. They stretch outward and outward as far as our best telescopes can see, with no visible sign of an end—perhaps a hundred billion of them in all.

And in more and more of those galaxies we are becoming aware of violence at the centers—of great explosions and outpourings of radiation, marking the death of perhaps millions of stars. Even at the center of our own galaxy there is incredible violence masked from our own solar system far in the outskirts by enormous clouds of dust and gas that lie between us and the heaving center.

Some galactic centers are so bright that they can be seen from distances of billions of light-years, distances from which the galaxies themselves cannot be seen and only the bright starlike centers of ravening energy show up—as quasars. Some of these have been detected from more than ten billion light-years away.

All these galaxies are hurrying outward from each other in a vast universal expansion that began fifteen billion years ago, when all the matter in the universe was in a tiny sphere that exploded in the hugest conceivable shatter to form the galaxies.

The universe may expand forever or the day may come when

the expansion slows and turns back into a contraction to re-form the tiny sphere and begin the game all over again so that the whole universe is exhaling and inhaling in breaths that are perhaps a trillion years long.

And all of this vision—far beyond the scale of human imaginings—was made possible by the works of hundreds of "learn'd" astronomers. All of it; *all* of it was discovered after the death of Whitman in 1892, and most of it in the past twenty-five years, so that the poor poet never knew what a stultified and limited beauty he observed when he "look'd up in perfect silence at the stars."

Nor can we know or imagine now the limitless beauty yet to be revealed in the future—by science.

ERNEST NAGEL

SECOND ONLY to atomic energy in revolutionary social consequences, as this revolutionary century draws to its close, is the coming of the robots—those astonishing feedback mechanisms that promise to eliminate jobs of mental drudgery as rapidly as inventions of the past century eliminated tasks of physical drudgery. No one knows of course exactly how this second industrial revolution is going to affect our lives. Will it burst upon us so quickly that millions of workers will lose their jobs, sending the economy into a disastrous tail spin? Or will we have the foresight to ease our way past catastrophe into an automated era of undreamed of leisure and abundance?

The following essay by Ernest Nagel (b. 1901) was originally an introduction to an issue of *Scientific American* devoted to automation. A long-time professor of philosophy at Columbia University, Dr. Nagel is one of the nation's most esteemed "philosophers of science." His concise, shrewdly reasoned article clarifies many of the complex issues that are certain to become more urgent as the robots continue to proliferate in the uncertain years ahead.

ERNEST NAGEL

Automation

AUTOMATIC CONTROL is not a new thing in the world. Self-regulative mechanisms are an inherent feature of innumerable processes in nature, living and nonliving. Men have long recognized the existence of such mechanisms in living forms, although, to be sure, they have often mistaken automatic regulation for the operation of some conscious design or vital force. Even the deliberate construction of self-regulating machines is no innovation: the history of such devices goes back at least several hundred years.

Nevertheless, the preacher's weary cry that there is nothing new under the sun is at best a fragment of the truth. The general notion of automatic control may be ancient, but the formulation of its principles is a very recent achievement. And the systematic exploitation of these principles—their subtle theoretical elaboration and far-reaching practical application—must be credited to the twentieth century. When human intelligence is disciplined by the analytical methods of modern science, and fortified by modern material resources and techniques, it can transform almost beyond recognition the most familiar aspects of the physical and social scene. There is surely a profound difference between a primitive recognition that some mechanisms are self-regulative while others are not, and the invention of analytic theory which not only accounts for the gross facts but guides the construction of new types of systems.

We now possess at least a first approximation to an

adequate theory of automatic control, and we are at a point of history when the practical application of that theory begins to be conspicuous and widely felt. The future of automatic control, and the significance for human weal or woe of its extension to fresh areas of modern life, are still obscure. But if the future is not to take us completely by surprise, we need to survey the principal content of automatic control theory, the problems that still face it and the role that automatic control is likely to play in our society.

The central ideas of the theory of self-regulative systems are simple. Every operating system, from a pump to a primate, exhibits a characteristic pattern of behavior, and requires a supply of energy and a favorable environment for its continued operation. A system will cease to function when variations in its intake of energy or changes in its external and internal environment become too large. What distinguishes an automatically controlled system is that it possesses working components which maintain at least some of its typical processes despite such excessive variations. As need arises, these components employ a small part of the energy supplied to the system to augment or diminish the total volume of that energy, or in other ways to compensate for environmental changes. Even these elementary notions provide fruitful clues for understanding not only inanimate automatically controlled systems, but also organic bodies and their interrelations. There is no longer any sector of nature in which the occurrence of self-regulating systems can be regarded as a theme for oracular mystery-mongering.

However, some systems permit a greater degree of automatic control than others. A system's susceptibility to control depends on the complexity of its behavior pattern and on the range of variations under which it can maintain that pattern. Moreover, responses of automatic controls to changes affecting the operation of a system are in practice rarely instantaneous, and never absolutely accurate. An adequate science of automatic control must therefore develop comprehensive ways of discriminating and measuring variations in quality; it must learn how signals (or information) may be transmitted and relayed; it must be familiar with the conditions under which self-excitations and oscillations may occur, and it must devise mechanisms which will anticipate the probable course and sequence of events. Such a science will use and develop current theories of fundamental physico-chemical processes. It is dependent upon the elaborate logico-mathematical analyses of statistical aggregates, and upon an integration of specialized researches which until recently have seemed only remotely related. Our present theory of self-regulative systems has sprung from the soil of contemporary theoretical science. Its future is contingent upon the continued advance of basic research—in mathematics, physics, chemistry, physiology and the sciences of human behavior.

Automatic controls have been introduced into modern industry only in part because of the desire to offset rising labor costs. They are in fact not primarily an economy measure but a necessity, dictated by the nature of modern services and manufactured products and by the large demand for goods of uniformly high quality. Many articles in current use must be processed under conditions of speed, temperature, pressure and chemical exchange which make human control impossible, or at least impracticable, on an extensive scale. Moreover, modern machines and instruments themselves must often satisfy unprecedentedly high standards of quality, and beyond certain limits the discrimination and control of qualitative differences elude human capacity. The automatic control of both the manufacturing process and the quality of the product manufactured is therefore frequently indispensable.

Once the pleasures of creating and contemplating the quasi-organic unity of self-regulative systems have been learned, it is only a short step to the extension of such controls to areas where they are not mandatory. Economic considerations undoubtedly play a role in this extension, but certain engineers are probably at least partly correct in their large claim that the modern development in automatic engineering is the consequence of a point of view which finds satisfaction in unified schemes for their own sake.

How likely is the total automatization of industry, and what are the broad implications for human welfare of present tendencies in that direction? Crystal-gazing is a natural and valuable pastime, even if the visions beheld are only infrequently accurate. Some things, at any rate, are seen more clearly and certainly than others. If it is safe to project recent trends into the future, and if fundamental research in relevant areas continues to prosper, there is every reason to believe that the self-regulation of industrial production, and even of industrial management, will steadily increase. On the other hand, in some areas automatization will never be complete—either because of the relatively high cost of conversion, or because we shall never be able to dispense with human ingenuity in coping with unforeseeable changes, or finally because of certain inherent limitations in the capacity of any machine which operates according to a closed system of rules. The dream of a productive system that entirely runs itself appears to be unrealizable.

Some consequences of large-scale automatic control in current technology are already evident. . . . Industrial productivity has increased out of proportion to the increase in capital outlay. Many products are now of finer quality than they have ever been before. Working hours have been generally reduced, and much brutalizing drudgery has been eliminated. In addition, there are signs of a new type of professional man—the automatic control system engineer. There has been considerable conversion and retraining of unskilled labor. A slow refashioning of educational facilities, in content as well as in organization, in engineering schools as well as in the research divisions of universities and industries, is in progress. In the main these developments contribute to human welfare.

However, commentators on automatic control also see it as a potential source of social evil, and express fears—not altogether illegitimate—concerning its ultimate effect. There is first the fear that continued expansion in this direction will be accompanied by large-scale technological unemployment, and in consequence by acute economic distress and social upheaval. The possibility of disastrous technological unemployment cannot be ruled out on purely theoretical grounds; special circumstances will determine whether or not it occurs. But the brief history of automatic control in the U. S. suggests that serious unemployment is not its inevitable concomitant, at least in this country. The U. S. appears to be capable of adjusting itself to a major industrial reorganization without uprooting its basic patterns of living. Large-scale technological unemployment may be a more acute danger in other countries, but the problem is not insurmountable, and measures to circumvent or to mitigate it can be taken.

There is next the fear that an automatic technology will impoverish the quality of human life, robbing it of opportunities for individual creation, for pride of workmanship and for sensitive qualitative discrimination. This fear is often associated with a condemnation of "materialism" and with a demand for a return to the "spiritual" values of earlier civilizations. All the available evidence shows, however, that great cultural achievements are attained only by societies in which at least part of the population possesses considerable worldly substance. There is a good empirical basis for the belief that automatic control, by increasing the material well-being of a greater fraction of mankind, will release fresh energies for the cultivation and flowering of human excellence. At any rate, though material abundance undoubtedly is not a sufficient condition for the appearance of great works of the human spirit, neither is material penury; the vices of poverty are surely more ignoble than those of wealth. Moreover, there is no reason why liberation from the unimaginative drudgery which has been the lot of so many men throughout the ages should curtail opportunities for creative thought and for satisfaction in work well done. For example, the history of science exhibits a steady tendency to eliminate intellectual effort in the solution of individualized problems, by developing comprehensive formulas which can resolve by rote a whole class of them. To paraphrase Alfred North Whitehead, acts of thought, like a cavalry charge in battle, should be introduced only at the critical junction of affairs.

There has been no diminution in opportunities for creative scientific activity, for there are more things still to be discovered than are dreamt of in many a discouraged philosophy. And there is no ground for supposing that the course of events will be essentially different in other areas of human activity. Why should the wide adoption of automatic control and its associated quantitative methods induce a general insensitivity to qualitative distinctions? It is precisely measurement that makes evident the distinctions between qualities, and it is by measurement that man has frequently refined his discriminations and gained for them a wider acceptance. The apprehension that the growth of automatic controls will deprive us of all that gives zest and value to our lives appears in the main to be baseless.

There is finally the fear that an automatic technology will encourage the concentration of political power; that authoritarian controls will be established for all social institutions—in the interest of the smooth operation of industry and of society but to the ruin of democratic freedom. This forecast is given some substance by the recent history of several nations, but the dictatorships differ so greatly from the Western democracies in political traditions and social stratifications that the prediction has dubious validity for us. Nevertheless, one element in this grim conjecture requires attention. Whatever the future of automatic control, governmental regulation of social institutions is certain to increase—population growth alone will make further regulation imperative. It does not necessarily follow that liberal civilizations must therefore disappear. To argue that it does is to commit a form of the pathetic fallacy. Aristotle argued that political democracy was possible only in small societies such as the Greek city-states. If our present complex governmental regulations in such matters as sanitation, housing, transportation and education could have been foreseen by our ancestors, many of them would doubtless have concluded that such regulations are incompatible with any sense of personal freedom. It is easy to confound what is merely peculiar to a given society with the indispensable conditions for democratic life.

The crucial question is not whether control of social transactions will be further centralized. The crucial question is whether, despite such a movement, freedom of inquiry, freedom of communication and freedom to participate actively in decisions affecting our lives will be preserved and enlarged. It is good to be jealous of these rights; they are the substance of a liberal society. The probable expansion of automatic technology does raise serious problems concerning them. But it also provides fresh opportunities for the exercise of creative ingenuity and extraordinary wisdom in dealing with human affairs.

JONATHAN NORTON LEONARD

SINCE THE beginning of civilization man has speculated about life on other worlds. Lucian, the Greek satirist, actually wrote an amusing science fantasy about a voyage to the moon and the curious inhabitants to be found thereon! Renaissance astronomy naturally stimulated interest in the topic, especially among theologians who debated the question of whether extra-terrestrial races were in a state of grace or under the curse of original sin. Milton's *Paradise Lost* takes for granted that there are rational creatures on other planets, but the angel Raphael advises Adam to:

> *Dream not of other worlds, what creatures there*
> *Live, in what state, condition, or degree—*

The sons of Adam have continued, however, to disregard Raphael's advice. Kant believed all the planets in the solar system to be populated, with the more admirable creatures in the most distant planets—"a view to be praised for its terrestrial modesty," comments Bertrand Russell, "but not supported by any scientific grounds." Carlyle, contemplating the starry heavens, was moved to exclaim, "A sad spectacle! If they be inhabited, what a scope for pain and folly; and if they be not inhabited, what a waste of space."

Today, most scientists and all science fiction writers assume that extra-terrestrial creatures are probable, but divide over how "humanoid" they are likely to be. This intriguing question is discussed in the following essay by Jonathan Norton Leonard (b. 1903), for many years science editor of *Time* magazine. It forms a chapter of his *Flight into Space,* in this editor's opinion the best of the many space travel books that were published in the early fifties.

JONATHAN NORTON LEONARD

Other-Worldly Life

THERE IS a curious blank in the scientific literature of space. So far as this observer knows, there is no full-scale, responsible and informed study of the kinds of life that might develop under circumstances different from those on earth. Plenty of amateurs have speculated wildly about it. They imagine planets with atmospheres of corrosive fluorine; they even describe in some detail the inhabitants of the sun, which is too hot for any chemical compound and which consists entirely of gases stirred by ferocious turbulence. But no fully competent scientist has made a serious attack on this interesting subject.

One reason may be that scientists have an inordinate respect for the jurisdictional boundaries between their specialties. A biologist feels like a prowler in the night when he ventures into psychology. Astronomers dive for their cyclone cellars when the conversation veers even slightly toward biology. Some sciences have become so finely subdivided that specialists in closely related fields hardly speak to one another. A protein chemist has little to say to a steroid chemist. Both may be trying to find a chemical cure for cancer, but neither would dream of commenting on the other's problems.

To write a successful book on the possibilities of extraterrestrial life would require good knowledge in many widely separated fields. The author would have to be a broad biologist, familiar with all the forms that have been explored by life on earth. He would have to know organic chemistry, which is concerned with carbon compounds, and also inorganic chemistry. He would have to know the many kinds of physics that deal with conditions in the atmospheres, the oceans and on the surfaces of other planets than the earth. He would have to know enough about astronomy to read the extremely difficult literature that professional astronomers circulate in their small, charmed circle.

Such a man does not exist, or if he does exist, he has kept his talents unknown to the general public. The subject of extraterrestrial life has been abandoned to the space-fiction writers, who usually imagine lovely girls with yellow eyes, antennae growing out of their foreheads and lungs full of fluorine. This is a crying shame. Life is the most interesting thing in the entire universe; it deserves better treatment.

The study of life in general starts with a strange blank. No one has succeeded in defining what life is. J.B.S. Haldane has called it "any self-perpetuating pattern of chemical reactions," but this is too inclusive to suit all of his colleagues. A flame, for instance, is chemical and self-perpetuating. It consumes fuel and oxygen and breathes out carbon dioxide just like the animals on earth. It "lives" as long as it can find both fuel and oxygen. It dies when one of them is exhausted, just as animals die of starvation or suffocation. But a flame is not alive in the sense of the biologist. It is not a living organism.

When biologists try to point to the simplest possible organisms, they run into difficulties too. Certain viruses, such as those that cause the mosaic disease of tobacco, behave from the point of view of the tobacco grower just like any other pathogenic organism. They infect his tobacco plants, grow inside them and spread to other plants throughout his field. But when the apparently living virus is isolated, it proves to be nothing but a large molecule. It forms regular crystals much like those of common salt and behaves in many other ways like a lifeless chemical compound.

To decide whether tobacco mosaic is really a living organism is a problem in semantics. To call it lifeless is dangerous. It certainly multiplies and replenishes the earth within its small field, just like the fruitful people of the Old Testament. To call it alive is dangerous too. When packed together in a crystal, its molecules show no signs of life. It is then a mere chemical compound, and although it is very complicated it might presumably be synthesized in the scientists' glassware. Then the scientists would have created life. Or wouldn't they?

For the present at least, the scientists have abandoned this problem of definition to the semanticists and philosophers, tribes for which they have little respect. A living organism, they say in effect, is anything that grows, reproduces, and perpetuates itself or its species. It needs some source of energy for this operation and it must have the ability to absorb substances of which to build its bodily mechanism. The energy can come from any available source and so can building materials.

When scientists try to figure out how life got started on earth, they are forced into many assumptions that they cannot prove. The simplest organism to be found on earth today—the viruses—are parasitic. They do not live independently, but prey on higher organisms. The tobacco virus for instance can live and grow only within the cells of the tobacco plant, and other viruses are restricted in the same way. Obviously, they cannot be the original forms of life, which must have lived independently before higher organisms developed.

Most of the bacteria, one step upward, are parasitic too. They live by digesting the organic materials elaborated by higher creatures. Even those few that live independently cannot be the original forms of life on earth; they are too highly developed.

A bacterium looks simple in the field of a microscope, but it is extremely complicated, both chemically and in its interior structure. It cannot have sprung full grown, like Athena from the forehead of Zeus, out of the inorganic chemicals in the lifeless, primitive earth. There must have been simpler organisms, now extinct, from which all higher forms evolved in the course of time. When the scientists make the assumption that such organisms existed, their task becomes simpler. They can give them any properties that seem to make sense.

At present the energy source of all, or nearly all, life on

earth is the light of the sun. It is absorbed by chlorophyll and similar compounds in the cells of plants and is used to combine water and carbon dioxide to form sugars and other organic compounds that the plant needs for its growth. This is an extremely complicated operation, and most of the experts on this subject believe that the earliest organism must have got along without it.

They picture the early earth as very different from the earth today. Its atmosphere contained many carbon compounds, such as the methane that can still be detected in the atmospheres of the outer planets. Carbon has the property of combining with itself to form large, complicated, "organic" molecules. Under the influence of sunlight and perhaps of cosmic rays, much of the methane in the early atmosphere combined into these large molecules and washed down into the sea. There the molecules grew larger by combining with one another and with other substances such as nitrogen, sulphur, phosphorus, iron, magnesium, oxygen and hydrogen.

This process continued slowly for hundreds of millions of years until the sea was full of a kind of organic soup. It probably contained examples of all the compounds that carbon will form with the other available elements. This stuff does not exist in the sea today; it could not exist because living organisms would attack it and destroy it immediately, but in the primitive sea there was no life. So the organic molecules could grow indefinitely.

At last the blind process of chemical combination, repeated hundreds of trillions of times in each microsecond, produced a molecule with an extraordinary property. It could grow by taking other molecules into its structure, and it could reproduce, probably by the simple maneuver of breaking in two. This molecule was "alive," and a new and powerful force had appeared on the earth.

Feeding on the lifeless foodstuffs dissolved in the water, the descendants of the Adam-and-Eve molecule quickly populated all of the primitive oceans. Some of them changed slightly so that they could utilize more of the available foodstuffs. Some of them became fierce molecular predators feeding on weaker fellows. How long this primitive kind of life had the earth to itself the scientists do not attempt to guess. It may have been several hundred million years, for the earth is four billion years old at least, and its surface has had a tolerable temperature for nearly as long.

Once life appeared, it was forced by its very nature to become more complicated. The simplest living molecules could not exploit all the possible modes of life available at the time, so when more complicated forms were created by chemical or physical accidents, they had an advantage over their more primitive relatives. They grew and multiplied faster, only to be replaced in turn by still more complicated forms. At last organisms appeared that did not depend on the carbon compounds dissolved in the ocean. They made their own by photosynthesis out of carbon dioxide and light.

This was a second great turning point. Now life was hitched to the unfailing power of the shining sun. The primitive photosynthetic organisms may have been red or purple or any other color, but they were real plants. They were so successful that they soon cleaned the carbon dioxide out of the atmosphere, replacing it with the oxygen that dominates it chemically today. Soon primitive parasites (animals) developed to devour the plants. These breathed oxygen and breathed out carbon dioxide, which flowed back to the plants.

Thus was established the celebrated carbon cycle. Plants dominate the earth today in a chemical sense, keeping the carbon dioxide in its atmosphere down to a mere trace. The animals and the parasitic bacteria and fungi try to keep up with the plants, returning the carbon to the atmosphere where it can be used again to build more plants. Once life had been established on this firm basis, the rest of evolution was only a matter of time. It probably took less than a billion years for evolution to produce the intelligent animal, man, that now dominates the planet.

Scientists point out that there is nothing miraculous or unrepeatable about the appearance of life on earth. They believe it would happen again, given the same sufficient time and the same set of circumstances. It would even happen under very different circumstances. There is no reason to believe that conditions in the atmosphere and oceans of the primitive earth were modified by any outside power to make them favorable for the development of life. They just happened that way, and it is likely that life would have appeared even if conditions had been considerably different.

Not very much skilled reasoning has been applied to the problem of what conditions are absolutely necessary for the appearance of life. Conservative theorists hold that life must be based on carbon compounds like those that form human and other living bodies. They claim that only carbon can join in the long chains, complicated rings, and other molecular patterns that are needed to make the life process work.

If this is true, the climate must be just right. At very low temperatures, carbon compounds do not react readily with one another, and at the temperature of boiling water many of them disintegrate. Another limitation insisted upon by the conservative theorists is that life must have large amounts of water. Complicated carbon compounds will dissolve in other liquids than water, but not as well and not in the same way. Besides liquid water, there must be a supply of simple carbon compounds in the early stages, and this rules out environments rich in chemical substances (such as free oxygen) that would destroy them. Another necessity is light, which is the activating agent that induces small molecules to combine into larger ones.

Less conservative theorists claim that the conditions demanded by the conservatives may be necessary for the production of life as it is known on earth. But other kinds of life, they say, may be possible, and they may demand or tolerate very different conditions. The chemistry of carbon has been studied more intensively than that of any other element, but the limits of its possibilities have not been approached. There may be carbon compounds that will react vigorously even when dissolved in some novel medium, such as liquid ammonia. There may be others that will tolerate extremely high temperatures.

Living organisms on earth have not had to synthesize such

carbon compounds, and human chemists have not tried very hard to do so. When they do try, they sometimes have rather surprising success. Synthetic rubber, which is made of typical organic carbon compounds, is now being formed deliberately at very low temperatures, and it turns out to be better than rubber made at high temperatures.

At the other end of the temperature scale, the silicones (compounds containing both carbon and silicon) have proved to be stable far above the boiling point of water. There seems to be an unlimited number of possible silicones, so a planet with a very dense atmosphere and an ocean of water above the normal boiling point of water on earth might possibly develop living organisms with bodies made of silicones. Some chemists think that this is impossible, but at the present state of their knowledge they cannot prove it.

Neither can they prove that life is impossible in a medium other than water. They know little about the chemistry of complicated substances dissolved, say, in liquid hydrocarbons at a very low temperature. Their reactions might be slow, but the universe has plenty of time at its disposal. No one can prove, for instance, that Jupiter does not have a cold hydrocarbon ocean, or that it does not contain a slow sort of life.

It is even possible, theoretically, that life can develop in a gaseous instead of a liquid medium. A theory advanced seriously by Dr. Heinz Haber of the University of California at Los Angeles suggests that the mysterious clouds in the atmosphere of Venus may be a "biological airsol," a fog of small living organisms supported at the most favorable altitude in respect to sunlight and temperature. They would be like the plankton that forms the bulk of life in the earth's oceans, and larger flying organisms may have developed to feed upon them, like the earth's fish. Perhaps the bodies of all these creatures rain down eventually to the surface of Venus, which is probably dark and may be rather hot. If not too hot, it may be populated by large scavenging animals that live on the nutritious rain, just like the crabs and mollusks on the bottoms of the earth's oceans.

Once life gets started, it seems to have an unlimited ability to fit itself to changing conditions. Life on earth has learned to thrive in unlikely places such as boiling volcanic springs and the cold, weather-beaten rocks that rise above the surfaces of the antarctic ice cap. The atmosphere of the earth and the conditions beneath it must have changed enormously during its long history, but life went on just the same.

Not only low forms of life can make such adaptations; high forms can make them too. The highest form of animal, the mammals, thrives under conditions of both heat and cold that would kill their less well-organized rivals. The highest mammal, man, protected by his clothing, fire and mechanical contrivances, can thrive where no other animal can. This adaptability of living organisms makes it conceivable that high types of life may be thriving today on planets whose present conditions would surely prevent the appearance of primitive life.

What such creatures would look like and act like is anyone's guess. One chain of reasoning suggests that they may look surprisingly like the familiar forms on earth. An internal skeleton of hard strong material, for instance, is a fine device, and other sequences of evolution may have hit on it. A brain (that is, a communication system with a central "telephone exchange") is a necessity too, and the best place to put a brain is in a movable, well-protected member that also contains the major senses, such as the eyes, ears, and organs of smell. So the inhabitants of unknown planets may have heads and skulls of some sort. They may have legs too, for movable supports to maneuver the animal's body are convenient devices in any place where gravitation is not too powerful.

If light is available, eyes will be developed to use it as a source of information, and since the laws of optics are presumably uniform throughout the universe, the eyes of extraterrestrial races will not look very different from human eyes. They will certainly have lenses and something resembling eyelids to keep their surfaces clean.

Other theorists pooh-pooh this sort of thing as mere anthropomorphism. The form of man and other high earthside animals, they say, is the product of a long series of accidents that reach back to the fish in the primitive seas. Man has four limbs because the primitive fish had four fins, and once this pattern had been established it could not be changed readily.

Man might do better with more limbs. Elephants are a successful form, and they have kept all four of their legs while making a manipulating "hand" out of their noses. Insects make good use of all six of their legs as well as various other specialized appendages. If insects had managed to get around their limitations, chiefly their external skeleton and their inefficient breathing system, the highest animals on earth might have six legs as well as antennae and tentacles. They might have brains in the small of their backs; they might lay eggs and nurse their young through the troublesome adolescence of metamorphosis. It is too much to expect, say these free-style evolutionists, that the intelligent inhabitants of other planets should look like earthside monkeys that have recently descended from the trees. Even on the earth, very slight changes of environment during the past two billion years would have modified the long chain of evolutionary accidents and produced a final product of very different appearance.

Imagining what kind of "people" may exist on other planets of the solar system is very different from determining what kinds do exist, or if any do. As an abode for intelligent life, Mercury looks impossible except to the most persistent optimist. Venus is a mystery under its thick clouds. If its atmosphere contains a biological airsol with large scavengers prowling the dark surface below, there is no reason why some of these should not be highly intelligent. But there is no evidence of it, only a possibility.

Nearly all astronomers admit that Mars has some sort of vegetation on it, and where plants live, there must be something equivalent to animal life. Animals (organisms that eat plants) are a necessary part of the carbon cycle. Without them the plants would soon extract all the carbon dioxide from the atmosphere. Then they would have to die.

The plants on Mars are not dead. They may be as low as the lichens that grow on earthside rocks (the sunlight reflected from them seems to resemble the light reflected from lichens) but they grow each season after their fashion, and perhaps they grow rather fast.

If Mars is cursed with great dust storms, as some observers believe, its plants must grow vigorously to keep above the dust that is deposited on them. If they are still growing, there must be animals on Mars that attack their bodies and return their carbon to the atmosphere as carbon dioxide. Perhaps these animals are no bigger than earthside bacteria or fungi, which perform this function too. But they may be large, even large enough to support well-developed brains. There is no conclusive evidence on this point.

The outer planets do not look favorable to life, but astronomers know almost nothing about conditions beneath the tops of their deep atmospheres. Life may have developed on them too, in some unimaginable form. They have plenty of fluids, either gaseous or liquid or both, and enough sunlight reaches them to keep life moving. It is weak, to be sure, compared to sunlight on earth, but earthside plants can grow in very dense shade where the energy arriving from the sun is a small fraction of what it is in the open. Space explorers are entitled to hope that some kind of life exists on the outer planets.

There is a good chance that life on other planets may be in some stage of social organization. Evolution on earth has adopted the social pattern many times over in many geological ages, and it is reasonable to suppose that on other planets it has its advantages, too. The process of socialization is at least as important a part of evolution as the process of developing the bodies of individual organisms.

Each one-celled organism, whether plant or animal, has in its nucleus a group of genes which control the growth and reproduction of the rest of the cell. These genes seem to be relics of the very early stage of life when the earth was populated only by living, reproducing molecules. In the course of time they banded together in co-operating groups and gathered around them subordinate molecules that did not possess the ability of reproduction. One-celled organisms such as the protozoa are, in effect, colonies of genes which have acquired a powerful competitive advantage by working together. These cells are now the dominant form of small-scale life. The living molecules (viruses) have been reduced to the status of dependent parasites.

The next step in socialization was for the cells to form colonies of their own. They banded together to form multi-celled plants and animals which vary in size from microscopic creatures no bigger than their one-celled rivals up to sequoias and whales. They are all built of cells whose individualities have been lost in the service of the larger unit to which they belong. Human bodies are societies too: colonies of trillions of cells, all of which resemble closely the one-celled organisms that were their distant ancestors. Inside each cell are the much smaller genes: the living molecules that banded together soon after the dawn of life.

There was an excellent reason for this banding together. The living molecules could grow only so large before they fetched up against limits to further increase in size. When they co-operated to form the nuclei of cells, they could grow larger and do many more things. But the cells also reached a limit eventually and were forced to co-operate too. Some of the patterns that they adopted were capable of attaining very great size, but some reached their limit soon and so had no recourse except a third stage of banding together.

About one hundred million years ago, the insects reached their size limit and did what the cells had done a billion years before. They formed large social groups with the individual insects subordinated to the welfare of the colony as a whole. These social insects (ants, wasps, bees, termites, etc.) were so successful that they have survived almost unchanged down to the present moment. There is hardly a square yard of land where life exists at all that does not contain either ants or termites. They inhabit the tropics and the temperate regions, both moist land and dry. An unprejudiced observer from another planet might well decide that the social insects are the most successful form on earth.

The socialization of the vertebrates was delayed for a long time. Their bodily pattern allowed them to grow much larger than the insects, and so they were not forced to use the device of co-operation. Their evolution tested many large experimental models (such as the dinosaurs) before it decided that mere giantism was not the path of progress. Then the evolution of the mammals, the highest type of vertebrate, abandoned giantism and turned at last to socialization. The result was man, the social mammal, who now shares control of the planet with the social insects.

The present state of things on the planet earth would be rather a puzzle to an observer from another planet. If he landed in the United States, the most conspicuous animals in sight would be automobiles, and if he examined these vigorous hard-shelled creatures, he would find that each contains one or more soft, feeble organisms that appear notably helpless when removed from their shells. He would decide, after talking with these defenseless creatures, that they had no independent existence. Few of them have anything to do with the production or transportation of food. They need clothing and shelter, but do not provide them for themselves. They are dependent on their distant fellows in thousands of complex ways. When isolated, they usually die—just like worker ants that wander helplessly and hopelessly if separated from their colony.

If the observer were intelligent (and extraterrestrial observers are always presumed to be intelligent) he would conclude that the earth is inhabited by a few very large organisms whose individual parts are subordinate to a central directing force. He might not be able to find any central brain or other controlling unit, but human biologists have the same difficulty when they try to analyze an ant hill. The individual ants are not impressive objects—in fact they are rather stupid, even for insects—but the colony as a whole behaves with striking intelligence.

When human observers descend on a foreign planet, they may find it inhabited by organisms in an even more advanced stage of social co-operation. Perhaps its moving and visible parts will be entirely secondary, like the machines of man. Perhaps the parts that are really alive will be even more helpless: mere clots of nerve tissue lying motionless and sedentary far underground. Perhaps this organic stuff, having served its creative purpose, will have withered away, leaving the machines that it has created in possession of the planet.

This state of things would not be much more extraordinary than the situation that evolution has already produced on the earth. In a sense, the cells of the human body were once independent. A few of them, the white corpuscles in the blood, retain some shreds of this independence; they wander around restlessly, acting and looking much like free-living amoebas. But most of the body's cells have lost all separateness. A higher organism has taken over, and the seat of its individuality cannot be found.

When imaginative men turn their eyes toward space and wonder whether life exists in any part of it, they may cheer themselves by remembering that life need not resemble closely the life that exists on earth. Mars looks like the only planet where life like ours could exist, and even this is doubtful. But there may be other kinds of life based on other kinds of chemistry, and they may be thriving on Venus or Jupiter. At least we cannot prove at present that they are not.

Even more interesting is the possibility that life on other planets may be in a more advanced stage of evolution. Present-day man is in a peculiar and probably fleeting stage. His individual units retain a strong sense of personality. They are, in fact, still capable under favorable circumstances of leading individual lives. But man's societies (analogous to ant colonies) are already sufficiently developed to have enormously more power and effectiveness than the individuals have.

It is not likely that this transitional situation will continue very long on the evolutionary time scale. Fifty thousand years ago man was a wild animal, living like wolves or beavers in small family groups. Fifty thousand years from now his societies may have become so close-knit that the individuals retain no sense of separate personality. Then little distinction will remain between the organic parts of the multiple organism and the inorganic parts (machines) that have been constructed by it. A million years further on—and a million years is a tick of the clock on the evolutionary time scale—man and his machines may have merged as closely as the muscles of the human body and the nerve cells that actuate them.

The explorers of space should be prepared for some such situation. If they arrive on a foreign planet when its living organisms are in an earlier stage of evolution, they may find the equivalent of dinosaurs or mollusks or even one-celled protozoa. If the planet has reached a later stage (and this is by no means impossible), it may be inhabited by a single large organism composed of many closely co-operating units.

The units may be "secondary"—machines created millions of years ago by a previous form of life and given the will and ability to survive and reproduce. They may be built entirely of metals, ceramics and other durable materials, like man's guided missiles. If this is the case, they may be much more tolerant of their environment, thriving under conditions that would destroy immediately any organism made of carbon compounds and dependent on the familiar carbon cycle.

They could live on very hot or very cold planets. They could breathe any atmosphere or none. They could build their bodies to any desirable size out of materials plentiful in their planet's crust. They could get their energy from sunlight or from nuclear reactions. Such creatures might be relics of a bygone age, many million years ago, when their planet was favorable to the origin of life, or they might be immigrants from a favored planet.

Man's space explorers are not exactly likely to find such a situation on any planet they can reach. But neither are they likely to find the equivalent of present-day man, whose semi-socialized stage of development, though interesting, can occupy only a tiny slice of evolutionary time.

J. ROBERT OPPENHEIMER

AT LOS ALAMOS, New Mexico, during the war years when work on the atom bomb was under way, the first siren of a working day sounded at 7:00 A.M. At that moment, Mrs. Enrico Fermi writes in her biography of her husband, Enrico would stretch out in his bed, yawn, and remark, "Oppie has whistled. It is time to get up."

"Oppie" was J. Robert Oppenheimer (1904-1967), director of the Los Alamos laboratories. A former physics professor, he proved to be, in Mrs. Fermi's words, "the real soul of the project," carrying "the burden of his responsibilities with an enthusiasm and a zeal bordering on religiosity." A decade later, in a nervous fit of mistrust, the Atomic Energy Commission withdrew security clearance from him. One is reminded of the remark reportedly made by a senator, that our nation's gravest mistake was to let the scientists in on the secret of the atom bomb.

It is not often one finds in the personality of a scientist as many contradictory facets as in Oppenheimer. A brilliant theoretician; a brilliant administrator. An expert mathematician; a man who learned Greek in high school and later studied Sanskrit as a hobby. A man deeply committed to the view that scientific truth should be pursued for its own sake; a man equally concerned with the effects of science on society. A man at one time so divorced from human affairs that he read no newspapers, owned no radio or telephone, never troubled to vote; a man who became a fellow-traveler, then like so many other fellow-travelers, broke with the Communist movement when he caught a glimpse of the dismal reality behind the pompous rhetoric of the comrades.

Perhaps the key to Oppenheimer's complicated personality can be found in quantum theory, in a principle he himself is fond of invoking—the principle of complementarity. In the middle ages it was called the doctrine of double truth—a view that there are truths of Revelation that contradict truths of philosophy, yet one can and must accept both sets of truths as valid. In our time it has been applied by Niels Bohr to an unavoidable experimental predicament. The "real" electron is a mystery. We can approach it as a particle, putting aside those experiments which prove it to be a wave. Or we can assume it is a wave, forgetting the data that prove it to be a particle. The two approaches are incommensurable, but both are valid. "To be touched with awe, or humor," Oppenheimer has written, "to be moved by beauty, to make a commitment or a determination, to understand some truth—these are complementary modes of the human spirit. All of them are part of man's spiritual life. None can replace the others, and where one is called for the others are in abeyance."

The essay by Oppenheimer chosen here was originally a lecture delivered at Massachusetts Institute of Technology in 1947. It is a stirring plea for that kind of revolution in thinking for which T. H. Huxley and John Dewey also plead in essays printed elsewhere in this volume, even though there is the "melancholy certainty" that while the sirens scream, few are getting up.

J. ROBERT OPPENHEIMER

Physics in the Contemporary World

If I have even in the title of this talk sought to restrict its theme, that does not imply an overestimate of physics among the sciences, nor a too great myopia for these contemporary days. It is rather that I must take my starting point in the science in which I have lived and worked, and a time through which my colleagues and I are living.

Nevertheless, I shall be talking tonight about things which are quite general for the relations between science and civilization. For it would seem that in the ways of science, its practice, the peculiarities of its discipline and universality, there are patterns which in the past have somewhat altered, and in the future may greatly alter, all that we think about the world and how we manage to live in it. What I shall be able to say of this will not be rich in exhortation, for this is ground that I know how to tread only very lightly.

But that I should be speaking of such general and such difficult questions at all reflects in the first instance a good deal of self-consciousness on the part of physicists. This self-consciousness is in part a result of the highly critical traditions which have grown up in physics in the last half century, which have shown in so poignant a way how much the applications of science determine our welfare and that of our fellows, and which have cast in doubt that traditional optimism, that confidence in progress, which have characterized Western culture since the Renaissance.

It is, then, *about* physics rather than *of* physics that I shall be speaking—and there is a great deal of difference. You know that when a student of physics makes his first acquaintance with the theory of atomic structure and of quanta, he must come to understand the rather deep and subtle notion which has turned out to be the clue to unraveling that whole domain of physical experience. This is the notion of complementarity, which recognizes that various ways of talking about physical experience may each have validity, and may each be necessary for the adequate description of the physical world, and may yet stand in a mutually exclusive relationship to each other, so that to a situation to which one applies, there may be no consistent possibility of applying the other. Teachers very often try to find illustrations, familiar from experience, for relationships of this kind; and one of the most apt is the exclusive relationship between the practicing of an art and the description of that practice. Both are a part of civilized life. But an analysis of what we do and the doing of it—these are hard to bed in the same bed.

As it did on everything else, the last world war had a great and at least a temporarily disastrous effect on the prosecution of pure science. The demands of military technology in this country and in Britain, the equally overriding demands of the Resistance in much of Europe, distracted the physicists from their normal occupations, as they distracted most other men.

We in this country, who take our wars rather spastically, perhaps witnessed a more total cessation of true professional activity in the field of physics, even in its training, than did any other people. For in all the doings of war we, as a country, have been a little like the young physicist who went to Washington to work for the National Defense Research Committee in 1940. There he met his first Civil Service questionnaire and came to the questions on drinking: "Never," "occasionally," "habitually," "to excess." He checked both "occasionally" and "to excess." So, in the past, we have taken war.

All over the world, whether because of the closing of universities, or the distractions of scientists called in one way or another to serve their countries, or because of devastation and terror and attrition, there was a great gap in physical science. It has been an exciting and an inspiring sight to watch the recovery—a recovery testifying to extraordinary vitality and vigor in this human activity. Today, barely two years after the end of hostilities, physics is booming.

One may have gained the impression that this boom derives primarily from the application of the new techniques developed during the war, such as the atomic reactor and microwave equipment; one may have gained the impression that in large part the flourishing of physics lies in exploitation of the eagerness of governments to promote it. These are indeed important factors. But they are only a small part of the story. Without in any way deprecating the great value of wartime technology, one nevertheless sees how much of what is today new knowledge can trace its origin directly, by an orderly yet imaginative extension, to the kind of things that physicists were doing in their laboratories and with their pencils almost a decade ago.

Let me try to give a little more substance to the physics that is booming. We are continuing the attempt to discover, to identify and characterize, and surely ultimately to order, our knowledge of what the elementary particles of physics really are. I need hardly say that in the course of this we are learning again how far our notion of *elementarity*, of what makes a particle elementary, is from the early atomic ideas of the Hindu and Greek atomists, or even from the chemical atomists of a century ago. We are finding out that what we are forced to call elementary particles retain neither permanence nor identity, and they are elementary only in the sense that their properties cannot be understood by breaking them down into subcomponents. Almost every month has surprises for us in the findings about these particles. We are meeting new ones for which we are not prepared. We are learning how poorly we had identified the properties even of our old friends among them. We are seeing what a challenging job the ordering of this experience is likely to be, and what a strange world we must enter to find that order.

In penetrating into this world perhaps our sharpest tool in the past has been the observation of the phenomena of the cosmic rays in interaction with matter. But the next years will see an important methodological improvement, when the great program of ultra-high-energy accelerators begins to get under way. This program is itself one of the expensive parts of physics. It has been greatly subsidized by the government, primarily through the Atomic Energy Commission and the Office of Naval Research. It is a superlative example, of which one could find so many, of the repayment that technology makes to basic science, in providing means whereby our physical experience can be extended and enriched.

Another progress is the definement of our knowledge of the behavior of electrons within atomic systems, a refinement which on the one hand is based on the microwave techniques, to the developments of which the Radiation Laboratory of the Massachusetts Institute of Technology made unique contributions, and which on the other hand has provided a newly vigorous criterion for the adequacy of our knowledge of the interactions of radiation and matter. Thus we are beginning to see in this field at least a partial resolution, and I am myself inclined to think rather more than that, of the paradoxes that have plagued the professional physical theorists for two decades.

A third advance in atomic physics is in the increasing understanding of those forces which give to atomic nuclei their great stability, and to their transmutations their great violence. It is the prevailing view that a true understanding of these forces may well not be separable from the ordering of our experience with regard to elementary particles, and that it may also turn on an extension to new fields of recent advances in electrodynamics.

However this may be, all of us who are physicists by profession know that we are embarked on another great adventure of exploration and understanding, and count ourselves happy for that.

In how far is this an account of physics in the United

States only? In how far does it apply to other parts of the world, more seriously ravaged and more deeply disturbed by the last war? That question may have a somewhat complex answer, to the varied elements of which one may pay respectful attention.

In much of Europe and in Japan, that part of physics which does not rest on the availability of elaborate and radical new equipment is enjoying a recovery comparable to our own. The traditional close associations of workers in various countries makes it just as difficult now to disentangle the contributions by nationality as it was in the past. But there can be little doubt that it is very much harder for a physicist in France, for instance, or the Low Countries, and very much more nearly impossible for him in Japan, to build a giant accelerator than it is for the workers in this country.

Yet in those areas of the world where science has not merely been disturbed or arrested by war and by terror, but where terror and its official philosophy have, in a deep sense, corrupted its very foundations, even the traditional fraternity of scientists has not proved adequate protection against decay. It may not be clear to us in what way and to what extent the spirit of scientific inquiry may come to apply to matters not yet and perhaps never to be part of the domain of science; but that it does apply, there is one very brutal indication. Tyranny, when it gets to be absolute, or when it tends so to become, finds it impossible to continue to live with science.

Even in the good ways of contemporary physics, we are reluctantly made aware of our dependence on things which lie outside our science. The experience of the war, for those who were called upon to serve the survival of their civilization through the Resistance, and for those who contributed more remotely, if far more decisively, by the development of new instruments and weapons of war, has left us with a legacy of concern. In these troubled times it is not likely that we shall be free of it altogether. Nor perhaps is it right that we should be.

Nowhere is this troubled sense of responsibility more

acute, and surely nowhere has it been more prolix, than among those who participated in the development of atomic energy for military purposes. I should think that most historians would agree that other technical developments, notably radar, played a more decisive part in determining the outcome of this last war. But I doubt whether that participation would have of itself created the deep trouble and moral concern which so many of us who were physicists have felt, have voiced, and have tried to get over feeling. It is not hard to understand why this should be so. The physics which played the decisive part in the development of the atomic bomb came straight out of war laboratories and our journals.

Despite the vision and the far-seeing wisdom of our wartime heads of state, the physicists felt a peculiarly intimate responsibility for suggesting, for supporting, and in the end, in large measure, for achieving the realization of atomic weapons. Nor can we forget that these weapons, as they were in fact used, dramatized so mercilessly the inhumanity and evil of modern war. In some sort of crude sense which no vulgarity, no humor, no overstatement can quite extinguish, the physicists have known sin; and this is a knowledge which they cannot lose.

Probably in giving expression to such feelings of concern most of us have belabored the influence of science on society through the medium of technology. This is natural, since the developments of the war years were almost exclusively technological, and since the participation of academic scientists forced them to be deeply aware of an activity of whose existence they had always known but which had been often remote from them.

When I was a student at Göttingen twenty years ago, there was a story current about the great mathematician Hilbert, who perhaps would have liked, had the world let him, to have thought of his science as something independent of worldly vicissitudes. Hilbert had a colleague, an equally eminent mathematician, Felix Klein, who was certainly aware, if not of the dependence of science generally on society, at least of the dependence of mathematics on the physical sciences which nourish it and give it application. Klein used to take some of his students to meet once a year with the engineers of the Technical High School in Hanover. One year he was ill and asked Hilbert to go in his stead, and urged him, in the little talk that he would give, to try to refute the then prevalent notion that there was a basic hostility between science and technology. Hilbert promised to do so; but when the time came a magnificent absent-mindedness led him instead to speak his own mind: "One hears a good deal nowadays of the hostility between science and technology. I don't think that is true, gentlemen. I am quite sure that it isn't true, gentlemen. It almost certainly isn't true. It really can't be true. *Sie haben ja gar nichts mit einander zu tun.* They have nothing whatever to do with one another." Today the wars and the troubled times deny us the luxury of such absent-mindedness.

The great testimony of history shows how often in fact the development of science has emerged in response to technological and even economic needs, and how in the economy of social effort, science, even of the most abstract and recondite kind, pays for itself again and again in providing the basis for radically new technological developments. In fact, most people—when they think of science as a good thing, when they think of it as worthy of encouragement, when they are willing to see their governments spend substance upon it, when they greatly do honor to men who in science have attained some eminence—have in mind that the conditions of their life have been altered just by such technology, of which they may be reluctant to be deprived.

The debt of science to technology is just as great. Even the most abstract researches owe their very existence to things that have taken place quite outside of science, and with the primary purpose of altering and improving the conditions of man's life. As long as there is a healthy physics, this mutual fructification will surely continue. Out of its work there will come in the future, as so often in the past, and with an apparently chaotic unpredictability, things which will improve man's health, ease his labor, and divert and edify him.

There will come things which, properly handled, will shorten his working day and take away the most burdensome part of his effort, which will enable him to communicate, to travel, and to have a wider choice both in the general question of how he is to spend his life and in the specific question of how he is to spend an hour of his leisure. There is no need to belabor this point, nor its obverse—that out of science there will come, as there has in this last war, a host of instruments of destruction which will facilitate that labor, even as they have facilitated all others.

But no scientist, no matter how aware he may be of these fruits of his science, cultivates his work, or refrains from it, because of arguments such as these. No scientist can hope to evaluate what his studies, his researches, his experiments may in the end produce for his fellow men, except in one respect—if they are sound, they will produce knowledge. And this deep complementarity between what may be conceived to be the social justification of science and what is for the individual his compelling motive in its pursuit makes us look for other answers to the question of the relation of science to society.

One of these is that the scientist should assume responsibility for the fruits of his work. I would not argue against this, but it must be clear to all of us how very modest such assumption of responsibility can be, how very ineffective it has been in the past, how necessarily ineffective it will surely be in the future. In fact, it appears little more than exhortation to the man of learning to be properly uncomfortable, and, in the worst instances, is used as a sort of screen to justify the most casual, unscholarly and, in the last analysis, corrupt intrusion of scientists into other realms of which they have neither experience nor knowledge, nor the patience to obtain them.

The true responsibility of a scientist, as we all know, is to the integrity and vigor of his science. And because most scientists, like all men of learning, tend in part also to be teachers, they have a responsibility for the communication of the truths they have found. This is at least a collective if not an individual responsibility. That we should see in this any insurance that the fruits of science will be used for man's benefit, or denied to man when they make for his distress or destruction, would be a tragic naïveté.

There is another side of the coin. This is the question of whether there are elements in the way of life of the scientist which need not be restricted to the professional, and which have hope in them for bringing dignity and courage and serenity to other men. Science is not all of the life of reason; it is a part of it. As such, what can it mean to man?

Perhaps it would be well to emphasize that I am talking neither of wisdom nor of an elite of scientists, but precisely of the kind of work and thought, of action and discipline, that makes up the everyday professional life of the scientist. It is not of any general insight into human affairs that I am talking. It is not the kind of thing we recognize in our greatest statesmen, after long service devoted to practical affairs and to the public interest. It is something very much more homely and robust than that. It has in it the kind of beauty that is inseparable from craftsmanship and form, but it has in it also the vigor that we rightly associate with the simple, ordered lives of artisans or of farmers, that we rightly associate with lives to which limitations of scope, and traditional ways, have given robustness and structure.

Even less would it be right to interpret the question of what there is in the ways of science that may be of general value to mankind in terms of the creation of an elite. The study of physics, and I think my colleagues in the other sciences will let me speak for them too, does not make philosopher-kings. It has not, until now, made kings. It almost never makes fit philosophers—so rarely that they must be counted as exceptions. If the professional pursuit of science makes good scientists, if it makes men with a certain serenity in their lives, who yield perhaps a little more slowly than others to the natural corruptions of their time, it is doing a great deal, and all that we may rightly ask of it. For if Plato believed that in the study of geometry, a man might prepare himself for wisdom and responsibility in the world of men, it was precisely because he thought so hopefully that the understanding of men could be patterned after the understanding of geometry. If we believe that today, it is in a much more recondite sense, and a much more cautious one.

Where, then, is the point? For one thing, it is to describe some of the features of the professional life of the scientist, which make of it one of the great phenomena of the contemporary world. Here again I would like to speak of physics; but I have enough friends in the other sciences to know how close their experience is to ours. And I know too that despite profound differences in method and technique, differences which surely are an appropriate reflection of the difference in the areas of the world under study, what I would say of physics will seem familiar to workers in other disparate fields, such as mathematics or biology.

What are some of these points? There is, in the first instance, a total lack of authoritarianism, which is hard to comprehend or to admit unless one has lived with it. This is accomplished by one of the most exacting of intellectual disciplines. In physics the worker learns the possibility of error very early. He learns that there are ways to correct his mistakes; he learns the futility of trying to conceal them. For it is not a field in which error awaits death and subsequent generations for verdict—the next issue of the journals will take care of it. The refinement of techniques for the prompt discovery of error serves as well as any other as a hallmark of what we mean by science.

In any case, it is an area of collective effort in which there is a clear and well-defined community whose canons of taste and order simplify the life of the practitioner. It is a field in which the technique of experiment has given an almost perfect harmony to the balance between thought and action. In it we learn, so frequently that we could almost become accustomed to it, how vast is the novelty of the world, and how much even the physical world transcends in delicacy and in balance the limits of man's prior imaginings. We learn that views may be useful and inspiriting although they are not complete. We come to have a great caution in all assertions of totality, of finality or absoluteness.

In this field quite ordinary men, using what are in the last analysis only the tools which are generally available in our society, manage to unfold for themselves and all others who wish to learn, the rich story of one aspect of the physical world, and of man's experience. We learn to throw away those instruments of action and those modes of description which are not appropriate to the reality we are trying to discern, and in this most painful discipline, find ourselves modest before the world.

The question which is so much in our mind is whether a comparable experience, a comparable discipline, a comparable community of interest, can in any way be available to mankind at large. I suppose that all the professional scientists together number some one one-hundredth of a per cent of the men of the world—even this will define rather generously what we mean by scientists. Scientists as professionals are, I suppose, rather sure to constitute a small part of our people.

Clearly, if we raise at all this question that I have raised, it must be in the hope that there are other areas of human experience that may be discovered or invented or cultivated, and to which the qualities which distinguish scientific life may be congenial and appropriate. It is natural that serious scientists, knowing of their own experience something of the quality of their profession, should just today be concerned about its possible extension. For it is a time when the destruction and the evil of the last quarter century make men everywhere eager to seek all that can contribute to their intellectual life, some of the order and freedom and purpose which we conceive the great days of the past to have. Of all intellectual activity, science alone has flourished in the last centuries, science alone has turned out to have the kind of universality among men which the times require. I shall be disputed in this; but it is near to truth.

If one looks at past history, one may derive some encouragement for the hope that science, as one of the forms of reason, will nourish all of its forms. One may note how integral the love and cultivation of science were with the whole awakening of the human spirit which characterized the Renaissance. Or one may look at the late seventeenth and eighteenth centuries in France and England and see what pleasure and what stimulation the men of that time derived from the growth of physics, astronomy and mathematics.

What perhaps characterizes these periods of the past, which we must be careful not to make more heroic because of their remoteness, was that there were many men who were able to combine in their own lives the activities of a scientist with activities of art and learning and politics, and were able to carry over from the one into the others this combination of courage and modesty which is the lesson that science always tries to teach to anyone who practices it.

And here we come to a point we touched earlier. It is very different to hear the results of science, as they may be descriptively or even analytically taught in a class or in a book or in the popular talk of the time; it is very different to hear these and to participate even in a modest way in the actual attainment of new knowledge. For it is just characteristic of all work in scientific fields that there is no authority to whom to refer, no one to give canon, no one to blame if the picture does not make sense.

Clearly these circumstances pose a question of great difficulty in the field of education. For if there is any truth in the views that I have outlined, there is all the difference in the world between hearing about science or its results and sharing in the experience of the scientist himself and of that of the scientific community. We all know that an awareness of this, and an awareness of the value of science as method, rather than science as doctrine, underlies the practices of teaching to scientist and layman alike. For surely the whole notion of incorporating a laboratory in a high school or college is a deference to the belief that not only what the scientist finds but how he finds it is worth learning and teaching and worth living through.

Yet there is something fake about all this. No one who

has had to do with elementary instruction can have escaped a sense of artificiality in the way in which students are led, by the calculations of their instructors, to follow paths which will tell them something about the physical world. Precisely that groping for what is the appropriate experiment, what are the appropriate terms in which to view subtle or complex phenomena, which are the substance of scientific effort, almost inevitably are distilled out of it by the natural patterns of pedagogy. The teaching of science to laymen is not wholly a loss; and here perhaps physics is an atypically bad example. But surely they are rare men who, entering upon a life in which science plays no direct part, remember from their early courses in physics what science is like or what it is good for. The teaching of science is at its best when it is most like an apprenticeship.

President Conant, in his sensitive and thoughtful book *On Understanding Science*, has spoken at length of these matters. He is aware of how false it is to separate scientific theory from the groping, fumbling, tentative efforts which lead to it. He is aware that it is science as method and not as doctrine which we should try to teach. His basic suggestion is that we attempt to find, in the history of our sciences, stories which can be re-created in the instruction and experiment of the student and which thus can enable him to see at firsthand how error may give way to less error, confusion to less confusion, and bewilderment to insight.

The problem that President Conant has here presented is indeed a deep one. Yet he would be quite willing, I think, that I express skepticism that one can re-create the experience of science as an artifact. And he would no doubt share my concern that science so taught would be corrupt with antiquarianism. It was not antiquarianism but a driving curiosity that inspired in the men of the Renaissance their deep interest in classical culture.

For it is in fact difficult, almost to the point of impossibility, to re-create the climate of opinion in which substantial errors about the physical world, now no longer entertained, were not only held but were held unquestioned as part of the obvious mode of thinking about reality. It is most difficult to do because in all human thought only the tiniest fraction of our experience is in focus, and because to this focus a whole vast unanalyzed account of experience must be brought to bear. Thus I am inclined to think that, with exceptions I hope will be many but fear will be few, the attempt to give the history of science as a living history will be far more difficult than either to tell of the knowledge that we hold today or to write externally of that history as it may appear in the learned books. It could easily lead to a sort of exercise of mental inventiveness on the part of teachers and students alike which is the very opposite of the candor, the "no holds barred" rules of Professor Bridgman, that characterize scientific understanding at its best.

If I am troubled by President Conant's suggestions, this is not at all because I doubt that the suggestions he makes are desirable. I do have a deep doubt as to the extent to which they may be practical. There is something irreversible about acquiring knowledge; and the simulation of the search for it differs in a most profound way from the reality. In fact, it would seem that only those who had some firsthand experience in the acquisition of new knowledge in some disciplined field would be able truly to appreciate how great the science of the past has been, and would be able to measure those giant accomplishments against their own efforts to penetrate a few millimeters farther into darkness that surrounds them.

Thus it would seem at least doubtful that the spiritual fruits of science could be made generally available, either by the communication of its results, or by the study of its history, or by the necessarily somewhat artificial re-enactment of its procedures. Rather it would seem that there are general features of the scientists' work the direct experience of which in any context could contribute more to this end. All of us, I suppose, would list such features and find it hard to define the words which we found it necessary to use in our lists. But on a few, a common experience may enable us to talk in concert.

In the first instance the work of science is co-operative; a scientist takes his colleagues as judges, competitors and collaborators. That does not mean, of course, that he loves his colleagues; but it gives him a way of living with them which would be not without its use in the contemporary world. The work of science is discipline in that its essential inventiveness is most of all dedicated to means for promptly revealing error. One may think of the rigors of mathematics and the virtuosity of physical experiment as two examples. Science is disciplined in its rejection of questions that cannot be answered and in its grinding pursuit of methods for answering all that can. Science is always limited, and is in a profound sense unmetaphysical, in that it necessarily bases itself upon the broad ground of common human experience, tries to refine it within narrow areas where progress seems possible and exploration fruitful. Science is novelty and change. When it closes it dies. These qualities constitute a way of life which of course does not make wise men from foolish, or good men from wicked, but which has its beauty and which seems singularly suited to man's estate on earth.

If there is to be any advocacy at all in this talk, it would be this: that we be very sensitive to all new possibilities of extending the techniques and the patterns of science into other areas of human experience. Even in saying this we must be aware how slow the past development of science has in fact been, how much error there has been, and how much in it that turned out to be contrary to intellectual health or honesty.

We become fully aware of the need for caution if we look for a moment at what are called the social problems of the day and try to think what one could mean by approaching them in the scientific spirit, of trying to give substance, for example, to the feeling that a society that could develop atomic energy could also develop the means of controlling it. Surely the establishment of a secure peace is very much in all our minds. It is right that we try to bring reason to bear on an understanding of this problem; but for that there are available to us no equivalents of the experimental techniques of science. Errors of conception can remain undetected and even undefined. No means of appropriately narrowing the focus of thinking is known to us. Nor have we found good avenues for extending or deepening our experience that bears upon this problem. In short, almost all the preconditions of scientific activity are missing, and in this case, at least, one may have a melancholy certainty that man's inventiveness will not rapidly provide them. All that we have from science in facing such great questions is a memory of our professional life, which makes us somewhat skeptical of other people's assertions, somewhat critical of enthusiasms so difficult to define and to control.

Yet the past century has seen many valid and inspiriting examples for the extension of science to new domains. As even in the case of physics, the initial steps are always controversial; probably we should not as a group be unanimous in saying which of these extensions were hopeful, and which not, for the science of the future. But one feature which I cannot fail to regard as sound—particularly in the fields of biology and psychology—is that they provide an appropriate means of correlating understanding and action, and involve new experimental procedures in terms of which a new conceptual apparatus can be defined; above all, they give us means of detecting error. In fact, one of the features which must arouse our suspicion of the dogmas some of Freud's followers have built up on the initial brilliant works of Freud is the tendency toward a self-sealing system, a system, that is, which has a way of almost automatically discounting evidence which might bear adversely on the doctrine. The whole point of science is to do just the opposite: to invite the detection of error and to welcome it. Some of you may think that in another field a comparable system has been developed by the recent followers of Marx.

Thus we may hope for an ever-widening and more diverse field of application of science. But we must be aware how slowly these things develop and how little their development is responsive to even the most desperate of man's needs. For me it is an open question, and yet not a trivial one, whether in a time necessarily limited by the threats of war and of chaos these expanding areas in which the scientific spirit can flourish may yet contribute in a decisive way to man's rational life.

I have had to leave this essential question unanswered: I am not at all proud of that. In lieu of apology perhaps I may tell a story of another lecturer, speaking at Harvard, a few miles from here, two decades ago. Bertrand Russell had given a talk on the then new quantum mechanics, of whose wonders he was most appreciative. He spoke hard and earnestly in the New Lecture Hall. And when he was done, Professor Whitehead, who presided, thanked him for his efforts, and not least for "leaving the vast darkness of the subject unobscured."

ALFRED NORTH WHITEHEAD

PRINCIPIA MATHEMATICA, one of the early classics of symbolic logic, was the outcome of ten years of close collaboration. The authors: Alfred North Whitehead (1861-1947), then a lecturer on mathematics at Cambridge, and his former pupil, Bertrand Russell. In subsequent decades the two men remained the best of friends, but their philosophical views diverged enormously. Russell remained allergic to metaphysics. Whitehead, after becoming a professor of philosophy at Harvard, proceeded to construct a vast edifice of thought that included such traditional concepts as God, free will, and even a kind of immortality.

Empiricists shocked by Whitehead's later metaphysics often forget that he never regarded these views as *a priori* in the classic sense, but merely tentative generalizations from experience. They were what he called "adventures of ideas," to be altered or discarded if they proved useless. Unfortunately, their usefulness is not a matter on which other philosophers agree. To idealists and Protestant theologians in sympathy with Whitehead's central vision, his ideas are as profound and exciting as they are bizarre and meaningless to Russell and the logical empiricists.

The following selection is a chapter from one of Whitehead's most readable books, *Science and the Modern World.* So much has been written about the conflict of science and religion that it would be folly to expect to find here a novel method of reconciliation. But, though his thesis be a platitude, Whitehead defends it with such freshness and zest that it acquires new persuasiveness. "A clash of doctrines is not a disaster—it is an opportunity."

ALFRED NORTH WHITEHEAD

Religion and Science

The difficulty in approaching the question of the relations between Religion and Science is, that its elucidation requires that we have in our minds some clear idea of what we mean by either of the terms, 'religion' and 'science.' Also I wish to speak in the most general way possible, and to keep in the background any comparison of particular creeds, scientific or religious. We have got to understand the type of connection which exists between the two spheres, and then to draw some definite conclusions respecting the existing situation which at present confronts the world.

The *conflict* between religion and science is what naturally occurs to our minds when we think of this subject. It seems as though, during the last half-century, the results of science and the beliefs of religion had come into a position of frank disagreement, from which there can be no escape, except by abandoning either the clear teaching of science, or the clear teaching of religion. This conclusion has been urged by controversialists on either side. Not by all controversialists, of course, but by those trenchant intellects which every controversy calls out into the open.

The distress of sensitive minds, and the zeal for truth, and the sense of the importance of the issues, must command our sincerest sympathy. When we consider what religion is for mankind, and what science is, it is no exaggeration to say that the future course of history depends upon the decision of this generation as to the relations between them. We have here the two strongest general forces (apart from the mere impulse of the various senses) which influence men, and they seem to be set one against the other—the force of our religious intuitions, and the force of our impulse to accurate observation and logical deduction.

A great English statesman once advised his countrymen to use large-scale maps, as a preservative against alarms, panics, and general misunderstanding of the true relations between nations. In the same way in dealing with the clash between permanent elements of human nature, it is well to map our history on a large scale, and to disengage ourselves from our immediate absorption in the present conflicts. When we do this, we immediately discover two great facts. In the first place, there has always been a conflict between religion and science; and in the second place, both religion and science have always been in a state of continual development. In the early days of Christianity, there was a general belief among Christians that the world was coming to an end in the lifetime of people then living. We can make only indirect inferences as to how far this belief was authoritatively proclaimed; but it is certain that it was widely held, and that it formed an impressive part of the popular religious doctrine. The belief proved itself to be mistaken, and Christian doctrine adjusted itself to the change. Again in the early Church individual theologians very confidently deduced from the Bible opinions concerning the nature of the physical universe. In the year A. D. 535, a monk named Cosmas wrote a book which he entitled, *Christian Topography*. He was a travelled man who had visited India and Ethiopia; and finally he lived in a monastery at Alexandria, which was then a great centre of culture. In this book, basing himself upon the direct meaning of Biblical texts as construed by him in a literal fashion, he denied the existence of the antipodes, and asserted that the world is a flat parallelogram whose length is double its breadth.

In the seventeenth century the doctrine of the motion of the earth was condemned by a Catholic tribunal. A hundred years ago the extension of time demanded by geological science distressed religious people, Protestant and Catholic. And to-day the doctrine of evolution is an equal stumbling-block. These are only a few instances illustrating a general fact.

But all our ideas will be in a wrong perspective if we think that this recurring perplexity was confined to contradictions between religion and science; and that in these controversies religion was always wrong, and that science was always right. The true facts of the case are very much more complex, and refuse to be summarised in these simple terms.

Theology itself exhibits exactly the same character of gradual development, arising from an aspect of conflict between its own proper ideas. This fact is a commonplace to theologians, but is often obscured in the stress of controversy. I do not wish to overstate my case; so I will confine myself to Roman Catholic writers. In the seventeenth century a learned Jesuit, Father Petavius, showed that the theologians of the first three centuries of Christianity made use of phrases and statements which since the fifth century would be condemned as heretical. Also Cardinal Newman devoted a treatise to the discussion of the development of doctrine. He wrote it before he became a great Roman Catholic ecclesiastic; but throughout his life, it was never retracted and continually reissued.

Science is even more changeable than theology. No man of science could subscribe without qualification to Galileo's beliefs, or to Newton's beliefs, or to all his own scientific beliefs of ten years ago.

In both regions of thought, additions, distinctions, and modifications have been introduced. So that now, even when the same assertion is made to-day as was made a thousand, or fifteen hundred years ago, it is made subject to limitations or expansions of meaning, which were not contemplated at the earlier epoch. We are told by logicians that a proposition must be either true or false, and that there is no middle term. But in practice, we may know that a proposition expresses an important truth, but that it is subject to limitations and qualifications which at present remain undiscovered. It is a general feature of our knowledge, that we are insistently aware of important truth; and yet that the only formulations of these truths which we are able to make presuppose a general standpoint of conceptions which may have to be modified. I will give you two illustrations, both from science: Galileo said that the earth moves and that the sun is fixed; the Inquisition said that the earth is fixed and the sun moves; and Newtonian astronomers, adopting an absolute theory of space, said that both the sun and the earth move. But now we say that any one of these three statements is equally true, provided that you have fixed your sense of 'rest' and 'motion' in the way required by the statement adopted. At the date of Galileo's controversy with the Inquisition, Galileo's way of stating the facts was, beyond question, the fruitful procedure for the sake of scientific research. But in itself it was not more true than the formulation of the Inquisition. But at that time the modern concepts of relative motion were in nobody's mind; so that the statements were made in ignorance of the qualifications required for their more perfect truth. Yet this question of the motions of the earth and the sun expresses a real fact in the universe; and all sides had got hold of important truths concerning it. But with the knowledge of those times, the truths appeared to be inconsistent.

Again I will give you another example taken from the state of modern physical science. Since the time of Newton and Huyghens in the seventeenth century there have been two theories as to the physical nature of light. Newton's theory was that a beam of light consists of a stream of very minute particles, or corpuscles, and that we have the sensation of light when these corpuscles strike the retinas of our eyes. Huyghens' theory was that light consists of very minute waves of trembling in an all-pervading ether, and that these waves are travelling along a beam of light. The two theories are contradictory. In the eighteenth century Newton's theory was believed, in the nineteenth century Huyghens' theory was believed. To-day there is one large group of phenomena which can be explained only on the wave theory, and another large group which can be explained only on the corpuscular theory. Scientists have to leave it at that, and wait for the future, in the hope of attaining some wider vision which reconciles both.

We should apply these same principles to the questions in which there is a variance between science and religion. We would believe nothing in either sphere of thought which does not appear to us to be certified by solid reasons based upon the critical research either of ourselves or of competent authorities. But granting that we have honestly taken this precaution, a clash between the two on points of detail where they overlap should not lead us hastily to abandon doctrines for which we have solid evidence. It may be that we are more interested in one set of doctrines than in the other. But, if we have any sense of perspective and of the history of thought, we shall wait and refrain from mutual anathemas.

We should wait: but we should not wait passively, or in despair. The clash is a sign that there are wider truths and finer perspectives within which a reconciliation of a deeper religion and a more subtle science will be found.

In one sense, therefore, the conflict between science and religion is a slight matter which has been unduly emphasised. A mere logical contradiction cannot in itself point to more than the necessity of some readjustments, possibly of a very minor character on both sides. Remember the widely different aspects of events which are dealt with in science and in religion respectively. Science is concerned with the general conditions which are observed to regulate physical phenomena; whereas religion is wholly wrapped up in the contemplation of moral and aesthetic values. On the one side there is the law of gravitation, and on the other the contemplation of the beauty of holiness. What one side sees, the other misses; and vice versa.

Consider, for example, the lives of John Wesley and of Saint Francis of Assisi. For physical science you have in these lives merely ordinary examples of the operation of the principles of physiological chemistry, and of the dynamics of nervous reactions: for religion you have lives of the most profound significance in the history of the world. Can you be surprised that, in the absence of a perfect and complete phrasing of the principles of science and of the principles of religion which apply to these specific cases, the accounts of these lives from these divergent standpoints should involve discrepancies? It would be a miracle if it were not so.

It would, however, be missing the point to think that we need not trouble ourselves about the conflict between science and religion. In an intellectual age there can be no active interest which puts aside all hope of a vision of the harmony of truth. To acquiesce in discrepancy is destructive of candour, and of moral cleanliness. It belongs to the self-respect of intellect to pursue every tangle of thought to its final unravelment. If you check that impulse, you will get no religion and no science from an awakened thoughtfulness. The important question is, In what spirit are we going to face the issue? There we come to something absolutely vital.

A clash of doctrines is not a disaster—it is an opportunity. I will explain my meaning by some illustrations from science. The weight of an atom of nitrogen was well known. Also it was an established scientific doctrine that the average weight of such atoms in any considerable mass will be always the same. Two experimenters, the late Lord Rayleigh and the late Sir William Ramsay, found that if they obtained nitrogen by two different methods, each equally effective for that purpose, they always observed a persistent slight difference between the average weights of the atoms in the two cases. Now I ask you, would it have been rational of these men to have despaired because of this conflict between chemical theory and scientific observation? Suppose that for some reason the chemical doctrine had been highly prized throughout some district as the foundation of its social order:—would it have been wise, would it have been candid, would it have been moral, to forbid the disclosure of the fact that the experiments produced discordant results? Or, on the other hand, should Sir William Ramsay and Lord Rayleigh have proclaimed that chemical theory was now a detected delusion? We see at once that either of these ways would have been a method of facing the issue in an entirely wrong spirit. What Rayleigh and Ramsay did was this: They at once perceived that they had hit upon a line of investigation which would disclose some subtlety of chemical theory that had hitherto eluded observation. The discrepancy was not a disaster: it was an opportunity to increase the sweep of chemical knowledge. You all know the end of the story: finally argon was discovered, a new chemical element which had lurked undetected, mixed with the nitrogen. But the story has a sequel which forms my second illustration. This discovery drew attention to the importance of observing accurately minute differences in chemical substances as obtained by different methods. Further researches of the most careful accuracy were undertaken. Finally another physicist, F. W. Aston, working in the Cavendish Laboratory at Cambridge in England, discovered that even the same element might assume two or more distinct forms, termed *isotopes,* and that the law of the constancy of average atomic weight holds for each of these forms, but as between the different isotopes differs slightly. The research has effected a great stride in the power of chemical theory, far transcending in importance the discovery of argon from which it originated. The moral of these stories lies on the surface, and I will leave to you their application to the case of religion and science.

In formal logic, a contradiction is the signal of a defeat: but in the evolution of real knowledge it marks the first step in progress towards a victory. This is one great reason for the utmost toleration of variety of opinion. Once and forever, this duty of toleration has been summed up in the words, 'Let both grow together until the harvest.' The failure of Christians to act up to this precept, of the highest authority, is one of the curiosities of religious history. But we have not yet exhausted the discussion of the moral temper required for the pursuit of truth. There are short cuts leading merely to an illusory success. It is easy enough to find a theory, logically harmonious and with important applications in the region of fact, provided that you are content to disregard half your evidence. Every age produces people with clear logical intellects, and with the most praiseworthy grasp of the importance of some sphere of human experience, who have elaborated, or inherited, a scheme of thought which exactly fits those experiences which claim their interest. Such people are apt resolutely to ignore, or to explain away, all evidence which confuses their scheme with contradictory instances; what they cannot fit in is for them nonsense. An unflinching determination to take the whole evidence into account is the only method of preservation against the fluctuating extremes of fashionable opinion. This advice seems so easy, and is in fact so difficult to follow.

One reason for this difficulty is that we cannot think first and act afterwards. From the moment of birth we are immersed in action, and can only fitfully guide it by taking thought. We have, therefore, in various spheres of experience to adopt those ideas which seem to work within those spheres. It is absolutely necessary to trust to ideas which are generally adequate, even though we know that there are subtleties and distinctions beyond our ken. Also apart from the necessities of action, we cannot even keep before our minds the whole evidence except under the guise of doctrines which are incompletely harmonised. We cannot think in terms of an indefinite multiplicity of detail; our evidence can acquire its proper importance only if it comes before us marshalled by general ideas. These ideas we inherit—they form the tradition of our civilisation. Such traditional ideas are never static. They are either fading into meaningless formulae, or are gaining power by the new lights thrown by a more delicate apprehension. They are transformed by the urge of critical reason, by the vivid evidence of emotional experience, and by the cold certainties of scientific perception. One fact is certain, you cannot keep them still. No generation can merely reproduce its ancestors. You may preserve the life in a flux of form, or preserve the form amid an ebb of life. But you cannot permanently enclose the same life in the same mould.

The present state of religion among the European races illustrates the statements which I have been making. The phenomena are mixed. There have been reactions and revivals. But on the whole, during many generations, there has been a gradual decay of religious influence in European civilisation. Each revival touches a lower peak than its predecessor, and each period of slackness a lower depth. The average curve marks a steady fall in religious tone. In some countries the interest in religion is higher than in others. But in those countries where the interest is relatively high, it still falls as the generations pass. Religion is tending to degenerate into a decent formula wherewith to embellish a comfortable life. A great historical movement on this scale results from the convergence of many causes. I wish to suggest two of them which lie within the scope of this chapter for consideration.

In the first place for over two centuries religion has been on the defensive, and on a weak defensive. The period has been one of unprecedented intellectual progress. In this way a series of novel situations have been produced for thought. Each such occasion has found the religious thinkers unprepared. Something, which has been proclaimed to be vital, has finally, after struggle, distress, and anathema, been modified and otherwise interpreted. The next generation of religious apologists then congratulates the religious world on the deeper insight which has been gained. The result of the continued repetition of this undignified retreat, during many generations, has at last almost entirely destroyed the intellectual authority of religious thinkers. Consider this contrast: when Darwin or Einstein proclaim theories which modify our ideas, it is a triumph for science. We do not go about saying that there is another defeat for science, because its old ideas have been abandoned. We know that another step of scientific insight has been gained.

Religion will not regain its old power until it can face change in the same spirit as does science. Its principles may be eternal, but the expression of those principles requires continual development. This evolution of religion is in the main a disengagement of its own proper ideas from the adventitious notions which have crept into it by reason of the expression of its own ideas in terms of the imaginative picture of the world entertained in previous ages. Such a release of religion from the bonds of imperfect science is all to the good. It stresses its own genuine message. The great point to be kept in mind is that normally an advance in science will show that statements of various religious beliefs require some sort of modification. It may be that they have to be expanded or explained, or indeed entirely restated. If the religion is a sound expression of truth, this modification will only exhibit more adequately the exact point which is of importance. This process is a gain. In so far, therefore, as any religion has any contact with physical facts, it is to be expected that the point of view of those facts must be continually modified as scientific knowledge advances. In this way, the exact relevance of these facts for religious thought will grow more and more clear. The progress of science must result in the unceasing codification of religious thought, to the great advantage of religion.

The religious controversies of the sixteenth and seventeenth centuries put theologians into a most unfortunate state of mind. They were always attacking and defending. They pictured themselves as the garrison of a fort surrounded by hostile forces. All such pictures express half-truths. That is why they are so popular. But they are dangerous. This particular picture fostered a pugnacious party spirit which really expresses an ultimate lack of faith. They dared not modify, because they shirked the task of disengaging their spiritual message from the associations of a particular imagery.

Let me explain myself by an example. In the early medieval times, Heaven was in the sky, and Hell was underground; volcanoes were the jaws of Hell. I do not assert that these beliefs entered into the official formulations: but they did enter into the popular understanding of the general doctrines of Heaven and Hell. These notions were what everyone thought to be implied by the doctrine of the future state. They entered into the explanations of the influential exponents of Christian belief. For example, they occur in the *Dialogues* of Pope Gregory, the Great, a man whose high official position is surpassed only by the magnitude of his services to humanity. I am not saying what we ought to believe about the future state. But whatever be the right doctrine, in this instance the clash between religion and science, which has relegated the earth to the position of a second-rate planet attached to a second-rate sun, has been greatly to the benefit of the spirituality of religion by dispersing these medieval fancies.

Another way of looking at this question of the evolution of religious thought is to note that any verbal form of statement which has been before the world for some time discloses ambiguities; and that often such ambiguities strike at the very heart of the meaning. The effective sense in which a doctrine has been held in the past cannot be determined by the mere logical analysis of verbal statements, made in ignorance of the logical trap. You have to take into account the whole reaction of human nature to the scheme of thought. This reaction is of a mixed character, including elements of emotion derived from our lower natures. It is here that the impersonal criticism of science and of philosophy comes to the aid of religious evolution. Example after example can be given of this motive force in development. For example, the logical difficulties inherent in the doctrine of the moral cleansing of human nature by the power of religion rent Christianity in the days of Pelagius and Augustine—that is to say, at the beginning of the fifth century. Echoes of that controversy still linger in theology.

So far, my point has been this: that religion is the expression of one type of fundamental experiences of mankind: that religious thought develops into an increasing accuracy of expression, disengaged from adventitious imagery: that the interaction between religion and science is one great factor in promoting this development.

I now come to my second reason for the modern fading of interest in religion. This involves the ultimate question which I stated in my opening sentences. We have to know what we mean by religion. The churches, in their presentation of their answers to this query, have put forward aspects of religion which are expressed in terms either suited to the emotional reactions of bygone times or directed to excite modern emotional interests of nonreligious character. What I mean under the first heading is that religious appeal is directed partly to excite that instinctive fear of the wrath of a tyrant which was inbred in the unhappy populations of the arbitrary empires of the ancient world, and in particular to excite that fear of an all-powerful arbitrary tyrant behind the unknown forces of nature. This appeal to the ready instinct of brute fear is losing its force. It lacks any directness of response, because modern science and modern conditions of life have taught us to meet occasions of apprehension by a critical analysis of their causes and conditions. Religion is the reaction of human nature to its search for God. The presentation of God under the aspect of power awakens every modern instinct of critical reaction. This is fatal; for religion collapses unless its main positions command immediacy of assent. In this respect the old phraseology is at variance with the psychology of modern civilisations. This change in psychology is largely due to science, and is one of the chief ways in which the advance of science has weakened the hold of the old religious forms of expression. The non-religious motive which has entered into modern religious thought is the desire for a comfortable organisation of modern society. Religion has been presented as valuable for the ordering of life. Its claims have been rested upon its function as a sanction to right conduct. Also the purpose of right conduct quickly degenerates into the formation of pleasing social relations. We have here a subtle degradation of religious ideas, following upon their gradual purification under the influence of keener ethical intuitions. Conduct is a by-product of religion—an inevitable by-product, but not the main point. Every great religious teacher has revolted against the presentation of religion as a mere sanction of rules of conduct. Saint Paul denounced the Law, and Puritan divines spoke of the filthy rags of righteousness. The insistence upon rules of conduct marks the ebb of religious fervour. Above and beyond all things, the religious life is not a research after comfort. I must now state, in all diffidence, what I conceive to be the essential character of the religious spirit.

Religion is the vision of something which stands beyond, behind, and within the passing flux of immediate things; something which is real, and yet waiting to be realised; something which is a remote possibility, and yet the greatest of present facts; something that gives meaning to all that passes, and yet eludes apprehension; something whose possession is the final good, and yet is beyond all reach; something which is the ultimate ideal, and the hopeless quest.

The immediate reaction of human nature to the religious vision is worship. Religion has emerged into human experience mixed with the crudest fancies of barbaric imagination. Gradually, slowly, steadily the vision recurs in history under nobler form and with clearer expression. It is the one element in human experience which persistently shows an upward trend. It fades and then recurs. But when it renews its force, it recurs with an added richness and purity of content. The fact of the religious vision, and its history of persistent expansion, is our one ground for optimism. Apart from it, human life is a flash of occasional enjoyments lighting up a mass of pain and misery, a bagatelle of transient experience.

The vision claims nothing but worship; and worship is a surrender to the claim for assimilation, urged with the motive force of mutual love. The vision never overrules. It is always there, and it has the power of love presenting the one purpose whose fulfilment is eternal harmony. Such order as we find in nature is never force—it presents itself as the one harmonious adjustment of complex detail. Evil is the brute motive force of fragmentary purpose, disregarding the eternal vision. Evil is overruling, retarding, hurting. The power of God is the worship He inspires. That religion is strong which in its ritual and its modes of thought evokes an apprehension of the commanding vision. The worship of God is not a rule of safety—it is an adventure of the spirit, a flight after the unattainable. The death of religion comes with the repression of the high hope of adventure.

JOHN DOS PASSOS

IN HIS GREAT TRILOGY, *U.S.A.*, novelist John Dos Passos (1896-1970) introduced three exceedingly novel literary devices. He splattered his narrative with "Newsreels" (compounded of headlines, news stories, advertisements, and snatches of popular songs), with "Camera Eyes" (impressionistic personal memories), and with short biographies of famous Americans. How well these devices succeed in giving the reader a *feel* of the period is a moot question, but most critics agree that the biographies are often small masterpieces of ironic, poetic prose. The following sketch of Charles Proteus Steinmetz, taken from *The 42nd Parallel*, was written long before Dos Passos softened his anger against capitalism and redirected it toward the views he himself had held in his early, pro-Communist days. Regardless of how one may react to its political overtones, it remains one of his most dramatic and moving pieces.

JOHN DOS PASSOS

Proteus

STEINMETZ was a hunchback,
son of a hunchback lithographer.

He was born in Breslau in eighteen-sixtyfive, graduated with highest honors at seventeen from the Breslau Gymnasium, went to the University of Breslau to study mathematics;

mathematics to Steinmetz was muscular strength and long walks over the hills and the kiss of a girl in love and big evenings spent swilling beer with your friends;

on his broken back he felt the topheavy weight of society the way workingmen felt it on their straight backs, the way poor students felt it, was a member of a Socialist club, editor of a paper called *The People's Voice*.

Bismarck was sitting in Berlin like a big paperweight to keep the new Germany feudal, to hold down the empire for his bosses the Hohenzollerns.

Steinmetz had to run off to Zurich for fear of going to jail; at Zurich his mathematics woke up all the professors at the Polytechnic;

but Europe in the eighties was no place for a penniless German student with a broken back and a big head filled with symbolic calculus and wonder about electricity that is mathematics made power

and a Socialist at that.

With a Danish friend he sailed for America steerage on an old French line boat *La Champagne*, lived in Brooklyn at first and commuted to Yonkers where he had a twelvedollar a week job with Rudolph Eichemeyer, who was a German exile from fortyeight, an inventor and electrician and owner of a factory where he made hatmaking machinery and electrical generators.

In Yonkers he worked out the theory of the Third Harmonics

and the law of hysteresis which states in a formula the hundredfold relations between the metallic heat, density, frequency, when the poles change places in the core of a magnet under an alternating current.

It is Steinmetz's law of hysteresis that makes possible all the transformers that crouch in little boxes and gableroofed houses in all the hightension lines all over everywhere. The mathematical symbols of Steinmetz's law are the patterns of all transformers everywhere.

In eighteen-ninetytwo, when Eichemeyer sold out to the corporation that was to form General Electric, Steinmetz was entered in the contract along with other valuable apparatus. All his life Steinmetz was a piece of apparatus belonging to General Electric.

First his laboratory was at Lynn, then it was moved and the little hunchback with it to Schenectady, the electric city.

General Electric humored him, let him be a Socialist, let him keep a greenhouseful of cactuses lit up by mercury lights, let him have alligators, talking crows, and a gila monster for pets, and the publicity department talked up the wizard, the medicine man who knew the symbols that opened up the doors of Ali Baba's cave.

Steinmetz jotted a formula on his cuff and next morning a thousand new powerplants had sprung up and the dynamos sang dollars and the silence of the transformers was all dollars,

and the publicity department poured oily stories into the

ears of the American public every Sunday and Steinmetz became the little parlor magician,

who made a toy thunderstorm in his laboratory and made all the toy trains run on time and the meat stay cold in the icebox and the lamp in the parlor and the great lighthouses and the searchlights and the revolving beams of light that guide airplanes at night towards Chicago, New York, St. Louis, Los Angeles,

and they let him be a Socialist and believe that human society could be improved the way you can improve a dynamo, and they let him be pro-German and write a letter offering his services to Lenin because mathematicians are so impractical who make up formulas by which you can build powerplants, factories, subway systems, light, heat, air, sunshine, but not human relations that affect the stockholders' money and the directors' salaries.

Steinmetz was a famous magician and he talked to Edison tapping with the Morse code on Edison's knee

because Edison was so very deaf

and he went out West

to make speeches that nobody understood

and he talked to Bryan about God on a railroad train

and all the reporters stood round while he and Einstein met face to face,

but they couldn't catch what they said

and Steinmetz was the most valuable piece of apparatus General Electric had

until he wore out and died.

JULIAN HUXLEY

BIOLOGISTS have long recognized that the singing of birds plays an important role in their mating rituals and also serves to identify the feeding territory over which each bird has staked a claim. Does this mean that the bird does not "enjoy" singing? One might as well say that a trumpet player does not care for the sound of his own horn because he earns a living by it or because it excites a girl friend. It is reassuring, therefore, to be told in the following essay by an eminent zoologist that the joy birds seem to experience is more than just an illusion in the minds of sentimental bird-watchers.

Julian Sorell Huxley's (1877-1975) resemblances to his grandfather Thomas Henry Huxley have often been singled out—a major interest in zoology supplemented by a broad background in all the sciences, a philosophic agnosticism coupled with a humanitarian faith in progress, a vigorous participation in political affairs (he was formerly director general of UNESCO), superb skill as teacher and lecturer, and above all, the ability to write popular books that are models of scientific accuracy and stylistic excellence.

The selection chosen here is from his *Essays of a Biologist*, a collection of early pieces dedicated to his colleagues at the Rice Institute, Houston, Texas, where he taught for several years. The reader does not have to be a member of the Audubon Society to be entranced by its insight into the minds of those bright-eyed, two-legged cousins who inhabit the atmosphere above our heads.

JULIAN HUXLEY

An Essay on Bird-Mind

"O Nightingale, thou surely art
A creature of a fiery heart."
—W. WORDSWORTH.

"The inferior animals, when the conditions of life are favourable, are subject to periodical fits of gladness, affecting them powerfully and standing out in vivid contrast to their ordinary temper. . . . Birds are more subject to this universal joyous instinct than mammals, and . . . as they are much freer than mammals, more buoyant and graceful in action, more loquacious, and have voices so much finer, their gladness shows itself in a greater variety of ways, with more regular and beautiful motions, and with melody."—W. H. HUDSON.

"How do you know but ev'ry Bird that cuts the airy way
Is an immense world of delight, clos'd by your senses five?"
—BLAKE.

"ILS N'ONT PAS *de cerveau—ils n'ont que de l'âme.*" [They have no brain—they have only a soul.] A dog was being described, with all his emotion, his apparent passion to make himself understood, his failure to reach comprehension; and that was how the French man of letters summed up the brute creation—"*pas de cerveau—que de l'âme.*"

Nor is it a paradox: it is a half-truth that is more than half true—more true at least than its converse, which many hold.

There is a large school to-day who assert that animals are "mere machines." Machines they may be: it is the qualification which does not fit. I suppose that by saying "mere" machines it is meant to imply that they have the soulless, steely quality of a machine which goes when it is set going, stops when another lever is turned, acts only in obedience to outer stimuli, and is in fact unemotional—a bundle of operations without any quality meriting the name of a self.

It is true that the further we push our analysis of animal behaviour, the more we find it composed of a series of automatisms, the more we see it rigorously determined by combination of inner constitution and outer circumstance, the more we have cause to deny to animals the possession of anything deserving the name of reason, ideals, or abstract thought. The more, in fact, do they appear to us as mechanisms (which is a much better word than machines, since this latter carries with it definite connotations of metal or wood, electricity or steam). They are mechanisms, because their mode of operation is regular; but they differ from any other type of mechanism known to us in that their working is—to put it in the most non-committal way—accompanied by emotion. It is, to be sure, a combination of emotion with reason that we attribute to a soul; but none the less, in popular parlance at least, the emotional side is predominant, and pure reason is set over against the emotional content which gives soul its essence. And this emotional content we most definitely find running through the lives of higher animals.

The objection is easily and often raised that we have no direct knowledge of emotion in an animal, no direct proof of the existence of any purely mental process in its life. But this is as easily laid as raised. We have no direct knowledge of emotion or any other conscious process in the life of any human being save our individual selves; and yet we feel no hesitation in deducing it from others' behaviour. Although it is an arguable point whether biological science may not for the moment be better served by confining the subject-matter and terms of analysis to behaviour alone, it is a very foolhardy "behaviorist" indeed who denies the *existence* of emotion and conscious process!

But the practical value of this method of thinking is, as I say, an arguable point; it is indeed clear that a great immediate advance, especially in non-human biology, has been and may still be made by translating the uncertain and often risky terms of subjective psychology into those based upon the objective description of directly observable behaviour. However, it is equally easy to maintain, and I for one maintain it, that to omit a whole category of phenomena from consideration is unscientific, and must in the long run lead to an unreal, because limited, view of things; and that, when great detail of analysis is not required, but only broad lines and general comparison, the psychological terminology, of memory, fear, anger, curiosity, affection, is the simpler and more direct tool, and should be used to supplement and make more real the cumbersome and less complete behavioristic terminology, of modification of behaviour, fright, aggression, and the rest.

It is at least abundantly clear that, if we are to believe in the principle of uniformity at all, we must ascribe emotion to animals as well as to men: the similarity of behaviour is so great that to assert the absence of a whole class of phenomena in one case, its presence in the other, is to make scientific reasoning a farce.

"Pas de cerveau—que de l'âme." Those especially who have studied birds will subscribe to this. The variety of their emotions is greater, their intensity more striking, than in four-footed beasts, while their power of modifying behaviour by experience is less, the subjection to instinct more complete. Those who are interested in the details can see from experiments, such as those recorded by Mr. Eliot Howard in his *Territory in Bird Life*, how limited is a bird's power of adjustment; but I will content myself with a single example, one of nature's experiments, recorded by Mr. Chance last year by the aid of the cinematograph—the behaviour of small birds when the routine of their life is upset by the presence of a young Cuckoo in the nest.

When, after prodigious exertions, the unfledged Cuckoo has ejected its foster-brothers and sisters from their home, it sometimes happens that one of them is caught on or close to the rim of the nest. One such case was recorded by Mr. Chance's camera. The unfortunate fledgling scrambled about on the branches below the nest; the parent Pipit flew back with food; the cries and open mouth of the ejected bird attracted attention, and it was fed; and the mother then settled down upon the nest as if all was in normal order. Meanwhile, the movements of the fledgling in the foreground grew feebler, and one could imagine its voice quavering off, fainter and fainter, as its vital warmth departed. At the next return of the parent with food the young one was dead.

It was the utter stupidity of the mother that was so impressive—its simple response to stimulus—of feeding to the stimulus of the young's cry and open mouth, of brooding to that of the nest with something warm and feathery contained in it—its neglect of any steps whatsoever to restore the fallen nestling to safety. It was almost as pitiable an exhibition of unreason as the well-attested case of the wasp attendant on a wasp-grub, who, on being kept without food for some time, grew more and more restless, and eventually bit off the hind end of the grub and offered it to what was left!

Birds in general are stupid, in the sense of being little able to meet unforeseen emergencies; but their lives are often emotional, and their emotions are richly and finely expressed. I have for years been interested in observing the courtship and the relations of the sexes in birds, and have in my head a number of pictures of their notable and dramatic moments. These seem to me to illustrate so well the emotional furnishing of birds, and to provide such a number of windows into that strange thing we call a bird's mind, that I shall simply set some of them down as they come to me.

First, then, the coastal plain of Louisiana; a pond, made and kept as a sanctuary by that public-spirited bird-lover Mr. E. A. McIlhenny, filled with noisy crowds of Egrets and little egret-like Herons. These, in great flocks, fly back across the "Mexique Bay" in the spring months from their winter quarters in South America. Arrived in Louisiana, they feed and roost in flocks for a time, but gradually split up into pairs. Each pair, detaching themselves from the flocks, choose a nesting-site (by joint deliberation) among the willows and maples of the breeding pond. And then follows a curious phenomenon. Instead of proceeding at once to biological business in the shape of nest-building and egg-laying, they indulge in what can only be styled a honeymoon. For three or four days both members of the pair are always on the chosen spot, save for the necessary visits which they alternately pay to the distant feeding grounds. When both are there, they will spend hours at a time sitting quite still, just touching one another. Generally the hen sits on a lower branch, resting her head against the cock bird's flanks; they look for all the world like one of those inarticulate but happy couples upon a bench in the park in spring. Now and again, however, this passivity of sentiment gives place to wild excitement. Upon some unascertainable cause the two birds raise their necks and wings, and, with loud cries, intertwine their necks. This is so remarkable a sight that the first time I witnessed it I did not fully credit it, and only after it had happened before my eyes on three or four separate occasions was I forced to admit it as a regular occurrence in their lives. The long necks are so flexible that they can and do make a complete single turn round each other—a real true-lover's-knot! This once accomplished, each bird then—most wonderful of all—runs its beak quickly and amorously through the just raised aigrettes of the other, again and again, nibbling and clappering them from base to tip. Of this I can only say that it seemed to bring such a pitch of emotion that I could have wished to be a Heron that I might experience it. This over, they would untwist their necks and subside once more into their usual quieter sentimentality.

This, alas! I never saw with the less common little White Egrets, but with the Louisiana Heron (which should, strictly speaking, be called an egret too); but since every other action of the two species is (in all save a few minor details) the same, I assume that the flashing white, as well as the slate and vinous and grey birds, behave thus.

The greeting ceremony when one bird of the pair, after having been away at the feeding grounds, rejoins its mate is also beautiful. Some little time before the human watcher notes the other's approach, the waiting bird rises on its branch, arches and spreads its wings, lifts its aigrettes into a fan and its head-plumes into a crown, bristles up the feathers of its neck, and emits again and again a hoarse cry. The other approaches, settles in the branches near by, puts itself into a similar position, and advances towards its mate; and after a short excited space they settle down close together. This type of greeting is repeated every day until the young leave the nest; for after the eggs are laid both sexes brood, and there is a nest-relief four times in every twenty-four hours. Each time the same attitudes, the same cries, the same excitement; only now at the end of it all, one steps off the nest, the other on. One might suppose that this closed the performance. But no: the bird that has been relieved is still apparently animated by stores of unexpended emotion; it searches about for a twig, breaks it off or picks it up, and returns with it in beak to present to the other. During the presentation the greeting ceremony is again gone through; after each relief the whole business of presentation and greeting may be repeated two, or four, or up even to ten or eleven times before the free bird flies away.

When there are numerous repetitions of the ceremony, it is extremely interesting to watch the progressive extinction of excitement. During the last one or two presentations the twig-bringing bird may scarcely raise his wings or plumes, and will often betray an absent air, turning his head in the direction in which he is proposing to fly off.

No one who has seen a pair of Egrets thus change places on the nest, bodies bowed forward, plumes a cloudy fan of lace, absolute whiteness of plumage relieved by gold of eye and lore and black of bill, and the whole scene animated by the repeated, excited cry, can ever forget it. But such unforgettable scenes are not confined to other countries. Here in

England you can see as good; I have seen them on the reservoirs of Tring, and within full view of the road by Frensham Pond—the courtship forms and dances of the Crested Grebe.

The Crested Grebe is happily becoming more familiar to bird-lovers in England. Its brilliant white belly, protective grey-brown back, rippleless and effortless diving, long neck, and splendid ruff and ear-tufts of black, chestnut, and white, conspire to make it a marked bird. In the winter the crest is small, and even when fully grown in spring it is usually held close down against the head, so as to be not at all conspicuous. When it is spread, it is almost, without exception, in the service of courtship or love-making. Ten years ago I spent my spring holiday watching these birds on the Tring reservoirs. I soon found out that their courtship, like the Herons', was mutual, not one-sidedly masculine as in Peacocks or fowls. It consisted most commonly in a little ceremony of head-shaking. The birds of a pair come close, face one another, raise their necks, and half-spread their ruffs. Then, with a little barking note, they shake their heads rapidly, following this by a slow swinging of them from side to side. This alternate shaking and swinging continues perhaps a dozen or twenty times; and the birds then lower their standards, become normal everyday creatures, and betake themselves to their fishing or resting or preening again. This is the commonest bit of love-making; but now and then the excitement evident even in these somewhat casual ceremonies is raised to greater heights and seems to reinforce itself. The little bouts of shaking are repeated again and again. I have seen over eighty succeed each other uninterruptedly. And at the close the birds do not relapse into ordinary life. Instead, they raise their ruffs still further, making them almost Elizabethan in shape. Then one bird dives; then the other: the seconds pass. At last, after perhaps half or three-quarters of a minute (half a minute is a long time when one is thus waiting for a bird's reappearance!) one after the other they emerge. Both hold masses of dark brownish-green weed, torn from the bottom of the pond, in their beaks, and carry their heads down and back on their shoulders, so that either can scarcely see anything of the other confronting it save the concentric colours of the raised ruff. In this position they swim together. It is interesting to see the eager looks of the first-emerged, and its immediate start towards the second when it too reappears. They approach, rapidly, until the watcher wonders what will be done to avert a collision. The answer is simple: there is no averting of a collision! But the collision is executed in a remarkable way: the two birds, when close to each other, leap up from the water and meet breast to breast, almost vertical, suddenly revealing the whole flashing white under-surface. They keep themselves in this position by violent splashings of the feet, rocking a little from side to side as if dancing, and very gradually sinking down (always touching with their breasts) towards the horizontal.

Meanwhile, they exchange some of the weed they are carrying; or at least nibbling and quick movements of the head are going on. And so they settle down on to the water, shake their heads a few times more, and separate, changing back from these performers of an amazing age-old rite—age-old but ever fresh—into the feeding- and sleeping-machines of every day, but leaving a vision of strong emotion, canalized into the particular forms of this dive and dance. The whole performance impresses the watcher not only with its strength, but as being apparently of very little direct (though possibly much indirect) biological advantage, the action being self-exhausting, not stimulating to further sexual relations, and carried out, it would seem, for its own sake.

Further acquaintance with the Grebe only deepened the interest and made clearer the emotional tinge underlying all the relations of the sexes. This bird, too, has its "greeting ceremony"; but since, unlike the colonial Herons and Egrets, it makes every effort to conceal its nest, this cannot take place at its most natural moment, that of nest-relief, but must be made to happen out on the open water where there are no secrets to betray. If the sitting bird wishes to leave the nest, and the other does not return, it flies off, after covering the eggs with weed, in search of its mate; it is common in the breeding season to see a Grebe in the "search-attitude," with neck stretched up and slightly forward and ear-tufts erected, emitting a special and far-carrying call. When this call is recognized and answered, the two birds do nothing so simple as to fly or swim to each other, but a special and obviously exciting ceremony is gone through. The bird that has been searched for and found puts itself into a very beautiful attitude, with wings half-spread and set at right angles to the body, ruff erected circularly, and head drawn back upon the shoulders, so that nothing is visible but the brilliant rosette of the spread ruff in the centre of the screen of wings, each wing showing a broad bar of brilliant white on its dusk-grey surface. In this position it swings restlessly back and forth in small arcs, facing towards its mate. The discoverer meanwhile has dived; but, swimming immediately below the surface of the water, its progress can be traced by the arrowy ripple it raises. Now and again it lifts its head and neck above the water, periscope-wise, to assure itself of its direction, and resumes its subaqueous course. Nor does it rise just in front of the other bird; but swims under and just beyond, and, as its mate swings round to the new orientation, emerges in a really extraordinary attitude. At the last it must have dived a little deeper; for now it appears perpendicularly from the water, with a slowish motion, slightly spiral, the beak and head pressed down along the front of the neck. I compared it in my notes of ten years ago with "the ghost of a Penguin," and that comparison is still the best I can think of to give some idea of the strange unreality of its appearance. It then settles down upon the water and the pair indulge in one of their never-failing bouts of head-shaking.

Two mated birds rejoin each other after a few hours' separation. Simple enough in itself—but what elaboration of detail, what piling on of little excitements, what purveying of thrills!

Other emotions too can be well studied in this bird, notably jealousy. Several times I have seen little scenes like the following enacted. A pair is floating idly side by side, necks drawn right down so that the head rests on the centre of the back. One—generally, I must admit, it has been the cock, but I think the hen may do so too on occasion—rouses himself from the pleasant lethargy, swims up to his mate, places himself in front of her, and gives a definite, if repressed, shake of the head. It is an obvious sign of his desire to "have a bit of fun"—to go through with one of those bouts of display and head-shaking in which pleasurable emotion clearly reaches its highest level in the birds' lives, as any one who has watched their habits with any thoroughness would agree. It also acts, by a simple extension of function, as an informative symbol. The other bird knows what is meant; it raises its head from beneath its wing, gives a sleepy, barely discernible shake—and replaces the head. In so doing it puts back the possibility of the ceremony and the thrill into its slumbers; for it takes two to make love, for Grebe as for human. The cock swims off; but he has a restless air, and in a minute or so is back again, and the same series of events is run through. This may be repeated three or four times.

If now another hen bird, unaccompanied by a mate, reveals herself to the eye of the restless and disappointed cock, he will make for her and try the same insinuating informative head-shake on her; and, in the cases that I have seen, she has responded, and a bout of shaking has begun. Flirtation—illicit love, if you will; for the Grebe, during each breeding season at least, is strictly monogamous, and the whole economics of its family life, if I may use the expression, are based on the co-operation of male and female in incubation and the feeding and care of the young. On the other hand, how natural and how human! and how harmless—for there is no evidence that the pretty thrills of the head-shaking display ever lead on to anything more serious.

But now observe. Every time that I have seen such a flirtation start, it has always been interrupted. The mate, so sleepy before, yet must have had one eye open all the time. She is at once aroused to action: she dives, and attacks the strange hen after the fashion of Grebes, from below, with an underwater thrust of the sharp beak in the belly. Whether the thrust ever goes home I do not know. Generally, I think, the offending bird becomes aware of the danger just in time, and, squawking, hastily flaps off. The rightful mate emerges. What does she do now? Peck the erring husband? Leave him in chilly disgrace? Not a bit of it! She approaches with an eager note, and in a moment the two are hard at it, shaking their heads; and, indeed, on such occasions you may see more vigour and excitement thrown into the ceremony than at any other time.

Again we exclaim, how human! And again we see to what a pitch of complexity the bird's emotional life is tuned.

It will have been observed that in the Grebe, whose chief skill lies in its wonderful powers of diving, these powers have been utilized as the raw material of several of the courtship ceremonies. This pressing of the everyday faculties of the bird into the service of emotion, the elevation and conversion of its useful powers of diving and underwater swimming into ceremonials of passion, is from an evolutionary point of view natural enough, and has its counterparts elsewhere. So in the Divers, not too distant relatives of the Grebes, swimming and diving have their rôle in courtship. Here too the thrilling, vertical emergence close to the mate takes place; and there is a strange ceremony in which two or three birds plough their way through the water with body set obliquely—hinder parts submerged, breast raised, and neck stretched forward and head downward with that strange look of rigidity or tension often seen in the courtship actions of birds.

Or, again, I once saw (strangely enough from the windows of the Headmaster's house at Radley!) the aerial powers of the Kestrel converted to the uses of courtship. The hen bird was sitting in a large bush beyond the lawn. A strong wind was blowing, and the cock again and again beat his way up against it, to turn when nearly at the house and bear down upon the bush in an extremity of speed. Just when it seemed inevitable that he would knock his mate off her perch and dash himself and her into the branches, he changed the angle of his wings to shoot vertically up the face of the bush; then turned and repeated the play. Sometimes he came so near to her that she would start back, flapping her wings, as if really fearing a collision. The wind was so strong—and blowing away from me—that I could not hear what cries may have accompanied the display.

A friend of mine who knows the Welsh mountains and is a watcher of birds as well, tells me that he has there seen the Peregrine Falcons do the same thing: the same thing—except that the speed was perhaps twice as great, and the background a savage rock precipice instead of a Berkshire garden.

Not only the activities of everyday life, but also those of nest-building, are taken and used to build up the ceremonies of courtship; but whereas in the former case the actions are simply those which are most natural to and best performed by the bird, in the latter there is, no doubt, actual association between the cerebral centres concerned with nest-building and with sexual emotion in general. Thus we almost invariably find the seizing of nest-material in the beak as a part of courtship, and this is often extended to a presentation of the material to the mate. This we see in the Grebes, with the dank weeds of which their sodden nest is built; the Divers use moss in the construction of theirs, and the mated birds repair to moss banks, where they nervously pluck the moss, only to drop it again or throw it over their shoulder. Among the Warblers, the males pluck or pick up a leaf or twig, and with this in their beak hop and display before the hens; and the Peewit plucks frenziedly at grass and straws. The Adelie Penguins, so well described by Dr. Levick, make their nests of stones, and use stones in their courtship.

A curious, unnatural transference of object may sometimes be seen in these Penguins. The normal course of things is for this brave but comic creature, having picked up a stone in its beak, to come up before another of opposite sex, and, with stiff bow and absurdly outstretched flippers, to deposit it at the other's feet. When, however, there are men near the rookery, the birds will sometimes in all solemnity come up to them with their stone offering and lay it at the feet of the embarrassed or amused human being.

The Adelies do not nest by their natural element the sea, but some way away from it on stony slopes and rock patches;

thus they cannot employ their brilliant dives and feats of swimming in courtship, but content themselves, apart from this presentation of household material, with what Dr. Levick describes as "going into ecstasy"—spreading their flippers sideways, raising their head quite straight upwards, and emitting a low humming sound. This a bird may do when alone, or the two birds of a pair may make a duet of it. In any case, the term applied to it by its observer well indicates the state of emotion which it suggests and no doubt expresses.

The depositing of courtship offerings before men by the Penguins shows us that there must be a certain freedom of mental connection in birds. Here an act, properly belonging to courtship, is performed as the outlet, as it were, of another and unusual emotion. The same is seen in many song-birds, who, like the Sedge Warbler, sing loudly for anger when disturbed near their nest; or in the Divers, who, when an enemy is close to the nest, express the violence of their emotion by short sharp dives which flip a fountain of spray into the air—a type of dive also used as a sign of general excitement in courtship.

Or, again, the actions may be performed for their own sake, as we may say: because their performance, when the bird is full of energy and outer conditions are favourable, gives pleasure. The best-known example is the song of song-birds. This, as Eliot Howard has abundantly shown, is in its origin and essential function a symbol of possession, of a nesting territory occupied by a male—to other males a notice that "trespassers will be prosecuted," to females an invitation to settle, pair, and nest. But in all song-birds, practically without exception, the song is by no means confined to the short period during which it actually performs these functions, but is continued until the young are hatched, often to be taken up again when they have flown, or after the moult, or even, as in the Song Thrush, on almost any sunny or warm day the year round.

And finally this leads on to what is perhaps the most interesting category of birds' actions—those which are not merely sometimes performed for their own sake, although they possess other and utilitarian function, but actually have no other origin or *raison d'être* than to be performed for their own sake. They represent, in fact, true play or sport among ourselves; and seem better developed among birds than among mammals, or at least than among mammals below the monkey. True that the cat plays with the mouse, and many young mammals, like kittens, lambs, and kids, are full of play; but the playing with the mouse is more like the singing of birds outside the mating season, a transference of a normal activity to the plane of play; and the play of young animals, as Groos successfully exerted himself to show, is of undoubted use. To be sure, the impulse to play must be *felt* by the young creature as an exuberance of emotion and spirits demanding expression; but a similar impulse must be felt for all instinctive actions. Psychologically and individually, if you like, the action is performed for its own sake; but from the standpoint of evolution and of the race it has been originated, or at least perfected, as a practice ground for immature limbs and a training and keeping ready of faculties that in the future will be needed in earnest.

We shall best see the difference between mammals' and birds' behaviour by giving some examples. A very strange one I saw in a pond near the Egret rookery in Louisiana. Here, among other interesting birds, were the Darters or Water Turkeys, curious-looking relatives of the Cormorants, with long, thin, flexible neck, tiny head, and sharp beak, who often swim with all the body submerged, showing nothing but the snake-like neck above water. One of these was sitting on a branch of swamp-cedar, solitary and apparently tranquil. But this tranquillity must have been the cloak of boredom. For suddenly the bird, looking restlessly about her (it was a hen), began to pluck at the little green twigs near by. She pulled one off in her beak, and then, tossing her head up, threw it into the air, and with dexterous twist caught it again in her beak as it descended. After five or six successful catches she missed the twig. A comic sideways and downward glance at the twig, falling and fallen, in meditative immobility; and then another twig was broken off, and the same game repeated. She was very clever at catching; the only bird that I have seen come up to her was a Toucan in the Zoo which could catch grapes thrown at apparently any speed. But then the Toucan had been specially trained—and had the advantage of a huge capacity of bill!

Here again it might, of course, be said that the catching of twigs is a practice for beak and eye, and helps keep the bird in training for the serious business of catching fish. This is no doubt true; but, as regards the evolution of the habit, I incline strongly to the belief that it must be quite secondary —that the bird, desirous of occupying its restless self in a satisfying way, fell back upon a modification of its everyday activities, just as these are drawn upon in other birds to provide much of the raw material of courtship. There is no evidence that young Darters play at catching twigs as preparation for their fishing, and until there is evidence of this it is simpler to think that the play habit here, instead of being rooted by the utilitarian dictates of natural selection in the behaviour of the species, as with kids or kittens, is a secondary outcome of leisure and restlessness combining to operate with natural aptitude—in other words, true sport, of however simple a kind.

The commonest form of play in birds is flying play. Any one who has kept his eyes open at the seaside will have seen the Herring Gulls congregate in soaring intersecting spirals where the cliff sends the wind upwards. But such flights are nothing compared with those of other birds. Even the staid black-coated Raven may sometimes be seen to go through a curious performance. One I remember, all alone, flying along the side of a mountain near Oban; but instead of progressing in the conventional way, he flew diagonally upwards for a short distance, then giving a special croak with something of gusto in it, turned almost completely over on to his back, and descended a corresponding diagonal in this position. Then with a strong flap of the wings he righted himself, and so continued until he disappeared round the shoulder of the hill half a mile on. It reminded me of a child who has learnt some new little trick of step or dance-rhythm, and tries it out happily all the way home along the road. Mr. Harold Massingham has seen the Ravens' games too, and set them down more vividly than I can. He also is clear that they play for the love of playing, and even believes that their love of sport has helped their downfall to rarity by rendering them too easy targets for the gunner.

Or again, at the Egret rookery in Louisiana, at evening when the birds returned in great numbers, they came back with steady wing-beats along an aerial stratum about two hundred feet up. Arrived over their nesting pond, they simply let themselves drop. Their plumes flew up behind like a comet's tail; they screamed aloud with excitement; and, not far above the level of the trees, spread the wings so that they caught the air again, and as result skidded and side-slipped in the wildest and most exciting-looking curves before recovering themselves with a brief upward glide and settling carefully on the branches. This certainly had no significance for courtship; and I never saw it done save over the pond at the birds' return. It seemed to be simply an entertaining bit of sport grafted on to the dull necessity of descending a couple of hundred feet.

Examples could be multiplied: Rooks and Crows, our solemn English Heron, Curlew, Swifts, Snipe—these and many others have their own peculiar flying sports. What is clear to the watcher is the emotional basis of these sports—a joy in controlled performance, and excitement in rapidity of motion, in all essentials like the pleasure to us of a well-hit ball at golf, or the thrill of a rapid descent on sledge skis.

For any one to whom the evolution theory is one of the master-keys to animate nature, there must be an unusual interest in tracing out the development of lines of life that, like the birds', have diverged comparatively early from the line which eventually and through many vicissitudes led to Man.

In the birds as in the mammals, and quite separately in the two groups, we see the evolution not only of certain structural characters such as division of heart, compactness of skeleton, increase of brain-size, not only of physiological characters like warm-bloodedness or efficiency of circulation, but also of various psychical characters. The power of profiting by experience becomes greater, as does that of distinguishing between objects; and there is most markedly an increase in the intensity of emotion. It has somehow been of advantage, direct or indirect, to birds to acquire a greater capacity for affection, for jealousy, for joy, for fear, for curiosity. In birds the advance on the intellectual side has been less, on the emotional side greater: so that we can study in them a part of the single stream of life where emotion, untrammelled by much reason, has the upper hand.

ARTHUR STANLEY EDDINGTON

ONE OF THE MOST STARTLING of all notions in modern nuclear physics is the notion that ultimate "particles" are not bound by rigid causal laws. Like tadpoles they scurry about anywhither, caught only in the inescapable net of probability equations. If this is true, as almost all physicists today believe, how should it affect our views on human will and destiny?

Proponents of determinism in human affairs usually point out that although chance may be a real factor on the submicroscopic level of quantum theory, on the macroscopic level of man's history, statistical laws of virtual certainty must prevail. This argument collapses when we reflect on the following possibility. Imagine a jet bomber moving above an enemy nation with supersonic speed. It carries a hydrogen bomb, the release of which is triggered by the click of a Geiger counter. If the timing of this click is a truly random matter, as nuclear physicists tell us, it follows that pure chance may determine whether city A or city B is vaporized. This in turn may alter the course of history.

Sir Arthur Stanley Eddington (1882-1944), a distinguished professor of astronomy at Cambridge, was a devout Quaker who saw in the new indeterminism the removal of a psychological stumbling block to belief in free will. If strict causality does not apply to the material world, he reasoned, then does it not become easier to accept our intuitive feeling that the will, in some equally mysterious fashion, may contain an element of creative spontaneity? His views were much ridiculed by determinists, especially in the Soviet Union where the essay chosen here has been held up as a notorious example of western science corrupted by "bourgeois idealism." Marx and Engels were determinists, and Soviet physicists against the extreme partiality for classical studies, the study of names instead of things, which has so long been shown in our educational system, this new cry is wholesome and good; but so far as it implies that science is capable of taking the place of the great literatures as an instrument of high culture, it is mischievous and misleading.

About the intrinsic value of science, its value as a factor in our civilization, there can be but one opinion; but about its value to the scholar, the thinker, the man of letters, there is room for very divergent views. It is certainly true that the great ages of the world have not been ages of exact science; nor have the great literatures, in which so much of the power and vitality of the race have been stored, sprung from minds which held correct views of the physical universe. Indeed, if the growth and maturity of man's moral and intellectual stature were a question of material appliances or conveniences, or of accumulated stores of exact knowledge, the world of to-day ought to be able to show more eminent achievements in all fields of human activity than ever before. But this it cannot do. Shakespeare wrote his plays for people who believed in witches, and probably believed in them himself; Dante's immortal poem could never have been produced in a scientific age. Is it likely that the Hebrew Scriptures would have been any more precious to the race, or their influence any deeper, had they been inspired by correct views of physical science?

It is not my purpose to write a diatribe against the physical sciences. I would as soon think of abusing the dictionary. But as the dictionary can hardly be said to be an end in itself, so I would indicate that the final value of physical science is its capability to foster in us noble ideals, and to lead us to new and larger views of moral and spiritual truths. The extent to which it is able to do this measures its value to the spirit,—measures its value to the educator.

That the great sciences can do this, that they are capable of becoming instruments of pure culture, instruments to refine and spiritualize the whole moral and intellectual nature, is no doubt true; but that they can ever usurp the place of the humanities or general literature in this respect is one of those

ARTHUR STANLEY EDDINGTON

The Decline of Determinism

Thus from the outset we can be quite clear about one very important fact, namely, that the validity of the law of causation for the world of reality is a question that cannot be decided on grounds of abstract reasoning.

MAX PLANCK, *Where Is Science Going?* p. 113.

The new theory appears to be well founded on observation, but one may ask whether *in the future,* by development or refinement, it may not be made deterministic again. As to this it must be said: It can be shown by rigorous mathematics that the accepted formal theory of quantum mechanics does not admit of any such extension. If anyone clings to the hope that determinism will ever return, he must hold the existed theory to be false in substance; it must be possible to disprove experimentally definite assertions of this theory. The determinist should therefore not protest but experiment.

MAX BORN, *Naturwissenschaften,* 1929, p. 117.

Whilst the feeling of free-will dominates the life of the spirit, the regularity of sensory phenomena lays down the demand for causality. But in both domains simultaneously the point in question is an idealisation, whose natural limitations can be more closely investigated, and which determine one another in the sense that the feeling of volition and the demand for causality are equally indispensable in the relation between Subject and Object which is the kernel of the problem of perception.

NIELS BOHR, *Naturwissenschaften,* 1930, p. 77.

We must await the further development of science, perhaps for centuries, before we can design a true and detailed picture of the interwoven texture of Matter, Life and Soul. But the old classical determinism of Hobbes and Laplace need not oppress us any longer.

HERMANN WEYL, *The Open World*, p. 55.

I

TEN YEARS AGO practically every physicist of repute was, or believed himself to be, a determinist, at any rate so far as inorganic phenomena are concerned. He believed he had come across a scheme of strict causality regulating the sequence of phenomena. It was considered to be the primary aim of science to fit as much of the universe as possible into such a scheme; so that, as a working belief if not as a philosophical conviction, the causal scheme was always held to be applicable in default of overwhelming evidence to the contrary. In fact, the methods, definitions and conceptions of physical science were so much bound up with the hypothesis of strict causality that the limits (if any) of the scheme of causal law were looked upon as the ultimate limits of physical science. No serious doubt was entertained that this determinism covered all inorganic phenomena. How far it applied to living or conscious matter or to consciousness itself was a matter of individual opinion; but there was naturally a reluctance to accept any restriction of an outlook which had proved so successful over a wide domain.

Then rather suddenly determinism faded out of theoretical physics. Its exit has been received in various ways. Some writers are incredulous and cannot be persuaded that determinism has really been eliminated from the present foundations of physical theory. Some think that it is no more than a domestic change in physics, having no reactions on general philosophic thought. Some decide cynically to wait and see if determinism fades in again.

The rejection of determinism is in no sense an abdication of scientific method. It is rather the fruition of a scientific method which had grown up under the shelter of the old causal method and has now been found to have a wider range. It has greatly increased the power and precision of the mathematical theory of observed phenomena. On the other hand I cannot agree with those who belittle the philosophical significance of the change. The withdrawal of physical science from an attitude it had adopted consistently for more than 200 years is not to be treated lightly; and it provokes a reconsideration of our views as to one of the most perplexing problems of our existence.

In a subject which arouses so much controversy it seems well to make clear at the outset certain facts regarding the extent of the change as to which there has frequently been a misunderstanding. Firstly, it is not suggested that determinism has been disproved. What we assert is that physical science is no longer based on determinism. Is it difficult to grasp this distinction? If I were asked whether astronomy has disproved the doctrine that "the moon is made of green cheese" I might have some difficulty in finding really conclusive evidence; but I could say unhesitatingly that the doctrine is not the basis of present-day selenography. Secondly, the denial of determinism, or as it is often called "the law of causality", does not mean that it is denied that effects may proceed from causes. The common regular association of cause and effect is a matter of experience; the law of causality is an extreme generalisation suggested by this experience. Such generalisations are always risky. To suppose that in doubting the generalisation we are denying the experience is like supposing that a person who doubts Newton's (or Einstein's) law of gravitation denies that apples fall to the ground. The first criterion applied to any theory, deterministic or indeterministic, is that it must account for the regularities in our sensory experience—notably our experience that certain effects regularly follow certain causes. Thirdly, the admission of indeterminism in the physical universe does not immediately clear up all the difficulties—not even all the physical difficulties—connected with Free Will. But it so far modifies the problem that the door is not barred and bolted for a solution less repugnant to our deepest intuitions than that which has hitherto seemed to be forced upon us.

Let us be sure that we agree as to what is meant by determinism. I quote three definitions or descriptions for your consideration. The first is by a mathematician (Laplace):

We ought then to regard the present state of the universe as the effect of its antecedent state and the cause of the state that is to follow. An intelligence, who for a given instant should be acquainted with all the forces by which Nature is animated and with the several positions of the entities composing it, if further his intellect were vast enough to submit those data to analysis, would include in one and the same formula the movements of the largest bodies in the universe and those of the lightest atom. Nothing would be uncertain for him; the future as well as the past would be present to his eyes. The human mind in the perfection it has been able to give to astronomy affords a feeble outline of such an intelligence. . . . All its efforts in the search for truth tend to approximate without limit to the intelligence we have just imagined.

The second is by a philosopher (C. D. Broad):

"Determinism" is the name given to the following doctrine. Let *S* be any substance, ψ any characteristic, and *t* any moment. Suppose that *S* is in fact in the state σ with respect to ψ at *t*. Then the compound supposition that everything else in the world should have been exactly as it in fact was, and that *S* should instead have been in one of the other two alternative states with respect to ψ is an impossible one. [The three alternative states (of which σ is one) are: to have the characteristic ψ, not to have it, and to be changing.]

The third is by a poet (Omar Khayyam):

With Earth's first Clay They did the Last Man's knead,
And then of the Last Harvest sow'd the Seed:

Yea, the first Morning of Creation wrote
What the Last Dawn of Reckoning shall read.

I regard the poet's definition as my standard. There is no doubt that his words express what is in our minds when we refer to determinism. In saying that the physical universe as now pictured is not a universe in which "the first morning of creation wrote what the last dawn of reckoning shall read", we make it clear that the abandonment of determinism is no technical quibble but is a fundamental change of outlook. The other two definitions need to be scrutinised suspiciously; we are afraid there may be a catch in them. In fact I think there is a catch in them.[1]

It is important to notice that all three definitions introduce the time element. Determinism postulates not merely causes but *pre-existing* causes. Determinism means predetermination. Hence in any argument about determinism the dating of the alleged causes is all-important; we must challenge them to produce their birth-certificates.

In the passage quoted from Laplace a definite aim of science is laid down. Its efforts "tend to approximate without limit to the intelligence we have just imagined", i.e. an intelligence who from the present state of the universe could foresee the whole of future progress down to the lightest atom. This aim was accepted without question until recent times. But the practical development of science is not always in a direct line with its ultimate aims; and about the middle of the nineteenth century there arose a branch of physics (thermodynamics) which struck out in a new direction. Whilst striving to perfect a system of law that would predict what *certainly* will happen, physicists also became interested in a system which predicts what *probably* will happen. Alongside the super-intelligence imagined by Laplace for whom "nothing would

[1] The catch that I suspect in Broad's definition is that it seems to convey no meaning without further elucidation of what is meant by the supposition being an *impossible* one. He does not mean impossible because it involves a logical contradiction. The supposition is not rejected as being contrary to logic nor as contrary to fact, but for a third reason undefined.

be uncertain" was placed an intelligence for whom nothing would be certain but some things would be exceedingly probable. If we could say of this latter being that for him *all* the events of the future were known with exceedingly high probability, it would be mere pedantry to distinguish him from Laplace's being who is supposed to know them with certainty. Actually, however, the new being is supposed to have glimpses of the future of varying degrees of probability ranging from practical certainty to entire indefiniteness according to his particular field of study. Generally speaking his predictions never approach certainty unless they refer to an average of a very large number of individual entities. Thus the aim of science to approximate to this latter intelligence is by no means equivalent to Laplace's aim. I shall call the aim defined by Laplace the *primary* aim, and the new aim introduced in the science of thermodynamics the *secondary* aim.

We must realise that the two aims are distinct. The prediction of what will probably occur is not a half-way stage in the prediction of what will certainly occur. We often solve a problem approximately, and subsequently proceed to second and third approximations, perhaps finally reaching an exact solution. But here the probable prediction is an end in itself; it is not an approximate attempt at a certain prediction. The methods differ fundamentally, just as the method of diagnosis of a doctor who tells you that you have just three weeks to live differs from that of a Life Insurance Office which tells you that your expectation of life is 18.7 years. We can, of course, occupy ourselves with the secondary aim without giving up the primary aim as an ultimate goal; but a survey of the present state of progress of the two aims produces a startling revelation.

The formulae given in modern textbooks on quantum theory—which are continually being tested by experiment and used to open out new fields of investigation—are exclusively concerned with probabilities and averages. This is quite explicit. The "unknown quantity" which is chased from formula to formula is a probability or averaging factor.

The quantum theory therefore contributes to the secondary aim, but adds nothing to the primary Laplacian aim which is concerned with causal certainty. But further it is now recognised that the classical laws of mechanics and electromagnetism (including the modifications introduced by relativity theory) are simply the limiting form assumed by the formulae of quantum theory when the number of individual quanta or particles concerned is very large. This connection is known as Bohr's Correspondence Principle. The classical laws are not a fresh set of laws, but are a particular adaptation of the quantum laws. So they also arise from the secondary scheme. We have already mentioned that it is when a very large number of individuals are concerned that the predictions of the secondary scheme have a high probability approaching certainty. Consequently the domain of the classical laws is just that part of the whole domain of secondary law in which the probability is so high as to be practically equivalent to certainty. That is how they came to be mistaken for causal laws whose operation is definitely certain. Now that their statistical character is recognised they are lost to the primary scheme. When Laplace put forward his ideal of a completely deterministic scheme he thought he already had the nucleus of such a scheme in the laws of mechanics and astronomy. That nucleus has now been transferred to the secondary scheme. Nothing is left of the old scheme of causal law, and we have not yet found the beginnings of a new one.

Measured by advance towards the secondary aim, the progress of science has been amazingly rapid. Measured by advance towards Laplace's aim its progress is just *nil*.

Laplace's aim has lapsed into the position of other former aims of science—the discovery of the elixir of life, the philosopher's stone, the North-West Passage—aims which were a fruitful inspiration in their time. We are like navigators on whom at last it has dawned that there are other enterprises worth pursuing besides finding the North-West Passage. I need hardly say that there are some old mariners who regard these new enterprises as a temporary diversion and predict an early return to the "true aim of geographical exploration".

II

Let us examine how the new aim of physics originated. We observe certain regularities in the course of phenomena and formulate these as laws of Nature. Laws can be stated positively or negatively, "Thou shalt" or "Thou shalt not". For the present purpose we shall formulate them negatively. Here are two regularities in the sensory experience of most of us:

(*a*) We never come across equilateral triangles whose angles are unequal.

(*b*) We never come across thirteen hearts in a hand dealt to us at Bridge.

In our ordinary outlook we explain these regularities in fundamentally different ways. We say that the first holds because a contrary experience is *impossible;* the second because a contrary experience is *too improbable.*

This distinction is theoretical. There is nothing in the observations themselves to suggest to which type a particular observed regularity belongs. We recognise that "impossible" and "too improbable" are both adequate explanations of any observed uniformity of experience; and formerly physics rather haphazardly explained some uniformities one way and others the other way. But now the whole of physical law (so far discovered) is found to be comprised in the secondary scheme which deals only with probabilities; and the only reason assigned for any regularity is that the contrary is too improbable. Our failure to find equilateral triangles with unequal angles is because such triangles are too improbable. Of course, I am not here referring to the theorem of pure geometry; I am speaking of a regularity of sensory experience and refer therefore to whatever measurement is supposed to confirm this property of equilateral triangles as being true of actual experience. Our measurements regularly confirm it to the highest accuracy attainable, and no doubt will always do so; but according to the present physical theory that is because a failure could only occur as the result of an extremely unlikely coincidence in the behaviour of the vast number of particles concerned in the apparatus of measurement.

The older view, as I have said, recognised two types of natural law. The earth keeps revolving round the sun because it is impossible that it should run away. That is the primary or deterministic type. Heat flows from a hot body to a cold body because it is too improbable that it should flow the other way. That is the secondary or statistical type. On the modern theory both regularities belong to the statistical type—it is too improbable that the earth should run away from the sun.[1]

So long as the aim of physics is to bring to light a deterministic scheme, the pursuit of secondary law is a blind alley since it leads only to probabilities. The determinist is not content with a law which ordains that, given reasonable luck, the fire will warm me; he agrees that that is the probable event, but adds that somewhere at the base of physics there are other laws which ordain just what the fire will do to me, luck or no luck.

To borrow an analogy from genetics, determinism is a *dominant character*. Granting a system of primary law, we can (and indeed must) have secondary indeterministic laws derivable from it stating what will probably happen under that system. So for a long time determinism watched with equanimity the development within itself of a subsidiary indeterministic system of law. What matter? Deterministic law remains dominant. It was not foreseen that the child would grow to supplant its parent. There is a game called "Think of a number". After doubling, adding, and other calculations, there comes the direction "Take away the number you first thought of". Determinism is now in the position of the number we first thought of.

---

[1] "Impossible" therefore disappears from our vocabulary except in the sense of involving a logical contradiction. But the logical contradiction or impossibility is in the description, not in the phenomenon which it attempts but (on account of the contradiction) fails to describe.

The growth of secondary law whilst still under the dominant deterministic scheme was remarkable, and whole sections of physics were transferred to it. There came a time when in the most progressive branches of physics it was used exclusively. The physicist might continue to profess allegiance to primary law but he ceased to use it. Primary law was the gold stored in the vaults; secondary law was the paper currency actually used. But everyone still adhered to the traditional view that paper currency needs to be backed by gold. As physics progressed the occasions when the gold was actually produced became rarer until they ceased altogether. Then it occurred to some of us to question whether there still was a hoard of gold in the vaults or whether its existence was a mythical tradition. The dramatic ending of the story would be that the vaults were opened and found to be empty. The actual ending is not quite so simple. It turns out that the key has been lost, and no one can say for certain whether there is any gold in the vaults or not. But I think it is clear that, with either termination, present-day physics is *off the gold standard.*

III

The nature of the indeterminism now admitted in the physical world will be considered in more detail in the next chapter. I will here content myself with an example showing its order of magnitude. Laplace's ideal intelligence could foresee the future positions of objects from the heaviest bodies to the lightest atoms. Let us then consider the lightest particle we know, viz. the electron. Suppose that an electron is given a clear course (so that it is not deflected by any unforeseen collisions) and that we know all that can be known about it at the present instant. How closely can we foretell its position one second later? The answer is that (in the most favourable circumstances) we can predict its position to within about 1½ inches—not closer. That is the nearest we can approximate to Laplace's super-intelligence. The error is not large if we recall that during the second covered by our prediction the electron may have travelled 10,000 miles or more.

The uncertainty would, however, be serious if we had to calculate whether the electron would hit or miss a small target such as an atomic nucleus. To quote Prof. Born: "If Gessler had ordered William Tell to shoot a hydrogen atom off his son's head by means of an α particle and had given him the best laboratory instruments in the world instead of a cross-bow, Tell's skill would have availed him nothing. Hit or miss would have been a matter of chance".

For contrast take a mass of .001 milligram—which must be nearly the smallest mass handled macroscopically. The indeterminacy is much smaller because the mass is larger. Under similar conditions we could predict the position of this mass a thousand years hence to within $\frac{1}{5000}$ of a millimetre.

This indicates how the indeterminism which affects the minutest constituents of matter becomes insignificant in ordinary mechanical problems, although there is no change in the basis of the laws. It may not at first be apparent that the indeterminacy of 1½ inches in the position of the electron after the lapse of a second is of any great practical importance either. It would not often be important for an electron pursuing a straight course through empty space; but the same indeterminism occurs whatever the electron is doing. If it is pursuing an orbit in an atom, long before the second has expired the indeterminacy amounts to atomic dimensions; that is to say, we have altogether lost track of the electron's position in the atom. Anything which depends on the relative location of electrons in an atom is unpredictable more than a minute fraction of a second ahead.

For this reason the break-down of an atomic nucleus, such as occurs in radio-activity, is not predetermined by anything in the existing scheme of physics. All that the most complete theory can prescribe is how frequently configurations favouring an explosion will occur on the average; the individual occurrences of such a configuration are unpredictable. In the solar system we can predict fairly accurately how many eclipses of the sun (i.e. how many recurrences of a special configuration of the earth, sun and moon) will happen in a thousand years; or we can predict fairly accurately the date and time of each particular eclipse. The theory of the second type of prediction is not an elaboration of the theory of the first; the occurrence of individual eclipses depends on celestial mechanics, whereas the frequency of eclipses is purely a problem of geometry. In the atom, which we have compared to a miniature solar system, there is nothing corresponding to celestial mechanics—or rather mechanics is stifled at birth by the magnitude of the indeterminacy—but the geometrical theory of frequency of configurations remains analogous.

The future is never entirely determined by the past, nor is it ever entirely detached. We have referred to several phenomena in which the future is *practically determined;* the break-down of a radium nucleus is an example of a phenomenon in which the future is *practically detached* from the past.

But, you will say, the fact that physics assigns no characteristic to the radium nucleus predetermining the date at which it will break up, only means that that characteristic has not yet been discovered. You readily agree that we cannot predict the future in all cases; but why blame Nature rather than our own ignorance? If the radium atom were an exception, it would be natural to suppose that there is a determining characteristic which, when it is found, will bring it into line with other phenomena. But the radio-active atom was not brought forward as an exception; I have mentioned it as an extreme example of that which applies in greater or lesser degree to all kinds of phenomena. There is a difference between explaining away an exception and explaining away a rule.

The persistent critic continues, "You are evading the point. I contend that there are characteristics unknown to you which completely predetermine not only the time of break-up of the radio-active atom but all physical phenomena. How do you know that there are not? You are not omniscient?" I can understand the casual reader raising this question; but when a man of scientific training asks it, he wants shaking up and waking. Let us try the effect of a story.

About the year 2000, the famous archaeologist Prof. Lambda discovered an ancient Greek inscription which recorded that a foreign prince, whose name was given as Κανδείκλης, came with his followers into Greece and established his tribe there. The Professor anxious to identify the prince, after exhausting other sources of information, began to look through the letters C and K in the *Encyclopaedia Athenica.* His attention was attracted by an article on Canticles who it appeared was the son of Solomon. Clearly that was the required identification; no one could doubt that Κανδείκλης was the Jewish Prince Canticles. His theory attained great notoriety. At that time the Great Powers of Greece and Palestine were concluding an Entente and the Greek Prime Minister in an eloquent peroration made touching reference to the newly discovered historical ties of kinship between the two nations. Some time later Prof. Lambda happened to refer to the article again and discovered that he had made an unfortunate mistake; he had misread "Son of Solomon" for "Song of Solomon". The correction was published widely, and it might have been supposed that the Canticles theory would die a natural death. But no; Greeks and Palestinians continued to believe in their kinship, and the Greek Minister continued to make perorations. Prof. Lambda one day ventured to remonstrate with him. The Minister turned on him severely, "How do you know that Solomon had not a son called Canticles? You are not omniscient." The Professor, having reflected on the rather extensive character of Solomon's matrimonial establishment, found it difficult to reply.

The curious thing is that the determinist who takes this line is under the illusion that he is adopting a more modest attitude in regard to our scientific knowledge than the indeterminist. The indeterminist is accused of claiming omniscience. I do not bring quite the same countercharge against the determinist; but surely it is only the man who thinks himself *nearly* omniscient who would have the audacity to enumerate the possibilities which (it occurs to him) might exist unknown to him. I suspect that some of the other chapters in this book will be criticised for including hypotheses and deductions for which the evidence is considered to be insufficiently conclusive; that is inevitable if one is to give a picture of physical science in the process of development and discuss the current problems which occupy our thoughts. I tremble to think what the critics would say if I included a conjecture solely on the ground that, not being omniscient, I do not know that it is false.

I have already said that determinism is not disproved by physics. But it is the determinist who puts forward a positive proposal and the onus of proof is on him. He wishes to base on our ordinary experience of the sequence of cause and effect a wide generalisation called the Principle of Causality. Since physics to-day represents this experience as the result of statistical laws without any reference to the principle of causality, it is obvious that the generalisation has nothing to commend it so far as observational evidence is concerned. The indeterminists therefore regard it as they do any other entirely unsupported hypothesis. It is part of the tactics of the advocate of determinism to turn our unbelief in his conjecture into a positive conjecture of our own—a sort of Principle of Uncausality. The indeterminist is sometimes said to postulate "something like free-will" in the individual atoms. *Something like* is conveniently vague; the various mechanisms used in daily life have their obstinate moods and may be said to display something like free-will. But if it is suggested that we postulate psychological characters in the individual atoms of the kind which appear in our minds as human free-will, I deny this altogether. We do not discard one rash generalisation only to fall into another equally rash.

IV

When determinism was believed to prevail in the physical world, the question naturally arose, how far did it govern human activities? The question has often been confused by assuming that human activity belongs to a totally separate sphere—a mental sphere. But man has a body as well as a mind. The movements of his limbs, the sound waves which issue from his lips, the twinkle in his eye, are all phenomena of the physical world, and unless expressly excluded would be predetermined along with other physical phenomena. We can, if we like, distinguish two forms of determinism: (1) The scheme of causal law predetermines all human thoughts, emotions and volitions; (2) it predetermines human actions but not human motives and volitions. The second seems less drastic and probably commends itself to the liberal-minded, but the concession really amounts to very little. Under it a man can think what he likes, but he can only say that which the laws of physics preordain.

The essential point is that, if determinism is to have any definable meaning, the domain of deterministic law must be a closed system; that is to say, all the data used in predicting must themselves be capable of being predicted. Whatever predetermines the future must itself be predetermined by the past. The movements of human bodies are part of the complete data of prediction of future states of the material universe; and if we include them for this purpose we must include them also as data which (it is asserted) can be predicted.

We must also note a semi-deterministic view, which asserts determinism for inorganic phenomena but supposes that it can be overridden by the interference of consciousness. Determinism in the material universe then applies only to phenomena in which it is assured that consciousness is not intervening directly or indirectly. It would be difficult to accept such a view nowadays. I suppose that most of those who expect determinism ultimately to reappear in physics do so from the feeling that there is some kind of logical necessity for it; but it can scarcely be a logical necessity if it is capable of being overridden. The hypothesis puts the scientific investigator in the position of being afraid to prove too much; he must show that effect is firmly linked to cause, but not so firmly that consciousness is unable to break the link. Finally we have to remember that physical law is arrived at from the analysis of conscious experience; it is the solution of the cryptogram contained in the story of consciousness. How then can we represent consciousness as being not only outside it but inimical to it?

The revolution of theory which has expelled determinism from present-day physics has therefore the important consequence that it is no longer necessary to suppose that human actions are completely predetermined. Although the door of human freedom is opened, it is not flung wide open; only a chink of daylight appears. But I think this is sufficient to justify a reorientation of our attitude to the problem. If our new-found freedom is like that of the mass of .001 mgm., which is only allowed to stray $\frac{1}{5000}$ mm. in a thousand years, it is not much to boast of. The physical results do not spontaneously suggest any higher degree of freedom than this. But it seems to me that philosophical, psychological, and in fact commonsense arguments for greater freedom are so cogent that we are justified in trying to prise the door further open now that it is not actually barred. How can this be done without violence to physics?

If we could attribute the large-scale movements of our bodies to the "trigger action" of the unpredetermined behaviour of a few key atoms in our brain cells the problem would be simple; for individual atoms have wide indeterminacy of behaviour. It is obvious that there is a great deal of trigger action in our bodily mechanism, as when the pent up energy of a muscle is released by a minute physical change in a nerve; but it would be rash to suppose that the physical controlling cause is contained in the configuration of a few dozen atoms. I should conjecture that the smallest unit of structure in which the physical effects of volition have their origin contains many billions of atoms. If such a unit behaved like an inorganic system of similar mass the indeterminacy would be insufficient to allow appreciable freedom. My own tentative view is that this "conscious unit" does in fact differ from an inorganic system in having a much higher indeterminacy of behaviour—simply because of the unitary nature of that which in reality it represents, namely the Ego.

We have to remember that the physical world of atoms, electrons, quanta, etc., is the abstract symbolic representation of something. Generally we do not know anything of the background of the symbols—we do not know the inner nature of what is being symbolised. But at a point of contact of the physical world with consciousness, we have acquaintance with the conscious unity—the self or mind—whose physical aspect and symbol is the brain cell. Our method of physical analysis leads us to dissect this cell into atoms similar to the atoms in any non-conscious region of the world. But whereas in other regions each atom (so far as its behaviour is indeterminate) is governed independently by chance, in the conscious cell the behaviour symbolises a single volition of the spirit and not a conflict of billions of independent impulses. It seems to me that we must attribute some kind of unitary behaviour to the physical terminal of consciousness, otherwise the physical symbolism is not an appropriate representation of the mental unit which is being symbolised.

We conclude then that the activities of consciousness do not violate the laws of physics, since in the present indeterministic scheme there is freedom to operate within them. But at first sight they seem to involve something which we previously described as worse than a violation of the laws of physics, namely an exceedingly improbable coincidence. That had reference to coincidences ascribed to chance. Here we do not suppose that the conspiracy of the atoms in a brain cell to bring about a certain physical result instead of all fighting against one another is due to a chance coincidence. The unanimity is rather the condition that the atoms form a legitimate representation of that which is itself a unit in the mental reality behind the world of symbols.

The two aspects of human freedom on which I would lay most stress are *responsibility* and *self-understanding*. The nature of responsibility brings us to a well-known dilemma which I am no more able to solve than hundreds who have tried before me. How can we be responsible for our own good or evil nature? We feel that we can to some extent change our nature; we can reform or deteriorate. But is not the reforming or deteriorating impulse also in our nature? Or, if it is not in us, how can we be responsible for it? I will not add to the many discussions of this difficulty, for I have no solution to suggest. I will only say that I cannot accept as satisfactory the solution sometimes offered, that responsibility is a self-contradictory illusion. The solution does not seem to me to fit the data. Just as a theory of matter has to correspond to our perceptions of matter so a theory of the human spirit has to correspond to our inner perception of our spiritual nature. And to me it seems that responsibility is one of the fundamental facts of our nature. If I can be deluded over such a matter of immediate knowledge—the very nature of the being that I myself am—it is hard to see where any trustworthy beginning of knowledge is to be found.

I pass on to another aspect of the freedom allowed under physical indeterminacy, which seems to be quite distinct from the question of Free Will. Suppose that I have hit on a piece of mathematical research which promises interesting results. The assurance that I most desire is that the result which I write down at the end shall be the work of a mind which respects truth and logic, not the work of a hand which respects Maxwell's equations and the conservation of energy. In this case I am by no means anxious to stress the fact (if it is a fact) that the operations of my mind are unpredictable. Indeed I often prefer to use a multiplying machine whose results are less unpredictable than those of my own mental arithmetic. But the truth of the result $7 \times 11 = 77$ lies in its character as a possible mental operation and not in the fact that it is turned out automatically by a special combination of cog-wheels. I attach importance to the physical unpredictability of the motion of my pen, because it leaves it free to respond to the thought evolved in my brain which may or may not have been predetermined by the mental characteristics of my nature. If the mathematical argument in my mind is compelled to reach the conclusion which a deterministic system of physical law has preordained that my hands shall write down, then reasoning must be explained away as a process quite other than that which I feel it to be. But my whole respect for reasoning is based on the hypothesis that it *is* what I feel it to be.

I do not think we can take liberties with that immediate self-knowledge of consciousness by which we are aware of ourselves as responsible, truth-seeking, reasoning, striving. The external world is not what it seems; we can transform our conception of it as we will provided that the system of signals passing from it to the mind is conserved. But as we draw nearer to the source of all knowledge the stream should run clearer. At least that is the hypothesis that the scientist is compelled to make, else where shall he start to look for truth? The Problem of Experience becomes unintelligible unless it is considered as the quest of a responsible, truth-seeking, reasoning spirit. These characteristics of the spirit therefore become the first datum of the problem.

The conceptions of physics are becoming difficult to understand. First relativity theory, then quantum theory and wave mechanics, have transformed the universe, making it seem fantastic to our minds. And perhaps the end is not yet. But there is another side to the transformation. Naïve realism, materialism, and mechanistic conceptions of phenomena were simple to understand; but I think that it was only by closing our eyes to the essential nature of conscious experience that they could be made to seem credible. These revolutions of scientific thought are clearing up the deeper contradictions between life and theoretical knowledge. The latest phase with its release from determinism is one of the greatest steps in the reconciliation. I would even say that in the present indeterministic theory of the physical universe we have reached something which a reasonable man might *almost* believe.

ALDOUS HUXLEY

ALDOUS LEONARD HUXLEY (1894-1963), unlike his grandfather T. H. Huxley and his older brother Julian, but like his maternal grand uncle Matthew Arnold, always viewed the progress of science with a mistrustful eye. In his greatest novel, *Point Counter Point,* a character modeled after D. H. Lawrence exhibits his sketches of two contrasting Outlines of History. One, in the mode of H. G. Wells, starts on the left with a small monkey. It proceeds through prehistoric man, then on through history, the figures growing larger and larger until they culminate in giant likenesses of Wells that spiral off the paper toward Utopia. The other Outline is history as Lawrence sees it. The largest figure is an ancient Greek, then the men grow steadily smaller. Victorians are dwarfish, twentieth century men still smaller. "Through the mists of the future one could see a diminishing company of little gargoyles and foetuses. . . ."

Brave New World, first published in 1932, is Huxley's sardonic version of the second Outline. It is the most noteworthy of "negative utopias," that disturbing genre of science fantasy in which undesirable current trends are projected into a future nightmare. Wells himself had done it earlier in *When the Sleeper Wakes,* and George Orwell was to do it later in *1984*. The three visions have much in common, but Huxley's has had the greatest shock effect in spite of the fact that titillated readers have been known to finish the novel without realizing it is satire! To understand the chapter reprinted here it is necessary to give a brief sketch of Huxley's brave new society.

The year is A.F. 632. "A.F." stands for "After Ford." As the prophet of mass production, by which the state now rears its children, Ford has replaced God in the government-sponsored religious mythology. Bokanovsky's Process enables the Hatchery Centers to split each fertilized human egg into 96 identical twins that are carefully incubated in rows of test tubes. A system of Neo-Pavlovian conditioning combined with hypnopaedia (sleep teaching) produces a rigid caste society ranging from the Alpha Plus administrators to the Epsilon Minus Semi-Morons who perform the lowest tasks. Citizens are kept young and healthy by science and happy by a combination of state controlled entertainment (especially the "feelies"—3D colored talking pictures with odor and tactile sensations), compulsory sex, and *soma*. *Soma* is a tranquilizing drug that provides escape from reality without unpleasant after-effects. "Mother" has become an obscene word. "Pneumatic" girls chew sex-hormone gum, wear Malthusian belts to carry contraceptives, dance the five-step to the wail of sexaphones, and worry about their mental health whenever they find themselves becoming overly fond of one man. The result is a completely stable society, its citizens happy and good in the way an infant or a bee is happy and good.

John, a "savage" from a Reservation where unconditioned men and women are kept as specimens, is brought to London by an insecure psychologist named Bernard Marx. Well-read in Shakespeare, the Savage expects to find Miranda's "brave new world," but finds instead a world in which a higher kind of happiness has been forgotten.

In the chapter selected here, a "world controller" named Mustapha Mond explains to the Savage why science as well as art must be censored by the state. In recent years we have seen two great examples of precisely this kind of government control. Hitler found it necessary to repress modern anthropology with its evidence for racial equality, and Stalin personally backed the Lysenko movement and its liquidation of modern genetics. *Brave New World* was a forward extrapolation of 600 years. "Today," Huxley wrote in a later introduction to his novel, "it seems possible that the horror may be upon us within a single century."

ALDOUS HUXLEY

Science in the Brave New World

THE ROOM into which the three were ushered was the Controller's study.

"His fordship will be down in a moment." The Gamma butler left them to themselves.

Helmholtz laughed aloud.

"It's more like a caffeine-solution party than a trial," he said, and let himself fall into the most luxurious of the pneumatic arm-chairs. "Cheer up, Bernard," he added, catching sight of his friend's green unhappy face. But Bernard would not be cheered; without answering, without even looking at Helmholtz, he went and sat down on the most uncomfortable chair in the room, carefully chosen in the obscure hope of somehow deprecating the wrath of the higher powers.

The Savage meanwhile wandered restlessly round the room, peering with a vague superficial inquisitiveness at the books in the shelves, at the sound-track rolls and the reading machine bobbins in their numbered pigeon-holes. On the table under the window lay a massive volume bound in limp black leather-surrogate, and stamped with large golden T's. He picked it up and opened it. MY LIFE AND WORK, BY OUR FORD. The book had been published at Detroit by the Society for the Propagation of Fordian Knowledge. Idly he turned the pages, read a sentence here, a paragraph there, and had just come to the conclusion that the book didn't interest him, when the door opened, and the Resident World Controller for Western Europe walked briskly into the room.

Mustapha Mond shook hands with all three of them; but it was to the Savage that he addressed himself. "So you don't much like civilization, Mr. Savage," he said.

The Savage looked at him. He had been prepared to lie, to bluster, to remain sullenly unresponsive; but, reassured by the good-humoured intelligence of the Controller's face, he decided to tell the truth, straightforwardly. "No." He shook his head.

Bernard started and looked horrified. What would the Controller think? To be labelled as the friend of a man who said that he didn't like civilization—said it openly and, of all people, to the Controller—it was terrible. "But, John," he began. A look from Mustapha Mond reduced him to an abject silence.

"Of course," the Savage went on to admit, "there are some very nice things. All that music in the air, for instance . . ."

"Sometimes a thousand twangling instruments will hum about my ears and sometimes voices."

The Savage's face lit up with a sudden pleasure. "Have you read it too?" he asked. "I thought nobody knew about that book here, in England."

"Almost nobody. I'm one of the very few. It's prohibited, you see. But as I make the laws here, I can also break them. With impunity, Mr. Marx," he added, turning to Bernard. "Which I'm afraid you *can't* do."

Bernard sank into a yet more hopeless misery.

"But why is it prohibited?" asked the Savage. In the excitement of meeting a man who had read Shakespeare he had momentarily forgotten everything else.

The Controller shrugged his shoulders. "Because it's old; that's the chief reason. We haven't any use for old things here."

"Even when they're beautiful?"

"Particularly when they're beautiful. Beauty's attractive, and we don't want people to be attracted by old things. We want them to like the new ones."

"But the new ones are so stupid and horrible. Those plays,

where there's nothing but helicopters flying about and you *feel* the people kissing." He made a grimace. "Goats and monkeys!" Only in Othello's words could he find an adequate vehicle for his contempt and hatred.

"Nice tame animals, anyhow," the Controller murmured parenthetically.

"Why don't you let them see *Othello* instead?"

"I've told you; it's old. Besides, they couldn't understand it."

Yes, that was true. He remembered how Helmholtz had laughed at *Romeo and Juliet*. "Well then," he said, after a pause, "something new that's like *Othello*, and that they could understand."

"That's what we've all been wanting to write," said Helmholtz, breaking a long silence.

"And it's what you never will write," said the Controller. "Because, if it were really like *Othello* nobody could understand it, however new it might be. And if it were new, it couldn't possibly be like *Othello*."

"Why not?"

"Yes, why not?" Helmholtz repeated. He too was forgetting the unpleasant realities of the situation. Green with anxiety and apprehension, only Bernard remembered them; the others ignored him. "Why not?"

"Because our world is not the same as Othello's world. You can't make flivvers without steel—and you can't make tragedies without social instability. The world's stable now. People are happy; they get what they want, and they never want what they can't get. They're well off; they're safe; they're never ill; they're not afraid of death; they're blissfully ignorant of passion and old age; they're plagued with no mothers or fathers; they've got no wives, or children, or lovers to feel strongly about; they're so conditioned that they practically can't help behaving as they ought to behave. And if anything should go wrong, there's *soma*. Which you go and chuck out of the window in the name of liberty, Mr. Savage. *Liberty!*" He laughed. "Expecting Deltas to know what liberty is! And now expecting them to understand *Othello!* My good boy!"

The Savage was silent for a little. "All the same," he insisted obstinately, "*Othello's* good, *Othello's* better than those feelies."

"Of course it is," the Controller agreed. "But that's the price we have to pay for stability. You've got to choose between happiness and what people used to call high art. We've sacrificed the high art. We have the feelies and the scent organ instead."

"But they don't mean anything."

"They mean themselves; they mean a lot of agreeable sensations to the audience."

"But they're . . . . they're told by an idiot."

The Controller laughed. "You're not being very polite to your friend, Mr. Watson. One of our most distinguished Emotional Engineers . . ."

"But he's right," said Helmholtz gloomily. "Because it *is* idiotic. Writing when there's nothing to say . . ."

"Precisely. But that requires the most enormous ingenuity. You're making flivvers out of the absolute minimum of steel—works of art out of practically nothing but pure sensation."

The Savage shook his head. "It all seems to me quite horrible."

"Of course it does. Actual happiness always looks pretty squalid in comparison with the over-compensations for misery. And, of course, stability isn't nearly so spectacular as instability. And being contented has none of the glamour of a good fight against misfortune, none of the picturesqueness of a struggle with temptation, or a fatal overthrow by passion or doubt. Happiness is never grand."

"I suppose not," said the Savage after a silence. "But need it be quite so bad as those twins?" He passed his hand over his eyes as though he were trying to wipe away the remembered image of those long rows of identical midgets at the assembling tables, those queued-up twin-herds at the entrance to the Brentford monorail station, those human maggots swarming round Linda's bed of death, the endlessly repeated face of his assailants. He looked at his bandaged left hand and shuddered. "Horrible!"

"But how useful! I see you don't like our Bokanovsky Groups; but, I assure you, they're the foundation on which everything else is built. They're the gyroscope that stabilizes the rocket plane of state on its unswerving course." The deep voice thrillingly vibrated; the gesticulating hand implied all space and the onrush of the irresistible machine. Mustapha Mond's oratory was almost up to synthetic standards.

"I was wondering," said the Savage, "why you had them at all—seeing that you can get whatever you want out of those bottles. Why don't you make everybody an Alpha Double Plus while you're about it?"

Mustapha Mond laughed. "Because we have no wish to have our throats cut," he answered. "We believe in happiness and stability. A society of Alphas couldn't fail to be unstable and miserable. Imagine a factory staffed by Alphas—that is to say by separate and unrelated individuals of good heredity and conditioned so as to be capable (within limits) of making a free choice and assuming responsibilities. Imagine it!" he repeated.

The Savage tried to imagine it, not very successfully.

"It's an absurdity. An Alpha-decanted, Alpha-conditioned man would go mad if he had to do Epsilon Semi-Moron work —go mad, or start smashing things up. Alphas can be completely socialized—but only on condition that you make them do Alpha work. Only an Epsilon can be expected to make Epsilon sacrifices, for the good reason that for him they aren't sacrifices; they're the line of least resistance. His conditioning has laid down rails along which he's got to run. He can't help himself; he's foredoomed. Even after decanting, he's still inside a bottle—an invisible bottle of infantile and embryonic fixations. Each one of us, of course," the Controller meditatively continued, "goes through life inside a bottle. But if we happen to be Alphas, our bottles are, relatively speaking, enormous. We should suffer acutely if we were confined in a narrower space. You cannot pour upper-caste champagne-surrogate into lower-caste bottles. It's obvious theoretically. But it has also been proved in actual practice. The result of the Cyprus experiment was convincing."

"What was that?" asked the Savage.

Mustapha Mond smiled. "Well, you can call it an experiment in rebottling if you like. It began in A.F. 473. The Controllers had the island of Cyprus cleared of all its existing inhabitants and re-colonized with a specially prepared batch of twenty-two thousand Alphas. All agricultural and industrial equipment was handed over to them and they were left to manage their own affairs. The result exactly fulfilled all the theoretical predictions. The land wasn't properly worked; there were strikes in all the factories; the laws were set at naught, orders disobeyed; all the people detailed for a spell of low-grade work were perpetually intriguing for high-grade jobs, and all the people with high-grade jobs were counter-intriguing at all costs to stay where they were. Within six years they were having a first-class civil war. When nineteen out of the twenty-two thousand had been killed, the survivors unanimously petitioned the World Controllers to resume the government of the island. Which they did. And that was the end of the only society of Alphas that the world has ever seen."

The Savage sighed, profoundly.

"The optimum population," said Mustapha Mond, "is modelled on the iceberg—eight-ninths below the water line, one-ninth above."

"And they're happy below the water line?"

"Happier than above it. Happier than your friend here, for example." He pointed.

"In spite of that awful work?"

"Awful? *They* don't find it so. On the contrary, they like it. It's light, it's childishly simple. No strain on the mind or the muscles. Seven and a half hours of mild, unexhausting labour, and then the *soma* ration and games and unrestricted copulation and the feelies. What more can they ask for? True," he added, "they might ask for shorter hours. And of course we could give them shorter hours. Technically, it would be perfectly simple to reduce all lower-caste working hours to three or four a day. But would they be any the happier for that? No, they wouldn't. The experiment was tried, more than a century and a half ago. The whole of Ireland was put on to the four-hour day. What was the result? Unrest and a large increase in the consumption of *soma;* that was all. Those three and a half hours of extra leisure were so far from being a source of happiness, that people felt constrained to take a holiday from them. The Inventions Office is stuffed with plans for labour-saving processes. Thousands of them." Mustapha Mond made a lavish gesture. "And why don't we put them into execution? For the sake of the labourers; it would be sheer cruelty to afflict them with excessive leisure. It's the same with agriculture. We could synthesize every morsel of food, if we wanted to. But we don't. We prefer to keep a third of the population on the land. For their own sakes—because it takes *longer* to get food out of the land than out of a factory. Besides, we have our stability to think of. We don't want to change. Every change is a menace to stability. That's another reason why we're so chary of applying new inventions. Every discovery in pure science is potentially subversive; even science must sometimes be treated as a possible enemy. Yes, even science."

Science? The Savage frowned. He knew the word. But what it exactly signified he could not say. Shakespeare and the old men of the pueblo had never mentioned science, and from Linda he had only gathered the vaguest hints: science was something you made helicopters with, something that caused you to laugh at the Corn Dances, something that prevented you from being wrinkled and losing your teeth. He made a desperate effort to take the Controller's meaning.

"Yes," Mustapha Mond was saying, "that's another item in the cost of stability. It isn't only art that's incompatible with happiness; it's also science. Science is dangerous; we have to keep it most carefully chained and muzzled."

"What?" said Helmholtz, in astonishment. "But we're always saying that science is everything. It's a hypnopædic platitude."

"Three times a week between thirteen and seventeen," put in Bernard.

"And all the science propaganda we do at the College . . ."

"Yes; but what sort of science?" asked Mustapha Mond sarcastically. "You've had no scientific training, so you can't judge. I was a pretty good physicist in my time. Too good—good enough to realize that all our science is just a cookery book, with an orthodox theory of cooking that nobody's allowed to question, and a list of recipes that mustn't be added to except by special permission from the head cook. I'm the head cook now. But I was an inquisitive young scullion once. I started doing a bit of cooking on my own. Unorthodox cooking, illicit cooking. A bit of real science, in fact." He was silent.

"What happened?" asked Helmholtz Watson.

The Controller sighed. "Very nearly what's going to happen to you young men. I was on the point of being sent to an island."

The words galvanized Bernard into a violent and unseemly activity. "Send *me* to an island?" He jumped up, ran across the room, and stood gesticulating in front of the Controller. "You can't send *me*. I haven't done anything. It was the others. I swear it was the others." He pointed accusingly to Helmholtz and the Savage. "Oh, please don't send me to Iceland. I promise I'll do what I ought to do. Give me another chance. Please give me another chance." The tears began to flow. "I tell you, it's their fault," he sobbed. "And not to Iceland. Oh please, your fordship, please . . ." And in a paroxysm of abjection he threw himself on his knees before the Controller. Mustapha Mond tried to make him get up; but Bernard persisted in his grovelling; the stream of words poured out inexhaustibly. In the end the Controller had to ring for his fourth secretary.

"Bring three men," he ordered, "and take Mr. Marx into a bedroom. Give him a good *soma* vaporization and then put him to bed and leave him."

The fourth secretary went out and returned with three

green-uniformed twin footmen. Still shouting and sobbing, Bernard was carried out.

"One would think he was going to have his throat cut," said the Controller, as the door closed. "Whereas, if he had the smallest sense, he'd understand that his punishment is really a reward. He's being sent to an island. That's to say, he's being sent to a place where he'll meet the most interesting set of men and women to be found anywhere in the world. All the people who, for one reason or another, have got too self-consciously individual to fit into community-life. All the people who aren't satisfied with orthodoxy, who've got independent ideas of their own. Every one, in a word, who's any one. I almost envy you, Mr. Watson."

Helmholtz laughed. "Then why aren't you on an island yourself?"

"Because, finally, I preferred this," the Controller answered. "I was given the choice: to be sent to an island, where I could have got on with my pure science, or to be taken on to the Controllers' Council with the prospect of succeeding in due course to an actual Controllership. I chose this and let the science go." After a little silence, "Sometimes," he added, "I rather regret the science. Happiness is a hard master—particularly other people's happiness. A much harder master, if one isn't conditioned to accept it unquestioningly, than truth." He sighed, fell silent again, then continued in a brisker tone, "Well, duty's duty. One can't consult one's own preferences. I'm interested in truth, I like science. But truth's a menace, science is a public danger. As dangerous as it's been beneficent. It has given us the stablest equilibrium in history. China's was hopelessly insecure by comparison; even the primitive matriarchies weren't steadier than we are. Thanks, I repeat, to science. But we can't allow science to undo its own good work. That's why we so carefully limit the scope of its researches—that's why I almost got sent to an island. We don't allow it to deal with any but the most immediate problems of the moment. All other enquiries are most sedulously discouraged. It's curious," he went on after a little pause, "to read what people in the time of Our

Ford used to write about scientific progress. They seemed to have imagined that it could be allowed to go on indefinitely, regardless of everything else. Knowledge was the highest good, truth the supreme value; all the rest was secondary and subordinate. True, ideas were beginning to change even then. Our Ford himself did a great deal to shift the emphasis from truth and beauty to comfort and happiness. Mass production demanded the shift. Universal happiness keeps the wheels steadily turning; truth and beauty can't. And, of course, whenever the masses seized political power, then it was happiness rather than truth and beauty that mattered. Still, in spite of everything, unrestricted scientific research was still permitted. People still went on talking about truth and beauty as though they were the sovereign goods. Right up to the time of the Nine Years' War. *That* made them change their tune all right. What's the point of truth or beauty or knowledge when the anthrax bombs are popping all around you? That was when science first began to be controlled—after the Nine Years' War. People were ready to have even their appetites controlled then. Anything for a quiet life. We've gone on controlling ever since. It hasn't been very good for truth, of course. But it's been very good for happiness. One can't have something for nothing. Happiness has got to be paid for. You're paying for it, Mr. Watson—paying because you happen to be too much interested in beauty. I was too much interested in truth; I paid too."

"But *you* didn't go to an island," said the Savage, breaking a long silence.

The Controller smiled. "That's how I paid. By choosing to serve happiness. Other people's—not mine. It's lucky," he added, after a pause, "that there are such a lot of islands in the world. I don't know what we should do without them. Put you all in the lethal chamber, I suppose. By the way, Mr. Watson, would you like a tropical climate? The Marquesas, for example; or Samoa? Or something rather more bracing?"

Helmholtz rose from his pneumatic chair. "I should like a thoroughly bad climate," he answered. "I believe one would write better if the climate were bad. If there were a lot of wind and storms, for example . . ."

The Controller nodded his approbation. "I like your spirit, Mr. Watson. I like it very much indeed. As much as I officially disapprove of it." He smiled. "What about the Falkland Islands?"

"Yes, I think that will do," Helmholtz answered. "And now, if you don't mind, I'll go and see how poor Bernard's getting on."

RACHEL CARSON

IN 1951 READERS of *The New Yorker* received something of a jolt. For several issues the magazine's profile department, normally devoted to persons in the public eye, ran a "profile" of the sea! The series of articles was taken from Rachel Carson's *The Sea Around Us.* When the book appeared later that year it became an immediate best-seller and the winner of many impressive awards.

The scientist and the poet, John Burroughs concludes in a previous essay of this collection, can and should be friends. On rare occasions, and the writing of Miss Carson's book is one of them, they become a single person. Novelists and poets before her had described the beauty, terror, and mystery of the sea; scientists before her had recorded in dry and perishable prose the "facts" about the sea. It remained for Miss Carson to fuse the science and the poetry into one magnificent, brilliantly executed volume.

Rachel Louise Carson (1907-1964) had been well trained for her task. After a period of research at the Marine Biological Laboratory, Woods Hole, Mass., and several years of university teaching, she joined the staff of what is now called the U.S. Fish and Wildlife Service, a branch of the Department of the Interior. From 1947 to 1952 she was the service's editor-in-chief.

In 1962 Miss Carson's book *Silent Spring* aroused the entire nation to the dangers inherent in the widespread, careless use of chemical sprays to kill insects. The ensuing struggle between proponents and opponents of the use of such sprays continues unabated today. The selection printed here, from Miss Carson's earlier book, *The Sea Around Us,* is one of that book's most dramatic chapters.

RACHEL CARSON

The Sunless Sea

Where great whales come sailing by,
Sail and sail, with unshut eye.

—MATTHEW ARNOLD.

BETWEEN THE sunlit surface waters of the open sea and the hidden hills and valleys of the ocean floor lies the least known region of the sea. These deep, dark waters, with all their mysteries and their unsolved problems, cover a very considerable part of the earth. The whole world ocean extends over about three-fourths of the surface of the globe. If we subtract the shallow areas of the continental shelves and the scattered banks and shoals, where at least the pale ghost of sunlight moves over the underlying bottom, there still remains about half the earth that is covered by miles-deep, lightless water, that has been dark since the world began.

This region has withheld its secrets more obstinately than any other. Man, with all his ingenuity, has been able to venture only to its threshold. Carrying tanks of compressed air, he can swim down to depths of about 300 feet. He can descend about 500 feet wearing a diving helmet and a rubberized suit. Only a few men in all the history of the world have had the experience of descending, alive, beyond the range of visible light. The first to do so were William Beebe and Otis Barton; in the bathysphere, they reached a depth of 3028 feet in the open ocean off Bermuda, in the year 1934. Barton alone, in the summer of 1949, descended to a depth of 4500 feet off California, in a steel sphere of somewhat different design; and in 1953 French divers penetrated depths greater than a mile, existing for several hours in a zone of cold and darkness where the presence of living man had never before been known.

Although only a fortunate few can ever visit the deep sea, the precise instruments of the oceanographer, recording light penetration, pressure, salinity, and temperature, have given us the materials with which to reconstruct in imagination these eerie, forbidding regions. Unlike the surface waters, which are sensitive to every gust of wind, which know day and night, respond to the pull of sun and moon, and change as the seasons change, the deep waters are a place where change comes slowly, if at all. Down beyond the reach of the sun's rays, there is no alternation of light and darkness. There is rather an endless night, as old as the sea itself. For most of its creatures, groping their way endlessly through its black waters, it must be a place of hunger, where food is scarce and hard to find, a shelterless place where there is no sanctuary from ever-present enemies, where one can only move on and on, from birth to death, through the darkness, confined as in a prison to his own particular layer of the sea.

They used to say that nothing could live in the deep sea. It was a belief that must have been easy to accept, for without proof to the contrary, how could anyone conceive of life in such a place?

A century ago the British biologist Edward Forbes wrote: 'As we descend deeper and deeper into this region, the inhabitants become more and more modified, and fewer and fewer, indicating our approach to an abyss where life is either extinguished, or exhibits but a few sparks to mark its lingering presence.' Yet Forbes urged further exploration of 'this vast deep-sea region' to settle forever the question of the existence of life at great depths.

Even then, the evidence was accumulating. Sir John Ross, during his exploration of the arctic seas in 1818, had brought up from a depth of 1000 fathoms mud in which there were worms, 'thus proving there was animal life in the bed of the ocean notwithstanding the darkness, stillness, silence, and immense pressure produced by more than a mile of superincumbent water.'

Then from the surveying ship *Bulldog*, examining a proposed northern route for a cable from Faroe to Labrador in 1860, came another report. The *Bulldog's* sounding line, which at one place had been allowed to lie for some time on the bottom at a depth of 1260 fathoms, came up with 13 starfish clinging to it. Through these starfish, the ship's naturalist wrote, 'the deep has sent forth the long coveted message.' But not all the zoologists of the day were prepared to accept the message. Some doubters asserted that the starfish had 'convulsively embraced' the line somewhere on the way back to the surface.

In the same year, 1860, a cable in the Mediterranean was raised for repairs from a depth of 1200 fathoms. It was found to be heavily encrusted with corals and other sessile animals that had attached themselves at an early stage of development and grown to maturity over a period of months or years. There was not the slightest chance that they had become entangled in the cable as it was being raised to the surface.

Then the *Challenger*, the first ship ever equipped for oceanographic exploration, set out from England in the year 1872 and traced a course around the globe. From bottoms lying under miles of water, from silent deeps carpeted with red clay ooze, and from all the lightless intermediate depths, net-haul after net-haul of strange and fantastic creatures came up and were spilled out on the decks. Poring over the weird beings thus brought up for the first time into the light of day, beings no man had ever seen before, the *Challenger* scientists realized that life existed even on the deepest floor of the abyss.

The recent discovery that a living cloud of some unknown creatures is spread over much of the ocean at a depth of several hundred fathoms below the surface is the most exciting thing that has been learned about the ocean for many years.

When, during the first quarter of the twentieth century, echo sounding was developed to allow ships while under way to record the depth of the bottom, probably no one suspected that it would also provide a means of learning something about deep-sea life. But operators of the new instruments soon discovered that the sound waves, directed downward from the ship like a beam of light, were reflected back from any solid object they met. Answering echoes were returned from intermediate depths, presumably from schools of fish, whales, or submarines; then a second echo was received from the bottom.

These facts were so well established by the late 1930's that fishermen had begun to talk about using their fathometers to search for schools of herring. Then the war brought the whole subject under strict security regulations, and little more was heard about it. In 1946, however, the United States Navy issued a significant bulletin. It was reported that several scientists, working with sonic equipment in deep water off the California coast, had discovered a widespread 'layer' of some sort, which gave back an answering echo to the sound waves. This reflecting layer, seemingly suspended between the surface and the floor of the Pacific, was found over an area 300 miles wide. It lay from 1000 to 1500 feet below the surface. The discovery was made by three scientists, C. F. Eyring, R. J. Christensen, and R. W. Raitt, aboard the U.S.S. *Jasper* in 1942, and for a time this mysterious phenomenon, of wholly unknown nature, was called the ECR layer. Then in 1945 Martin W. Johnson, marine biologist of the Scripps Institution of Oceanography, made a further discovery which gave the first clue to the nature of the layer. Working aboard the vessel, *E. W. Scripps*, Johnson found that whatever sent back the echoes moved upward and downward in rhythmic fashion, being found near the surface at night, in deep water during the day. This discovery disposed of speculations that the reflections came from something inanimate, perhaps a mere physical discontinuity in the water, and showed that the layer is composed of living creatures capable of controlled movement.

From this time on, discoveries about the sea's 'phantom bottom' came rapidly. With widespread use of echo-sounding instruments, it has become clear that the phenomenon is not something peculiar to the coast of California alone. It occurs almost universally in deep ocean basins—drifting by day at a depth of several hundred fathoms, at night rising to the surface, and again, before sunrise, sinking into the depths.

On the passage of the U.S.S. *Henderson* from San Diego to the Antarctic in 1947, the reflecting layer was detected during the greater part of each day, at depths varying from 150 to 450 fathoms, and on a later run from San Diego to Yokosuka, Japan, the *Henderson's* fathometer again recorded the layer every day, suggesting that it exists almost continuously across the Pacific.

During July and August 1947, the U.S.S. *Nereus* made a continuous fathogram from Pearl Harbor to the Arctic and found the scattering layer over all deep waters along this course. It did not develop, however, in the shallow Bering and Chuckchee seas. Sometimes in the morning, the *Nereus'* fathogram showed two layers, responding in different ways to the growing illumination of the water; both descended into deep water, but there was an interval of twenty miles between the two descents.

Despite attempts to sample it or photograph it, no one is sure what the layer is, although the discovery may be made any day. There are three principal theories, each of which has its group of supporters. According to these theories, the sea's phantom bottom may consist of small planktonic shrimps, of fishes, or of squids.

As for the plankton theory, one of the most convincing arguments is the well-known fact that many plankton creatures make regular vertical migrations of hundreds of feet, rising toward the surface at night, sinking down below the zone of light penetration very early in the morning. This is, of course, exactly the behavior of the scattering layer. Whatever composes it is apparently strongly repelled by sunlight. The creatures of the layer seem almost to be held prisoner at the end—or beyond the end—of the sun's rays throughout the hours of daylight, waiting only for the welcome return of darkness to hurry upward into the surface waters. But what is the power that repels; and what the attraction that draws them surfaceward once the inhibiting force is removed? Is it comparative safety from enemies that makes them seek darkness? Is it more abundant food near the surface that lures them back under cover of night?

Those who say that fish are the reflectors of the sound waves usually account for the vertical migrations of the layer as suggesting that the fish are feeding on planktonic shrimp and are following their food. They believe that the air bladder of a fish is, of all structures concerned, most likely from its construction to return a strong echo. There is one outstanding difficulty in the way of accepting this theory: we have no other evidence that concentrations of fish are universally present in the oceans. In fact, almost everything else we know suggests that the really dense populations of fish live over the continental shelves or in certain very definite determined zones of the open ocean where food is particularly abundant. If the reflecting layer is eventually proved to be composed of fish, the prevailing views of fish distribution will have to be radically revised.

The most startling theory (and the one that seems to have the fewest supporters) is that the layer consists of concentrations of squid, 'hovering below the illuminated zone of the sea and awaiting the arrival of darkness in which to resume their raids into the plankton-rich surface waters.' Proponents of this theory argue that squid are abundant enough, and of wide enough distribution, to give the echoes that have been picked up almost everywhere from the equator to the two poles. Squid are known to be the sole food of the sperm whale, found in the open oceans in all temperate and tropical waters. They also form the exclusive diet of the bottlenosed whale and are eaten extensively by most other toothed whales, by seals, and by many sea birds. All these facts argue that they must be prodigiously abundant.

It is true that men who have worked close to the sea surface at night have received vivid impressions of the abundance and activity of squids in the surface waters in darkness. Long ago Johan Hjort wrote:

'One night we were hauling long lines on the Faroe slope, working with an electric lamp hanging over the side in order to see the line, when like lightning flashes one squid after another shot towards the light . . . In October 1902 we were one night steaming outside the slopes of the coast banks of Norway, and for many miles we could see the squids moving in the surface waters like luminous bubbles, resembling large milky white electric lamps being constantly lit and extinguished.'[1]

Thor Heyerdahl reports that at night his raft was literally bombarded by squids; and Richard Fleming says that in his oceanographic work off the coast of Panama it was common to see immense schools of squid gathering at the surface at night and leaping upward toward the lights that were used by the men to operate their instruments. But equally spectacular surface displays of shrimp have been seen, and most people find it difficult to believe in the ocean-wide abundance of squid.

Deep-water photography holds much promise for the solution of the mystery of the phantom bottom. There are technical difficulties, such as the problem of holding a camera still as it swings at the end of a long cable, twisting and turning, suspended from a ship which itself moves with the sea. Some of the pictures so taken look as though the photographer has pointed his camera at a starry sky and swung it in an arc as he exposed the film. Yet the Norwegian biologist Gunnar Rollefson had an encouraging experience in correlating photography with echograms. On the research ship *Johan Hjort* off the Lofoten Islands, he persistently got reflection of sound from schools of fish in 20 to 30 fathoms. A specially constructed camera was lowered to the depth indicated by the echogram. When developed, the film showed moving shapes of fish at a distance, and a large and clearly recognizable

---

[1] From *The Depths of the Ocean*, by Sir John Murray and Johan Hjort, 1912 edition, Macmillan & Co., p. 649.

cod appeared in the beam of light and hovered in front of the lens.

Direct sampling of the layer is the logical means of discovering its identity, but the problem is to develop large nets that can be operated rapidly enough to capture swift-moving animals. Scientists at Woods Hole, Massachusetts, have towed ordinary plankton nets in the layer and have found that euphausiid shrimps, glassworms, and other deep-water plankton are concentrated there; but there is still a possibility that the layer itself may actually be made up of larger forms feeding on the shrimps—too large or swift to be taken in the presently used nets. New nets may give the answer. Television is another possibility.

Shadowy and indefinite though they be, these recent indications of an abundant life at mid-depths agree with the reports of the only observers who have actually visited comparable depths and brought back eyewitness accounts of what they saw. William Beebe's impressions from the bathysphere were of a life far more abundant and varied than he had been prepared to find, although, over a period of six years, he had made many hundreds of net hauls in the same area. More than a quarter of a mile down, he reported aggregations of living things 'as thick as I have ever seen them.' At half a mile—the deepest descent of the bathysphere—Dr. Beebe recalled that 'there was no instant when a mist of plankton . . . was not swirling in the path of the beam.'

The existence of an abundant deep-sea fauna was discovered, probably millions of years ago, by certain whales and also, it now appears, by seals. The ancestors of all whales, we know by fossil remains, were land mammals. They must have been predatory beasts, if we are to judge by their powerful jaws and teeth. Perhaps in their foragings about the deltas of great rivers or around the edges of shallow seas, they discovered the abundance of fish and other marine life and over the centuries formed the habit of following them farther and farther into the sea. Little by little their bodies took on a form more suitable for aquatic life; their hind limbs were reduced to rudiments, which may be discovered in a modern whale by dissection, and the forelimbs were modified into organs for steering and balancing.

Eventually the whales, as though to divide the sea's food resources among them, became separated into three groups: the plankton-eaters, the fish-eaters, and the squid-eaters. The plankton-eating whales can exist only where there are dense masses of small shrimp or copepods to supply their enormous food requirements. This limits them, except for scattered areas, to arctic and antarctic waters and the high temperate latitudes. Fish-eating whales may find food over a somewhat wider range of ocean, but they are restricted to places where there are enormous populations of schooling fish. The blue water of the tropics and of the open ocean basins offers little to either of these groups. But that immense, square-headed, formidably toothed whale known as the cachalot or sperm whale discovered long ago what men have known for only a short time—that hundreds of fathoms below the almost untenanted surface waters of these regions there is an abundant animal life. The sperm whale has taken these deep waters for his hunting grounds; his quarry is the deep-water population of squids including the giant squid Architeuthis, which lives pelagically at depths of 1500 feet or more. The head of the sperm whale is often marked with long stripes, which consist of a great number of circular scars made by the suckers of the squid. From this evidence we can imagine the battles that go on, in the darkness of the deep water, between these two huge creatures—the sperm whale with its 70-ton bulk, the squid with a body as long as 30 feet, and writhing, grasping arms extending the total length of the animal to perhaps 50 feet.

The greatest depth at which the giant squid lives is not definitely known, but there is one instructive piece of evidence about the depth to which sperm whales descend, presumably in search of the squids. In April 1932, the cable repair ship *All America* was investigating an apparent break in the submarine cable between Balboa in the Canal Zone and Esmeraldas, Ecuador. The cable was brought to the surface off the coast of Colombia. Entangled in it was a dead

45-foot male sperm whale. The submarine cable was twisted around the lower jaw and was wrapped around one flipper, the body, and the caudal flukes. The cable was raised from a depth of 540 fathoms, or 3240 feet.

Some of the seals also appear to have discovered the hidden food reserves of the deep ocean. It has long been something of a mystery where, and on what, the northern fur seals of the eastern Pacific feed during the winter, which they spend off the coast of North America from California to Alaska. There is no evidence that they are feeding to any great extent on sardines, mackerel, or other commercially important fishes. Presumably four millon seals could not compete with commercial fishermen for the same species without the fact being known. But there is some evidence on the diet of the fur seals, and it is highly significant. Their stomachs have yielded the bones of a species of fish that has never been seen alive. Indeed, not even its remains have been found anywhere except in the stomachs of seals. Ichthyologists say that this 'seal fish' belongs to a group that typically inhabits very deep water, off the edge of the continental shelf.

How either whales or seals endure the tremendous pressure changes involved in dives of several hundred fathoms is not definitely known. They are warm-blooded mammals like ourselves. Caisson disease, which is caused by the rapid accumulation of nitrogen bubbles in the blood with sudden release of pressure, kills human divers if they are brought up rapidly from depths of 200 feet or so. Yet, according to the testimony of whalers, a baleen whale, when harpooned, can dive straight down to a depth of half a mile, as measured by the amount of line carried out. From these depths, where it has sustained a pressure of half a ton on every inch of body, it returns almost immediately to the surface. The most plausible explanation is that, unlike the diver, who has air pumped to him while he is under water, the whale has in its body only the limited supply it carries down, and does not have enough nitrogen in its blood to do serious harm. The plain truth is, however, that we really do not know, since it is obviously impossible to confine a living whale and experiment on it, and almost as difficult to dissect a dead one satisfactorily.

At first thought it seems a paradox that creatures of such great fragility as the glass sponge and the jellyfish can live under the conditions of immense pressure that prevail in deep water. For creatures at home in the deep sea, however, the saving fact is that the pressure inside their tissues is the same as that without, and as long as this balance is preserved, they are no more inconvenienced by a pressure of a ton or so than we are by ordinary atmospheric pressure. And most abyssal creatures, it must be remembered, live out their whole lives in a comparatively restricted zone, and are never required to adjust themselves to extreme changes of pressure.

But of course there are exceptions, and the real miracle of sea life in relation to great pressure is not the animal that lives its whole life on the bottom, bearing a pressure of perhaps five or six tons, but those that regularly move up and down through hundreds or thousands of feet of vertical change. The small shrimps and other planktonic creatures that descend into deep water during the day are examples. Fish that possess air bladders, on the other hand, are vitally affected by abrupt changes of pressure, as anyone knows who has seen a trawler's net raised from a hundred fathoms. Apart from the accident of being captured in a net and hauled up through waters of rapidly diminishing pressures, fish may sometimes wander out of the zone to which they are adjusted and find themselves unable to return. Perhaps in their pursuit of food they roam upward to the ceiling of the zone that is theirs, and beyond whose invisible boundary they may not stray without meeting alien and inhospitable conditions. Moving from layer to layer of drifting plankton as they feed, they may pass beyond the boundary. In the lessened pressure of these upper waters the gas enclosed within the air bladder expands. The fish becomes lighter and more buoyant. Perhaps he tries to fight his way down again, opposing the upward lift with all the power of his muscles. If he does not succeed, he 'falls' to the surface, injured and dying, for the abrupt release of pressure from without causes distension and rupture of the tissues.

The compression of the sea under its own weight is relatively slight, and there is no basis for the old and picturesque belief that, at the deeper levels, the water resists the downward passage of objects from the surface. According to this belief, sinking ships, the bodies of drowned men, and presumably the bodies of the larger sea animals not consumed by hungry scavengers, never reach the bottom, but come to rest at some level determined by the relation of their own weight to the compression of the water, there to drift forever. The fact is that anything will continue to sink as long as its specific gravity is greater than that of the surrounding water, and all large bodies descend, in a matter of a few days, to the ocean floor. As mute testimony to this fact, we bring up from the deepest ocean basins the teeth of sharks and the hard ear bones of whales.

Nevertheless the weight of sea water—the pressing down of miles of water upon all the underlying layers—does have a certain effect upon the water itself. If this downward compression could suddenly be relaxed by some miraculous suspension of natural laws, the sea level would rise about 93 feet all over the world. This would shift the Atlantic coastline of the United States westward a hundred miles or more and alter other familiar geographic outlines all over the world.

Immense pressure, then, is one of the governing conditions of life in the deep sea; darkness is another. The unrelieved darkness of the deep waters has produced weird and incredible modifications of the abyssal fauna. It is a blackness so divorced from the world of the sunlight that probably only the few men who have seen it with their own eyes can visualize it. We know that light fades out rapidly with descent below the surface. The red rays are gone at the end of the first 200 or 300 feet, and with them all the orange and yellow warmth of the sun. Then the greens fade out, and at 1000 feet only a deep, dark, brilliant blue is left. In very clear waters the violet rays of the spectrum may penetrate another thousand feet. Beyond this is only the blackness of the deep sea.

In a curious way, the colors of marine animals tend to be related to the zone in which they live. Fishes of the surface waters, like the mackerel and herring, often are blue or green; so are the floats of the Portuguese men-of-war and the azure-tinted wings of the swimming snails. Down below the diatom meadows and the drifting sargassum weed, where the water becomes ever more deeply, brilliantly blue, many creatures are crystal clear. Their glassy, ghostly forms blend with their surroundings and make it easier for them to elude the ever-present, ever-hungry enemy. Such are the transparent hordes of the arrowworms or glassworms, the comb jellies, and the larvae of many fishes.

At a thousand feet, and on down to the very end of the sun's rays, silvery fishes are common, and many others are red, drab brown, or black. Pteropods are a dark violet. Arrow-worms, whose relatives in the upper layers are colorless, are here a deep red. Jellyfish medusae, which above would be transparent, at a depth of 1000 feet are a deep brown.

At depths greater than 1500 feet, all the fishes are black, deep violet, or brown, but the prawns wear amazing hues of red, scarlet, and purple. Why, no one can say. Since all the red rays are strained out of the water far above this depth, the scarlet raiment of these creatures can only look black to their neighbors.

The deep sea has its stars, and perhaps here and there an eerie and transient equivalent of moonlight, for the mysterious phenomenon of luminescence is displayed by perhaps half of all the fishes that live in dimly lit or darkened waters, and by many of the lower forms as well. Many fishes carry luminous torches that can be turned on or off at will, presumably helping them find or pursue their prey. Others have rows of lights over their bodies, in patterns that vary from species to species and may be a sort of recognition mark or badge by which the bearer can be known as friend or enemy. The deep-sea squid ejects a spurt of fluid that becomes a luminous cloud, the counterpart of the 'ink' of his shallow-water relative.

Down beyond the reach of even the longest and strongest of the sun's rays, the eyes of fishes become enlarged, as though to make the most of any chance illumination of whatever sort, or they may become telescopic, large of lens, and protruding. In deep-sea fishes, hunting always in dark waters, the eyes tend to lose the 'cones' or color-perceiving cells of the retina, and to increase the 'rods,' which perceive dim light. Exactly the same modification is seen on land among the strictly nocturnal prowlers which, like abyssal fish, never see the sunlight.

In their world of darkness, it would seem likely that some of the animals might have become blind, as has happened to some cave fauna. So, indeed, many of them have, compensating for the lack of eyes with marvelously developed feelers and long, slender fins and processes with which they grope their way, like so many blind men with canes, their whole knowledge of friends, enemies, or food coming to them through the sense of touch.

The last traces of plant life are left behind in the thin upper layer of water, for no plant can live below about 600 feet even in very clear water, and few find enough sunlight for their food-manufacturing activities below 200 feet. Since no animal can make its own food, the creatures of the deeper waters live a strange, almost parasitic existence of utter dependence on the upper layers. These hungry carnivores prey fiercely and relentlessly upon each other, yet the whole community is ultimately dependent upon the slow rain of descending food particles from above. The components of this never-ending rain are the dead and dying plants and animals from the surface, or from one of the intermediate layers. For each of the horizontal zones or communities of the sea that lie, in tier after tier, between the surface and the sea bottom, the food supply is different and in general poorer than for the layer above. There is a hint of the fierce and uncompromising competition for food in the saber-toothed jaws of some of the small, dragonlike fishes of the deeper waters, in the immense mouths and in the elastic and distensible bodies that make it possible for a fish to swallow another several times its size, enjoying swift repletion after a long fast.

Pressure, darkness, and—we should have added only a few years ago—silence, are the conditions of life in the deep sea. But we know now that the conception of the sea as a silent place is wholly false. Wide experience with hydrophones and other listening devices for the detection of submarines has proved that, around the shore lines of much of the world, there is the extraordinary uproar produced by fishes, shrimps, porpoises and probably other forms not yet identified. There has been little investigation as yet of sound in the deep, offshore areas, but when the crew of the *Atlantis* lowered a hydrophone into deep water off Bermuda, they recorded strange mewing sounds, shrieks, and ghostly moans, the sources of which have not been traced. But fish of shallower zones have been captured and confined in aquaria, where their voices have been recorded for comparison with sounds heard at sea, and in many cases satisfactory identification can be made.

During the Second World War the hydrophone network set up by the United States Navy to protect the entrance to Chesapeake Bay was temporarily made useless when, in the spring of 1942, the speakers at the surface began to give forth, every evening, a sound described as being like 'a pneumatic drill tearing up pavement.' The extraneous noises that came over the hydrophones completely masked the sounds of the passage of ships. Eventually it was discovered that the sounds were the voices of fish known as croakers, which in the spring move into Chesapeake Bay from their offshore wintering grounds. As soon as the noise had been identified and analyzed, it was possible to screen it out with an electric filter, so that once more only the sounds of ships came through the speakers.

Later in the same year, a chorus of croakers was discovered off the pier of the Scripps Institution at La Jolla. Every year from May until late September the evening chorus begins about sunset, and 'increases gradually to a steady uproar of harsh froggy croaks, with a background of soft drumming.

This continues unabated for two to three hours and finally tapers off to individual outbursts at rare intervals.' Several species of croakers isolated in aquaria gave sounds similar to the 'froggy croaks,' but the authors of the soft background drumming—presumably another species of croaker—have not yet been discovered.

One of the most extraordinarily widespread sounds of the undersea is the crackling, sizzling sound, like dry twigs burning or fat frying, heard near beds of the snapping shrimp. This is a small, round shrimp, about half an inch in diameter, with one very large claw which it uses to stun its prey. The shrimp are forever clicking the two joints of this claw together, and it is the thousands of clicks that collectively produce the noise known as shrimp crackle. No one had any idea the little snapping shrimps were so abundant or so widely distributed until their signals began to be picked up on hydrophones. They have been heard over a broad band that extends around the world, between latitudes 35° N and 35° S (for example, from Cape Hatteras to Buenos Aires) in ocean waters less than 30 fathoms deep.

Mammals as well as fishes and crustaceans contribute to the undersea chorus. Biologists listening through a hydrophone in an estuary of the St. Lawrence River heard 'high-pitched resonant whistles and squeals, varied with the ticking and clucking sounds slightly reminiscent of a string orchestra tuning up, as well as mewing and occasional chirps.' This remarkable medley of sounds was heard only while schools of the white porpoise were seen passing up or down the river, and so was assumed to be produced by them.

The mysteriousness, the eeriness, the ancient unchangingness of the great depths have led many people to suppose that some very old forms of life—some 'living fossils'—may be lurking undiscovered in the deep ocean. Some such hope may have been in the minds of the *Challenger* scientists. The forms they brought up in their nets were weird enough, and most of them had never before been seen by man. But basically they were modern types. There was nothing like the trilobites of Cambrian time or the sea scorpions of the

Silurian, nothing reminiscent of the great marine reptiles that invaded the sea in the Mesozoic. Instead, there were modern fishes, squids, and shrimps, strangely and grotesquely modified, to be sure, for life in the difficult deep-sea world, but clearly types that have developed in rather recent geologic time.

Far from being the original home of life, the deep sea has probably been inhabited for a relatively short time. While life was developing and flourishing in the surface waters, along the shores, and perhaps in the rivers and swamps, two immense regions of the earth still forbade invasion by living things. These were the continents and the abyss. As we have seen, the immense difficulties of surviving on land were first overcome by colonists from the sea about 300 million years ago. The abyss, with its unending darkness, its crushing pressures, its glacial cold, presented even more formidable difficulties. Probably the successful invasion of this region—at least by higher forms of life—occurred somewhat later.

Yet in recent years there have been one or two significant happenings that have kept alive the hope that the deep sea may, after all, conceal strange links with the past. In December 1938, off the southeast tip of Africa, an amazing fish was caught alive in a trawl—a fish that was supposed to have been dead for at least 60 million years! This is to say, the last known fossil remains of its kind date from the Cretaceous, and no living example had been recognized in historic time until this lucky net-haul.

The fishermen who brought it up in their trawl from a depth of only 40 fathoms realized that this five-foot, bright blue fish, with its large head and strangely shaped scales, fins, and tail, was different from anything that they ever caught before, and on their return to port they took it to the nearest museum, where it was christened Latimeria. It was identified as a coelacanth, or one of an incredibly ancient group of fishes that first appeared in the seas some 300 million years ago. Rocks representing the next 200 million and more years of earth history yielded fossil coelacanths; then, in the Cretaceous, the record of these fishes came to an end. After 60 million years of mysterious oblivion, one of the group, Latimeria, then appeared before the eyes of the South African fishermen, apparently little changed in structure from its ancient ancestors. But where had these fishes been in the meantime?

The story of the coelacanths did not end in 1938. Believing there must be other such fish in the sea, an ichthyologist in South Africa, Professor J. L. B. Smith, began a patient search that lasted 14 years before it was successful. Then, in December 1952, a second coelacanth was captured near the island of Anjouan, off the northwestern tip of Madagascar. It differed enough from Latimeria to be placed in a separate genus, but like the first coelacanth known in modern times, it can tell us much of a shadowy chapter in the evolution of living things.

Occasionally a very primitive type of shark, known from its puckered gills as a 'frillshark,' is taken in waters between a quarter of a mile and half a mile down. Most of these have been caught in Norwegian and Japanese waters—there are only about 50 preserved in the museums of Europe and America—but recently one was captured off Santa Barbara, California. The frillshark has many anatomical features similar to those of the ancient sharks that lived 25 to 30 million years ago. It has too many gills and too few dorsal fins for a modern shark, and its teeth, like those of fossil sharks, are three-pronged and briarlike. Some ichthyologists regard it as a relic derived from very ancient shark ancestors that have died out in the upper waters but, through this single species, are still carrying on their struggle for earthly survival, in the quiet of the deep sea.

Possibly there are other such anachronisms lurking down in these regions of which we know so little, but they are likely to be few and scattered. The terms of existence in these deep waters are far too uncompromising to support life unless that life is plastic, molding itself constantly to the harsh conditions, seizing every advantage that makes possible the survival of living protoplasm in a world only a little less hostile than the black reaches of interplanetary space.

MAURICE MAETERLINCK

THE TREE of animal evolution divides into two enormous trunks—the vertebrates who have spinal columns and the arthropods who do not. Vertebrate life reaches its climax in man. Arthropod behavior culminates in the social insects. So radically different are these two ways of meeting problems of survival that one would almost expect the two groups to have evolved on different planets. All the actions we consider "human" are learned reactions, acquired by each individual through long years of training by his elders. Insects, on the other hand, are virtually unteachable. They come into existence fully equipped with an elaborate pattern of inborn behavior. In the social insects—such as bees, wasps, ants, and termites—these instincts reach unbelievable levels of specialization. The colony itself takes on the characteristics of a super-organism—a totalitarian state in which there is not even the possibility of "anti-social" behavior. It is little wonder that of the two million or so different insect species, the social insects—the honeybee in particular—have excited the greatest interest of zoologists and laymen alike.

As a young man, Belgian poet and playwright Maurice Maeterlinck (1862-1949) made beekeeping his principal hobby. Inspired by the essays of Fabre, he began a period of observation and experiment with his apiary that resulted in 1901 in the publication of *The Life of the Bee.* Written in a highly poetic vein, with a skillful blending of fact, fancy, and mystical speculation, it became far and away the most popular book ever written about insect life.

Subsequent research has corrected some of Maeterlinck's errors and added new and even more fantastic data. In a series of beautifully devised experiments the Bavarian zoologist Karl von Frisch proved that bee scouts, after returning to the hive, execute a series of rapid gyrations that tell the workers how far away the new-found nectar is and in exactly what direction! More recent experiments have shown that bees possess a built-in method of measuring time irrespective of external factors such as sunlight.

The selection presented here is Maeterlinck's unrivaled description of the nuptial flight of the queen. The dramatic meeting in the sky, followed by the death of the pursuer, provides a magnificent background for the Belgian poet's reflections on the meaning of good and evil, love and death.

MAURICE MAETERLINCK

The Nuptial Flight

WE WILL now consider the manner in which the impregnation of the queen-bee comes to pass. Here again nature has taken extraordinary measures to favor the union of males with females of a different stock; a strange law, whereto nothing would seem to compel her; a caprice, or initial inadvertence, perhaps, whose reparation calls for the most marvelous forces her activity knows.

If she had devoted half the genius she lavishes on crossed fertilization and other arbitrary desires to making life more certain, to alleviating pain, to softening death and warding off horrible accidents, the universe would probably have presented an enigma less incomprehensible, less pitiable, than the one we are striving to solve. But our consciousness, and the interest we take in existence, must grapple, not with what might have been, but with what is.

Around the virgin queen, and dwelling with her in the hive, are hundreds of exuberant males, forever drunk on honey; the sole reason for their existence being one act of love. But, notwithstanding the incessant contact of two desires that elsewhere invariably triumph over every obstacle, the union never takes place in the hive, nor has it been possible to bring about the impregnation of a captive queen. While she lives in their midst the lovers about her know not what she is. They seek her in space, in the remote depths of the horizon, never suspecting that they have but this moment quitted her, have shared the same comb with her, have brushed against her, perhaps, in the eagerness of their departure. One might almost believe that those wonderful eyes of theirs, that cover their head as though with a glittering helmet, do not recognize or desire her save when she soars in the blue. Each day, from noon till three, when the sun shines resplendent, this plumed horde sallies forth in search of the bride, who is indeed more royal, more difficult of conquest, than the most inaccessible princess of fairy legend; for twenty or thirty tribes will hasten from all the neighboring cities, her court thus consisting of more than ten thousand suitors; and from these ten thousand one alone will be chosen for the unique kiss of an instant that shall wed him to death no less than to happiness; while the others will fly helplessly round the intertwined pair, and soon will perish without ever again beholding this prodigious and fatal apparition.

I am not exaggerating this wild and amazing prodigality of nature. The best-conducted hives will, as a rule, contain four to five hundred males. Weaker or degenerate ones will often have as many as four or five thousand; for the more a hive inclines to its ruin, the more males will it produce. It may be said that, on an average, an apiary composed of ten colonies will at a given moment send an army of ten thousand males into the air, of whom ten or fifteen at most will have the occasion of performing the one act for which they were born.

In the meanwhile they exhaust the supplies of the city; each one of the parasites requiring the unceasing labor of five or six workers to maintain it in its abounding and voracious idleness, its activity being indeed solely confined to its jaws. But nature is always magnificent when dealing with the privileges and prerogatives of love. She becomes miserly only when doling out the organs and instruments of labor. She is especially severe on what men have termed virtue, whereas she strews the path of the most uninteresting lovers with innumerable jewels and favors. "Unite and multiply; there is no other law, or aim, than love," would seem to be her constant cry on all sides, while she mutters to herself, perhaps:

"and exist afterward if you can; that is no concern of mine." Do or desire what else we may, we find, everywhere on our road, this morality that differs so much from our own. And note, too, in these same little creatures, her unjust avarice and insensate waste. From her birth to her death, the austere forager has to travel abroad in search of the myriad flowers that hide in the depths of the thickets. She has to discover the honey and pollen that lurk in the labyrinths of the nectaries and in the most secret recesses of the anthers. And yet her eyes and olfactory organs are like the eyes and organs of the infirm, compared with those of the male. Were the drones almost blind, had they only the most rudimentary sense of smell, they scarcely would suffer. They have nothing to do, no prey to hunt down; their food is brought to them ready prepared, and their existence is spent in the obscurity of the hive, lapping honey from the comb. But they are the agents of love; and the most enormous, most useless gifts are flung with both hands into the abyss of the future. Out of a thousand of them, one only, once in his life, will have to seek, in the depths of the azure, the presence of the royal virgin. Out of a thousand one only will have, for one instant, to follow in space the female who desires not to escape. That suffices. The partial power flings open her treasury, wildly, even deliriously. To every one of these unlikely lovers, of whom nine hundred and ninety-nine will be put to death a few days after the fatal nuptials of the thousandth, she has given thirteen thousand eyes on each side of their head, while the worker has only six thousand. According to Cheshire's calculations, she has provided each of their antennæ with thirty-seven thousand eight hundred olfactory cavities, while the worker has only five thousand in both. There we have an instance of the almost universal disproportion that exists between the gifts she rains upon love and her niggardly doles to labor; between the favors she accords to what shall, in an ecstasy, create new life, and the indifference wherewith she regards what will patiently have to maintain itself by toil. Whoever would seek faithfully to depict the character of nature, in accordance with the traits we discover here, would design an extraordinary figure, very foreign to our ideal, which nevertheless can only emanate from her. But too many things are unknown to man for him to essay such a portrait, wherein all would be deep shadow save one or two points of flickering light.

Very few, I imagine, have profaned the secret of the queen-bee's wedding, which comes to pass in the infinite, radiant circles of a beautiful sky. But we are able to witness the hesitating departure of the bride-elect and the murderous return of the bride.

However great her impatience, she will yet choose her day and her hour, and linger in the shadow of the portal till a marvelous morning fling open wide the nuptial spaces in the depths of the great azure vault. She loves the moment when drops of dew still moisten the leaves and the flowers, when the last fragrance of dying dawn still wrestles with burning day, like a maiden caught in the arms of a heavy warrior; when through the silence of approaching noon is heard, once and again, a transparent cry that has lingered from sunrise.

Then she appears on the threshold—in the midst of indifferent foragers, if she have left sisters in the hive; or surrounded by a delirious throng of workers, should it be impossible to fill her place.

She starts her flight backwards, returns twice or thrice to the alighting-board; and then, having definitely fixed in her mind the exact situation and aspect of the kingdom she has never yet seen from without, she departs like an arrow to the zenith of the blue. She soars to a height, a luminous zone, that other bees attain at no period of their life. Far away, caressing their idleness in the midst of the flowers, the males have beheld the apparition, have breathed the magnetic perfume that spreads from group to group till every apiary near is instinct with it. Immediately crowds collect, and follow her into the sea of gladness, whose limpid boundaries ever recede. She, drunk with her wings, obeying the magnificent law of the race that chooses her lover, and enacts that the strongest alone shall attain her in the solitude of the ether, she rises still; and, for the first time in her life, the blue morning air rushes into her stigmata singing its song, like the blood of heaven, in the myriad tubes of the tracheal sacs, nourished on space, that fill the center of her body. She rises still. A region must be found unhaunted by birds, that else might profane the mystery. She rises still; and already the ill-assorted troop below are dwindling and falling asunder. The feeble, infirm, the aged, unwelcome, ill-fed, who have flown from inactive or impoverished cities, these renounce the pursuit and disappear in the void. Only a small, indefatigable cluster remain, suspended in infinite opal. She summons her wings for one final effort; and now the chosen of incomprehensible forces has reached her, has seized her, and bounding aloft with united impetus, the ascending spiral of their intertwined flight whirls for one second in the hostile madness of love.

Most creatures have a vague belief that a very precarious hazard, a kind of transparent membrane, divides death from love; and that the profound idea of nature demands that the giver of life should die at the moment of giving. Here this idea, whose memory lingers still over the kisses of man, is realized in its primal simplicity. No sooner has the union been accomplished than the male's abdomen opens, the organ detaches itself, dragging with it the mass of the entrails; the wings relax, and, as though struck by lightning, the emptied body turns and turns on itself and sinks into the abyss.

The same idea that, before, in parthenogenesis, sacrificed the future of the hive to the unwonted multiplication of males, now sacrifices the male to the future of the hive.

This idea is always astounding; and the further we penetrate into it, the fewer do our certitudes become. Darwin, for instance, to take the man of all men who studied it the most methodically and most passionately, Darwin, though scarcely confessing it to himself, loses confidence at every step, and retreats before the unexpected and the irreconcilable. Would you have before you the nobly humiliating spectacle of human genius battling with infinite power, you have but to follow Darwin's endeavors to unravel the strange, incoherent, inconceivably mysterious laws of the sterility and fecundity of hybrids, or of the variations of specific and generic characters. Scarcely has he formulated a principle when numberless exceptions assail him; and this very principle, soon completely overwhelmed, is glad to find refuge in some corner, and preserve a shred of existence there under the title of an exception.

For the fact is that in hybridity, in variability (notably in the simultaneous variations known as correlations of growth), in instinct, in the processes of vital competition, in geologic succession and the geographic distribution of organized beings, in mutual affinities, as indeed in every other direction, the idea of nature reveals itself, in one and the same phenomenon and at the very same time, as circumspect and shiftless, niggard and prodigal, prudent and careless, fickle and stable, agitated and immovable, one and innumerable, magnificent and squalid. There lay open before her the immense and virgin fields of simplicity; she chose to people them with trivial errors, with petty contradictory laws that stray through existence like a flock of blind sheep. It is true that our eye, before which these things happen, can only reflect a reality proportionate to our needs and our stature; nor have we any warrant for believing that nature ever loses sight of her wandering results and causes.

In any event she will rarely permit them to stray too far, or approach illogical or dangerous regions. She disposes of two forces that never can err; and when the phenomenon shall have trespassed beyond certain limits, she will beckon to life or to death—which arrives, re-establishes order, and unconcernedly marks out the path afresh.

She eludes us on every side; she repudiates most of our rules and breaks our standards to pieces. On our right she sinks far beneath the level of our thoughts, on our left she towers mountain-high above them. She appears to be constantly blundering, no less in the world of her first experiments than in that of her last, of man. There she invests with her sanction the instincts of the obscure mass, the unconscious injustice of the multitude, the defeat of intelligence and virtue, the uninspired morality which urges on the great wave of the race, though manifestly inferior to the morality that could be conceived or desired by the minds composing the small and the clearer wave that ascends the other. And yet, can such a mind be wrong if it ask itself whether the whole truth—moral truths, therefore, as well as non-moral—had not better be sought in this chaos than in itself, where these truths would seem comparatively clear and precise?

The man who feels thus will never attempt to deny the reason or virtue of his ideal, hallowed by so many heroes and sages; but there are times when he will whisper to himself that this ideal has perhaps been formed at too great a distance from the enormous mass whose diverse beauty it would fain represent. He has, hitherto, legitimately feared that the attempt to adapt his morality to that of nature would risk the destruction of what was her masterpiece. But today he understands her a little better; and from some of her replies, which, though still vague, reveal an unexpected breadth, he has been enabled to seize a glimpse of a plan and an intellect vaster than could be conceived by his unaided imagination; wherefore he has grown less afraid, nor feels any longer the same imperious need of the refuge his own special virtue and reason afford him. He concludes that what is so great could surely teach nothing that would tend to lessen itself. He wonders whether the moment may not have arrived for submitting to a more judicious examination his convictions, his principles, and his dreams.

Once more, he has not the slightest desire to abandon his human ideal. That even which at first diverts him from this ideal teaches him to return to it. It were impossible for nature to give ill advice to a man who declines to include in the great scheme he is endeavoring to grasp, who declines to regard as sufficiently lofty to be definitive, any truth that is not at least as lofty as the truth he himself desires. Nothing shifts its place in his life save only to rise with him; and he knows he is rising when he finds himself drawing near to his ancient image of good. But all things transform themselves more freely in his thoughts; and he can descend with impunity, for he has the presentiment that numbers of successive valleys will lead him to the plateau that he expects. And, while he thus seeks for conviction, while his researches even conduct him to the very reverse of that which he loves, he directs his conduct by the most humanly beautiful truth, and clings to the one that provisionally seems to be highest. All that may add to beneficent virtue enters his heart at once; all that would tend to lessen it remaining there in suspense, like insoluble salts that change not till the hour for decisive experiment. He may accept an inferior truth, but before he will act in accordance therewith he will wait, if need be for centuries, until he perceive the connection this truth must possess with truths so infinite as to include and surpass all others.

In a word, he divides the moral from the intellectual order, admitting in the former that only which is greater and more beautiful than was there before. And blameworthy as it may be to separate the two orders in cases, only too frequent in life, where we suffer our conduct to be inferior to our thoughts, where, seeing the good, we follow the worse—to see the worse and follow the better, to raise our actions high over our idea, must ever be reasonable and salutary; for human experience renders it daily more clear that the highest thought we can attain will long be inferior still to the mysterious truth we seek. Moreover, should nothing of what goes before be true, a reason more simple and more familiar would counsel him not yet to abandon his human ideal. For the more strength he accords to the laws which would seem to set egoism, injustice, and cruelty as examples for men to follow, the more strength does he at the same time confer on the others that ordain generosity, justice, and pity; and these last laws are found to contain something as profoundly natural as the first, the moment he begins to equalize, or allot more methodically, the share he attributes to the universe and to himself.

Let us return to the tragic nuptials of the queen. Here it is evidently nature's wish, in the interests of crossed fertilization, that the union of the drone and the queen-bee should be possible only in the open sky. But her desires blend network-fashion, and her most valued laws have to pass through the meshes of other laws, which, in their turn, the moment after, are compelled to pass through the first.

In the sky she has planted so many dangers—cold winds, storm-currents, birds, insects, drops of water, all of which also obey invincible laws—that she must of necessity arrange for this union to be as brief as possible. It is so, thanks to the startlingly sudden death of the male. One embrace suffices; the rest all enacts itself in the very flanks of the bride.

She descends from the azure heights and returns to the hive, trailing behind her, like an oriflamme, the unfolded entrails of her lover. Some writers pretend that the bees manifest great joy at this return so big with promise—Büchner, among others, giving a detailed account of it. I have many a time lain in wait for the queen-bee's return, and I confess that I have never noticed any unusual emotion except in the case of a young queen who had gone forth at the head of a swarm, and represented the unique hope of a newly founded and still empty city. In that instance the workers were all wildly excited, and rushed to meet her. But as a rule they appear to forget her, even though the future of their city will often be no less imperiled. They act with consistent prudence in all things, till the moment when they authorize the massacre of the rival queens. That point reached, their instinct halts; and there is, as it were, a gap in their foresight.—They appear to be wholly indifferent. They raise their heads; recognize, probably, the murderous tokens of impregnation; but, still mistrustful, manifest none of the gladness our expectation had pictured. Being positive in their ways, and slow at illusion, they probably need further proofs before permitting themselves to rejoice. Why endeavor to render too logical, or too human, the feelings of little creatures so different from ourselves? Neither among the bees nor among any other animals that have a ray of our intellect, do things happen with the precision our books record. Too many circumstances remain unknown to us. Why try to depict the bees as more perfect than they are, by saying that which is not? Those who would deem them more interesting did they resemble ourselves, have not yet truly realized what it is that should awaken the interest of a sincere mind. The aim of the observer is not to surprise, but to comprehend; and to point out the gaps existing in an intellect, and the signs of a cerebral organization different from our own, is more curious by far than the relating of mere marvels concerning it.

But this indifference is not shared by all; and when the breathless queen has reached the alighting-board, some groups will form and accompany her into the hive; where the sun, hero of every festivity in which the bees take part, is entering with little timid steps, and bathing in azure and shadow the waxen walls and curtains of honey. Nor does the new bride, indeed, show more concern than her people, there being not room for many emotions in her narrow, barbarous, practical brain. She has but one thought, which is to rid herself as quickly as possible of the embarrassing souvenirs her consort has left her, whereby her movements are hampered. She seats herself on the threshold, and carefully strips off the useless organs, that are borne far away by the workers; for the male has given her all he possessed, and much more than she requires. She retains only, in her spermatheca, the seminal liquid where millions of germs are floating, which, until her last day, will issue one by one, as the eggs pass by, and in the obscurity of her body accomplish the mysterious union of the male and female element, whence the worker-bees are born. Through a curious inversion, it is she who furnishes the male principle, and the drone who provides the female. Two days after the union she lays her first eggs, and her people immediately surround her with the most particular care. From that moment, possessed of a dual sex, having within her an inexhaustible male, she begins her veritable life; she will never again leave the hive, unless to accompany a swarm; and her fecundity will cease only at the approach of death.

Prodigious nuptials these, the most fairylike that can be conceived, azure and tragic, raised high above life by the impetus of desire; imperishable and terrible, unique and bewildering, solitary and infinite. An admirable ecstasy, wherein death supervening in all that our sphere has of most limpid and loveliest, in virginal, limitless space, stamps the instant of happiness in the sublime transparence of the great sky; purifying in that immaculate light the something of wretchedness that always hovers around love, rendering the kiss one that can never be forgotten; and, content this time with moderate tithe, proceeding herself, with hands that are almost maternal, to introduce and unite, in one body, for a long and inseparable future, two little fragile lives.

Profound truth has not this poetry, but possesses another that we are less apt to grasp, which, however, we should end, perhaps, by understanding and loving. Nature has not gone out of her way to provide these two "abbreviated atoms," as Pascal would call them, with a resplendent marriage, or an ideal moment of love. Her concern, as we have said, was merely to improve the race by means of crossed fertilization. To ensure this she has contrived the organ of the male in such a fashion that he can make use of it only in space. A prolonged flight must first expand his two great tracheal sacs; these enormous receptacles being gorged on air will throw back the lower part of the abdomen, and permit the exsertion of the organ. There we have the whole physiological secret—which will seem ordinary enough to some, and almost vulgar to others—of this dazzling pursuit and these magnificent nuptials.

"But must we always, then," the poet will wonder, "rejoice in regions that are loftier than the truth?"

Yes, in all things, at all times, let us rejoice, not in regions loftier than the truth, for that were impossible, but in regions higher than the little truths that our eyes can seize. Should a chance, a recollection, an illusion, a passion,—in a word, should any motive whatever cause an object to reveal itself to us in a more beautiful light than to others, let that motive be first of all dear to us. It may only be error, perhaps; but this error will not prevent the moment wherein we are likeliest to perceive its real beauty. The beauty we lend it directs our attention to its veritable beauty and grandeur, which, derived as they are from the relation wherein every object must of necessity stand to general, eternal, forces and laws, might otherwise escape observation. The faculty of admiring which an illusion may have created within us will serve for the truth that must come, be it sooner or later. It is with the words, the feelings, and ardor created by ancient and imaginary beauties, that humanity welcomes today truths which perhaps would have never been born, which might not have been able to find so propitious a home, had these sacrificed illusions not first of all dwelt in, and kindled, the heart and the reason whereinto these truths should descend. Happy the eyes that need no illusion to see that the spectacle is great! It is illusion that teaches the others to look, to admire, and rejoice. And look as high as they will, they never can look too high. Truth rises as they draw nearer; they draw nearer when they admire. And whatever the heights may be whereon they rejoice, this rejoicing can never take place in the void, or above the unknown and eternal truth that rests over all things like beauty in suspense.

Does this mean that we should attach ourselves to falsehood, to an unreal and factitious poetry, and find our gladness therein for want of anything better? Or that in the example before us—in itself nothing, but we dwell on it because it stands for a thousand others, as also for our entire attitude in face of divers orders of truths—that here we should ignore the physiological explanation, and retain and taste only the emotions of this nuptial flight, which is yet, and whatever the cause, one of the most lyrical, most beautiful acts of that suddenly disinterested, irresistible force which all living creatures obey and are wont to call love? That were too childish; nor is it possible, thanks to the excellent habits every loyal mind has today acquired.

The fact being incontestable, we must evidently admit that the exsertion of the organ is rendered possible only by the expansion of the tracheal vesicles. But if we, content with this fact, did not let our eyes roam beyond it; if we deduced therefrom that every thought that rises too high or wanders too far must be of necessity wrong, and that truth must be looked for only in the material details; if we did not seek, no matter where, in uncertainties often far greater than the one this little explanation has solved, in the strange mystery of crossed fertilization for instance, or in the perpetuity of the race and life, or in the scheme of nature; if we did not seek in these for something beyond the current explanation, something that should prolong it, and conduct us to the beauty and grandeur that repose in the unknown, I would almost venture to assert that we should pass our existence further away from the truth than those, even, who in this case willfully shut their eyes to all save the poetic and wholly imaginary interpretation of these marvelous nuptials. They evidently misjudge the form and color of the truth, but they live in its atmosphere and its influence far more than the others, who complacently believe that the entire truth lies captive within their two hands. For the first have made ample preparations to receive the truth, have provided most hospitable lodging within them; and even though their eyes may not see it, they are eagerly looking toward the beauty and grandeur where its residence surely must be.

We know nothing of nature's aim, which for us is the truth that dominates every other. But for the very love of this truth, and to preserve in our soul the ardor we need for its search, it behooves us to deem it great. And if we should find one day that we have been on a wrong road, that this aim is incoherent and petty, we shall have discovered its pettiness by means of the very zeal its presumed grandeur had created within us; and this pettiness once established, it will teach us what we have to do. In the meanwhile it cannot be unwise to devote to its search the most strenuous, daring efforts of our heart and our reason. And should the last word of all this be wretched, it will be no little achievement to have laid bare the inanity and the pettiness of the aim of nature.

H. G. WELLS

WHEN THE FIRST atom bomb fell on Japan in August, 1945, an old man dying in London must have read the headlines with a strange emotion. The old man was H. G. Wells. He had already completed his final book in which he spoke of a "frightful queerness" descending on the world, and he had projected his own sentence of death into the prediction that *homo sapiens* would have to "give place to some other animal better adapted to face the fate that closes in more and more swiftly upon mankind." But more than this, he saw in the headlines a dramatic confirmation of one of his most accurate prophecies.

In 1914 Herbert George Wells (1866-1946) had written a science fiction novel titled *The World Set Free.* It opens with a prelude called "The Sun Snarers," tracing the history of man's conquest of power from the first crude use of tools and domesticated animals to the steam and electrical power of modern times. The prelude closes with a lecture by a professor of physics at the University of Edinburgh. The professor discusses the possibility of quickening radio-active decay of uranium, thereby releasing its immense energy and starting a new chapter in the history of mankind. A Scottish student, stimulated by the lecture, later watches an evening sun drop behind distant hills. " 'Ye auld thing,' he said, and his eyes were shining and he made a kind of grabbing gesture with his hand; 'ye auld red thing. . . . We'll have ye *yet.*' "

The novel's first chapter opens in 1933 when a young scientist named Holsten succeeds in inducing artificial radioactivity—the first step in the tapping of atomic energy. (It was in January, 1934, that Frédéric Joliot-Curie and his wife first did exactly this by bombarding aluminum with beta particles. In that same year Enrico Fermi achieved similar results by bombarding fluorine and other elements with neutrons.) Twenty years pass in Wells' novel before controlled chain reactions are achieved. (Fermi achieved the first chain reaction in 1942, eleven years ahead of Wells' schedule.) Shortly after 1956 the first "atomic bombs," as Wells called them, are dropped from airplanes in the world's first full-scale atomic war.

In the following selection we listen to the thoughts that race through Holsten's mind as he reflects on the terrible consequences of his discovery. Such thoughts must surely have troubled the minds of the Joliot-Curies and Fermi. "To moral questions there are no universal answers," Mrs. Fermi writes in her biography of her husband. ". . . Some men said the atomic bomb should never have been built. . . . Enrico did not think this would have been a sensible solution. It is no good trying to stop knowledge from going forward. Whatever Nature has in store for mankind, unpleasant as it may be, men must accept, for ignorance is never better than knowledge. Besides, if they had not built an atomic bomb, if they had destroyed all the data they had found and collected, others would come in the near future who in their quest for truth would proceed on the same path and rediscover what had been obliterated." It is almost a paraphrase of sentences recorded by Holsten in his diary!

The second selection is taken from Wells' little-read survey of the social sciences, *The Work, Wealth, and Happiness of Mankind.* It should remind us, in a day when it is not fashionable to praise Wells too highly, that in addition to the soundness of his science and the richness of his ideas, he was also a writer of extraordinary power and skill.

H. G. WELLS

The New Source of Energy

THE PROBLEM which was already being mooted by such scientific men as Ramsay, Rutherford, and Soddy, in the very beginning of the twentieth century, the problem of inducing radio-activity in the heavier elements and so tapping the internal energy of atoms, was solved by a wonderful combination of induction, intuition and luck by Holsten so soon as the year 1933. From the first detection of radio-activity to its first subjugation to human purpose measured little more than a quarter of a century. For twenty years after that, indeed, minor difficulties prevented any striking practical application of his success, but the essential thing was done, this new boundary in the march of human progress was crossed, in that year. He set up atomic disintegration in a minute particle of bismuth, it exploded with great violence into a heavy gas of extreme radio-activity, which disintegrated in its turn in the course of seven days, and it was only after another year's work that he was able to show practically that the last result of this rapid release of energy was gold. But the thing was done,—at the cost of a blistered chest and an injured finger, and from the moment when the invisible speck of bismuth flashed into riving and rending energy, Holsten knew that he had opened a way for mankind, however narrow and dark it might still be, to worlds of limitless power. He recorded as much in the strange diary biography he left the world, a diary that was up to that particular moment a mass of speculations and calculations, and which suddenly became for a space an amazingly minute and human record of sensations and emotions that all humanity might understand.

He gives, in broken phrases and often single words, it is true, but none the less vividly for that, a record of the twenty-four hours following the demonstration of the correctness of his intricate tracery of computations and guesses. "I thought I should not sleep," he writes—the words he omitted are supplied in brackets—(on account of) "pain in (the) hand and chest and (the wonder of) what I had done. . . . Slept like a child."

He felt strange and disconcerted the next morning; he had nothing to do, he was living alone in apartments in Bloomsbury, and he decided to go up to Hampstead Heath, which he had known when he was a little boy as a breezy playground. He went up by the underground tube that was then the recognised means of travel from one part of London to another, and walked up Heath Street from the tube station to the open heath. He found it a gully of planks and scaffoldings between the hoardings of housewreckers. The spirit of the times had seized upon that narrow, steep and winding thoroughfare, and was in the act of making it commodious and interesting according to the remarkable ideals of Neo-Georgian æstheticism. Such is the illogical quality of humanity that Holsten, fresh from work that was like a petard under the seat of the current civilisation, saw these changes with regret. He had come up Heath Street perhaps a thousand times, had known the windows of all the little shops, spent hours in the vanished cinematograph theatre, and marvelled at the high-flung early Georgian houses upon the westward bank of that old gully of a thoroughfare; he felt strange with all these familiar things gone. He escaped at last with a feeling of relief from this choked alley of trenches and holes and cranes, and emerged upon the old familiar scene about the White Stone Pond. That at least was very much as it used to be.

There were still the fine old red-brick houses to left and right of him; the reservoir had been improved by a portico of marble, the white-fronted inn with the clustering flowers above its portico still stood out at the angle of the ways, and the blue view to Harrow Hill and Harrow spire, a view of hills and trees and shining waters and wind-driven cloud-shadows, was like the opening of a great window to the ascending Londoner. All that was very reassuring. There were the same strolling crowd, the same perpetual miracle of motors dodging through it harmlessly, escaping headlong into the country from the Sabbatical stuffiness behind and below them. There was a band still, a women's suffrage meeting—for the suffrage women had won their way back to the tolerance, a trifle derisive, of the populace again—Socialist orators, politicians, a band, and the same wild uproar of dogs, frantic with the gladness of their one blessed weekly release from the back-yard and the chain. And away along the road to the "Spaniards" strolled a vast multitude, saying as ever that the view of London was exceptionally clear that day.

Young Holsten's face was white. He walked with that uneasy affectation of ease that marks an overstrained nervous system and an under-exercised body. He hesitated at the White Stone Pond whether to go to the left of it or the right, and again at the fork of the roads. He kept shifting his stick in his hand, and every now and then he would get in the way of people on the footpath or be jostled by them because of the uncertainty of his movements. He felt, he confesses, "inadequate to ordinary existence." He seemed to himself to be something inhuman and mischievous. All the people about him looked fairly prosperous, fairly happy, fairly well adapted to the lives they had to lead,—a week of work and a Sunday of best clothes and mild promenading—and he had launched something that would disorganise the entire fabric that held their contentments and ambitions and satisfactions together. "Felt like an imbecile who has presented a box of loaded revolvers to a Crêche," he notes.

He met a man named Lawson, an old school-fellow, of whom history now knows only that he was red-faced and had a terrier. He and Holsten walked together, and Holsten was sufficiently pale and jumpy for Lawson to tell him he overworked and needed a holiday. They sat down at a little table outside the County Council house of Golders Hill Park and sent one of the waiters to the "Bull and Bush" for a couple of bottles of beer, no doubt at Lawson's suggestion. The beer warmed Holsten's rather dehumanised system. He began to tell Lawson as clearly as he could to what his great discovery amounted. Lawson feigned attention, but indeed he had neither the knowledge nor the imagination to understand. "In the end, before many years are out, this must eventually change war, transit, lighting, building, and every sort of manufacture, even agriculture, every material human concern——"

Then Holsten stopped short. Lawson had leapt to his feet. "Damn that dog!" cried Lawson. "Look at it now. Hi! Here! *Phewoo-phewoo-phewoo!* Come *here, Bobs!* Come *here!*"

The young scientific man with his bandaged hand sat at the green table, too tired to convey the wonder of the thing he had sought so long, his friend whistled and bawled for his dog, and the Sunday people drifted about them through the spring sunshine. For a moment or so Holsten stared at Lawson in astonishment, for he had been too intent upon what he had been saying to realise how little Lawson had attended.

Then he remarked, "*Well!*" and smiled faintly and finished the tankard of beer before him.

Lawson sat down again. "One must look after one's dog," he said, with a note of apology. "What was it you were telling me?"

In the evening Holsten went out again. He walked to Saint Paul's Cathedral and stood for a time near the door listening to the evening service. The candles upon the altar reminded him in some odd way of the fireflies at Fiesole. Then he walked back through the evening lights to Westminster. He was oppressed, he was indeed scared, by his sense of the immense consequences of his discovery. He had a vague idea that night that he ought not to publish his results, that they were premature, that some secret association of wise men should take care of his work and hand it on from generation to generation until the world was riper for its practical application. He felt that nobody in all the thousands of people he passed had really awakened to the fact of change; they trusted the world for what it was, not to alter too rapidly, to respect their trusts, their assurances, their habits, their little accustomed traffics and hard-won positions.

He went into those little gardens beneath the overhanging, brightly-lit masses of the Savoy Hotel and the Hotel Cecil. He sat down on a seat and became aware of the talk of the two people next to him. It was the talk of a young couple evidently on the eve of marriage. The man was congratulating himself on having regular employment at last. "They like me," he said, "and I like the job. If I work up—in'r dozen years or so I ought to be gettin' somethin' pretty comfortable. That's the plain sense of it, Hetty. There ain't no reason whatsoever why we shouldn't get along very decently—very decently, indeed."

The desire for little successes amidst conditions securely fixed! So it struck upon Holsten's mind. He added in his diary: "I had a sense of all this globe as that. . . ."

By that phrase he meant a kind of clairvoyant vision of this populated world as a whole, of all its cities and towns and villages, its high roads and the inns beside them, its gardens and farms and upland pastures, its boatmen and sailors, its ships coming along the great circles of the ocean, its timetables and appointments and payments and dues, as it were one unified and unprogressive spectacle. Sometimes such visions came to him; his mind, accustomed to great generalisations and yet acutely sensitive to detail, saw things far more comprehensively than the minds of most of his contemporaries. Usually the teeming sphere moved on to its predestined ends and circled with a stately swiftness on its path about the sun. Usually it was all a living progress that altered under his regard. But now fatigue a little deadened him to that incessancy of life, it seemed just now an eternal circling. He lapsed to the commoner persuasion of the great fixities and recurrencies of the human routine. The remoter past of wandering savagery, the inevitable changes of to-morrow were veiled, and he saw only day and night, seed-time and harvest, loving and begetting, births and deaths, walks in the summer sunlight and tales by the winter fireside, the ancient sequence of hope and acts and age perennially renewed, eddying on for ever and ever,—save that now the impious hand of research was raised to overthrow this drowsy, gently humming, habitual, sunlit spinning-top of man's existence. . . .

For a time he forgot wars and crimes and hates and persecutions, famine and pestilence, the cruelties of beasts, weariness and the bitter wind, failure and insufficiency and retrocession. He saw all mankind in terms of the humble Sunday couple upon the seat beside him, who schemed their inglorious outlook and improbable contentments. "I had a sense of all this globe as that."

His intelligence struggled against this mood and struggled for a time in vain. He reassured himself against the invasion of this disconcerting idea that he was something strange and inhuman, a loose wanderer from the flock returning with evil gifts from his sustained unnatural excursions amidst the darknesses and phosphorescences beneath the fair surfaces of life. Man had not been always thus; the instincts and desires of the little home, the little plot, was not all his nature; also he was an adventurer, an experimenter, an unresting curiosity, an insatiable desire. For a few thousand generations, indeed, he had tilled the earth and followed the seasons, saying his prayers, grinding his corn and trampling the October winepress, yet not for so long but that he was still full of restless stirrings. . . .

"If there have been home and routine and the field," thought Holsten, "there have also been wonder and the sea."

He turned his head and looked up over the back of the seat at the great hotels above him, full of softly shaded lights and the glow and colour and stir of feasting. Might his gift to mankind mean simply more of that? . . .

He got up and walked out of the garden, surveyed a passing tramcar, laden with warm light against the deep blues of evening, dripping and trailing long skirts of shining reflection; he crossed the Embankment and stood for a time watching the dark river and turning ever and again to the lit buildings and bridges. His mind began to scheme conceivable replacements of all those clustering arrangements. . . .

"It has begun," he writes in the diary in which these things are recorded. "It is not for me to reach out to consequences I cannot foresee. I am a part, not a whole; I am a little instrument in the armoury of Change. If I were to burn all these papers, before a score of years had passed some other man would be doing this. . . ."

H. G. WELLS

Science and Ultimate Truth

In recent years very extensive readjustments have been made in the general formulæ which the man of science has used to simplify and systematize his facts. These readjustments have occurred mainly in the world of physical science; they have affected the steady advance of biological and social science very little. It is the professor of physics who is most concerned. The philosophical concepts that have served to guide and sustain his enquiries hitherto have been, so to speak, under repair. He has had to alter his general diagrams.

The reader will have heard endless echoes and repercussions from these enquiries into philosophico-scientific technique, even if he has not deliberately studied them, and so it is well to explain how far they concern us and how far they do not concern us in this work.

Some recent experiments and observations have jarred heavily with the general philosophical ideas that have hitherto satisfied and served the scientific worker. His diagrams have had to undergo a considerable revision. They were much too naïve and "obvious." In certain fields he has had to question the essential reality of that framework of space and time in which he—in common with the man in the street—has been wont to arrange his facts. He has had to scrutinize the ideas of time and eternity afresh. He has been brought to consider Euclidean space as only one of a great number of theoretical spaces, and to replace it by other and subtler concepts of space that seem more compatible with these recently observed facts. The old issue between predestination and free-will has in effect been revived in terms of mathematical physics. Is the universe a fixed, rigid time-space system, or has it movement in still other dimensions? Is it a continuous or an intermittent universe? The mere asking of such strange questions is very exciting to the speculative mind. But they do not affect the practical everyday life, either of the individual or of mankind, and we note these interesting developments of modern thought here as fascinating exercises for the intelligence outside our subject altogether.

It may be that we exist and cease to exist in alternations, like the minute dots in some forms of toned printing or the succession of pictures on a cinema film. It may be that consciousness is an illusion of movement in an eternal, static, multidimensional universe. We may be only a story written on a ground of inconceivable realities, the pattern of a carpet beneath the feet of the incomprehensible. We may be, as Sir James Jeans seems to suggest, part of a vast idea in the meditation of a divine circumambient mathematician. It is wonderful exercise for the mind to peer at such possibilities. It brings us to the realization of the entirely limited nature of our intelligence, such as it is, and of existence as we know it. It leads plainly towards the belief that with minds such as ours the ultimate truth of things is forever inconceivable and unknowable. It brings us to the realization that these theories, the working diagrams of modern science are in the end less provisional only in the measure of their effective working than the mythologies and symbols of barbaric religions.

But it does not give us any present escape from this world of work and wealth and war. For us, while we live, there must always be a to-morrow and choice, and no play of logic and formulæ can ever take us out of these necessities. To be taken out of these necessities would be to be taken out of existence as we know it altogether.

It is impossible to dismiss mystery from life. Being is altogether mysterious. Mystery is all about us and in us, the Inconceivable permeates us, it is "closer than breathing and nearer than hands and feet." For all we know, that which we are may rise at death from living, as an intent player wakes up from his absorption when a game comes to an end, or as a spectator turns his eyes from the stage as the curtain falls, to look at the auditorium he has for a time forgotten. These are pretty metaphors, that have nothing to do with the game or the drama of space and time. Ultimately the mystery may be the only thing that matters, but *within the rules and limits of the game of life,* when you are catching trains or paying bills or earning a living, the mystery does not matter at all.

It is this sense of an unfathomable reality to which not only life but all present being is but a surface, it is this realization "of the gulf beneath all seeming and of the silence above all sounds," which makes a modern mind impatient with the tricks and subterfuges of those ghost-haunted metaphysicians and creed-entangled apologists who are continually asserting and clamouring that science is dogmatic—with would-be permanent dogmas that are forever being overthrown. They try to degrade science to their own level. But she has never pretended to that finality which is the quality of religious dogmas. Science pits no dogmas against the dogmas of the ghost worshippers. Only sometimes, when perforce science touches their dogmas, do these latter dissolve away. Science is of set intention superficial. It touches religious dogma only in so far as religious dogma is materialistic, only in so far as religious dogma is a jumble of impossible stories about origins and destinies in space and time, a story pretending to a "spirituality" that is merely a dreamy, crazy attenuation of things material. And even then does it touch these dogmas only because they involve magic irrational distractions, interferences and limitations of the everyday life of man.

I wish that there was a plain and popular book in existence upon the history of scientific ideas. It would be fascinating to reconstruct the intellectual atmosphere that surrounded Galileo and show the pre-existing foundations on which his ideas were based. Or ask what did Gilbert, the first student of magnetism, know, and what was the ideology with which the natural philosophers of the Stuart period had to struggle? It would be very interesting and illuminating to trace the rapid modification of these elementary concepts as the scientific process became vigorous and spread into general thought.

Few people realize how recent that invasion is, how new the current diagram of the universe is, and how recently the ideas of modern science have reached the commoner sort of people. The present writer is sixty-five. When he was a little boy his mother taught him out of a book she valued very highly, *Magnell's Questions*. It had been her own school book. It was already old-fashioned, but it was still in use and on sale. It was a book on the eighteenth-century plan of question and answer, and it taught that there were four elements, earth, air, fire and water.

These four elements are as old at least as Aristotle. It never occurred to me in my white-sock and plaid-petticoat days to ask in what proportion these fundamental ingredients were mixed in myself or the tablecloth or my bread and milk. I just swallowed them as I swallowed the bread and milk.

From Aristotle I made a stride to the eighteenth century. The two elements of the Arabian alchemists, sulphur and mercury, I never heard of then, nor of Paracelsus and his universe of salt, sulphur, mercury, water, and the vital elixir. None of that ever got through to me. I went to a boys' school, and there I learnt, straightaway, that I was made up of hard, definite molecules, built up of hard definite indestructible atoms of carbon, oxygen, hydrogen, nitrogen, phosphorus, calcium, sodium, chlorine, and a few others. These were the real elements. They were shown plainly in my textbook like peas or common balls suitably grouped. That also I accepted for a time without making any fuss about it. I do not remember parting with the Four Elements: they got lost and I went on with the new lot.

At another school, and then at the Royal College of Science I learnt of a simple eternity of atoms and force. But the atoms now began to be less solid and simple. We talked very much of ether and protyle at the Royal College, but protons and electrons were still to come, and atoms, though taking on strange shapes and movements, were intact. Atoms could neither be transformed nor destroyed, but forces, though they could not be destroyed, could be transformed. This indestructibility of the chameleon of force, was the celebrated Conservation of Energy, which has since lost prestige, though it remains as a sound working generalization for the everyday engineer.

But in those days, when I debated and philosophized with my fellow students, I was speedily made aware that these atoms and molecules were not realities at all; they were, it was explained to me, essentially mnemonics; they satisfied, in the simplest possible arrangement of material models and images, what was needed to assemble and reconcile the known phenomena of matter. That was all they were. That I grasped without much difficulty. There was no shock to me, therefore, when presently new observations necessitated fresh elaborations of the model. My schoolmaster had been a little too crude in his instructions. He had not been a scientific man, but only a teacher of science. He had been an unredeemed Realist, teaching science in a dogmatic Realist way. Science, I now understood, never contradicts herself absolutely, but she is always busy in revising her classifications and touching up and rephrasing her earlier cruder statements. Science never professes to present more than a working diagram of fact. She does not *explain*, she *states the relations and associations of facts as simply as possible.*

Her justification for her diagrams lies in her increasing power to change matter. The test of all her theories is that they work. She has always been true, and continually she becomes truer. But she never expects to reach Ultimate Truth. At their truest her theories are not, and never pretend to be, more than diagrams to fit, not even all possible facts, but simply the known facts.

In my student days, forty-five years ago now, we were already quite aware that the *exact* equivalence of cause and effect was no more than a convenient convention, and that it was possible to represent the universe as a system of unique events in a spacetime framework. These are not new ideas. They were then common student talk. When excited journalists announce that such intellectualists as Professor Eddington and Professor Whitehead have made astounding discoveries to overthrow the "dogmas of science," they are writing in sublime ignorance of the fact that there are no dogmas of science, and that these ideas that seem such marvellous "discoveries" to them have been in circulation for more than half a century.

No engineer bothers about these considerations of marginal error and the relativity of things, when he plans out the making of a number of machines "in series" with replaceable parts. Every part is unique indeed and a little out of the straight, but it is near enough and straight enough to serve. The machines work. And no appreciable effect has been produced upon the teaching of machine drawing by the possibility that space is curved and expanding. In this book, let the reader bear in mind, we are always down at the level of the engineer and the machine drawing. From cover to cover we are dealing with practical things on the surface of the earth, where gravitation is best represented as a centripetal pull, and where a pound of feathers weighs equal to a pound of lead, and things are what they seem. We deal with the daily life of human beings now and in the ages immediately ahead. We remain in the space and time of ordinary experience throughout this book, at an infinite distance from ultimate truth.

LAURA FERMI

Scientific research is becoming more and more a cooperative enterprise. Great inventions and discoveries rest upon fragments of information that issue from a thousand laboratories. We know Edison invented the phonograph, but who is the inventor of television? This cooperative trend is augmented by the fact that tools of modern research are often enormously expensive. The individual inventor, puttering in his basement, can scarce afford an electron microscope or cyclotron. To gain access to these tools he must join the staff of a great university or the research department of a giant corporation.

It is not surprising, therefore, that the story of the atom bomb involves many scientists in many different countries. There is no "inventor" of the bomb. Yet many times it is possible to survey a vast mosaic of interlocking research and single out one man who more than any other contributed to the final dramatic result. In the story of atomic energy that man is Enrico Fermi. And the day on which it became certain that atomic energy could be harnessed, for peace as well as war, was December 2, 1942, when Fermi directed the operation of the world's first atomic pile.

The experiment itself was the outcome of much previous research in which the Italian scientist had played an important role. In 1938 he was awarded the Nobel Prize for his work with slow neutron bombardment, a work that led to the German discovery of uranium fission. Later he and four co-workers were granted Italian and American patents on their method of retarding neutrons, an essential aspect of early atomic piles. Four of the five men were eventually compensated by the Atomic Energy Commission for their patent rights, but the fifth man was unable to collect. He was Bruno Pontecorvo who had slipped behind the Iron Curtain to become an important figure in Soviet nuclear research. After the war, Fermi received the Congressional Medal of Merit for his work as associate director of the Los Alamos laboratories where the first bomb was constructed.

Shortly before Fermi's untimely death in 1954, his wife Laura (b. 1908) wrote *Atoms in the Family,* a graceful, entertaining biography of her famous husband. The following selection is a chapter of this book. It tells the story of that cold gray afternoon in Chicago when a group of scientists gathered in an abandoned squash court beneath an abandoned football stadium to open the atomic age.

LAURA FERMI

Success

MEANWHILE Herbert Anderson and his group at the Met. Lab. had also been building small piles and gathering information for a larger pile from their behavior. The best place Compton had been able to find for work on the pile was a squash court under the West Stands of Stagg Field, the University of Chicago stadium. President Hutchins had banned football from the Chicago campus, and Stagg Field was used for odd purposes. To the west, on Ellis Avenue, the stadium is closed by a tall gray-stone structure in the guise of a medieval castle. Through a heavy portal is the entrance to the space beneath the West Stands. The Squash Court was part of this space. It was 30 feet wide, twice as long, and over 26 feet high.

The physicists would have liked more space, but places better suited for the pile, which Professor Compton had hoped he could have, had been requisitioned by the expanding armed forces stationed in Chicago. The physicists were to be contented with the Squash Court, and there Herbert Anderson had started assembling piles. They were still "small piles," because material flowed to the West Stands at a very slow, if steady, pace. As each new shipment of crates arrived, Herbert's spirits rose. He loved working and was of impatient temperament. His slender, almost delicate, body had unsuspected resilience and endurance. He could work at all hours and drive his associates to work along with his same intensity and enthusiasm.

A shipment of crates arrived at the West Stands on a Saturday afternoon, when the hired men who normally unpack them were not working. A university professor, older by several years than Herbert, gave a look at the crates and said lightly: "Those fellows will unpack them Monday morning."

"Those fellows, Hell! We'll do them now," flared up Herbert, who had never felt inhibited in the presence of older men, higher up in the academic hierarchy. The professor took off his coat, and the two of them started wrenching at the crates.

Profanity was freely used at the Met. Lab. It relieved the tension built up by having to work against time. Would Germany get atomic weapons before the United States developed them? Would these weapons come in time to help win the war? These unanswered questions constantly present in the minds of the leaders in the project pressed them to work faster and faster, to be tense, and to swear.

Success was assured by the spring. A small pile assembled in the Squash Court showed that all conditions—purity of materials, distribution of uranium in the graphite lattice—were such that a pile of critical size would chain-react.

"It could be May, or early June at latest," Enrico told me, as we recently reminisced about the times of the Met. Lab. "I remember I talked about that experiment on the Indiana dunes, and it was the first time I saw the dunes. You were still in Leonia. I went with a group from the Met. Lab. I liked the dunes: it was a clear day, with no fog to dim colors. . . ."

"I don't want to hear about the dunes," I said. "Tell me about that experiment."

"I like to swim in the lake. . . ." Enrico paid no attention to my remark. I knew that he enjoyed a good swim, and I could well imagine him challenging a group of younger people, swimming farther and for a longer time than any of them, then emerging on the shore with a triumphant grin.

"Tell me about that experiment," I insisted.

"We came out of the water, and we walked along the beach."

I began to feel impatient. He did not have to mention the walk. He always walks after swimming, dripping wet, water streaming from his hair. In 1942 there was certainly much more hair on his head to shed water, not just the little fringe on the sides and on the back that there is now, and it was much darker.

". . . and I talked about the experiment with Professor Stearns. The two of us walked ahead of the others on the beach. I remember our efforts to speak in such a way that the others would not understand. . . ."

"Why? Didn't everyone at the Met. Lab. know that you were building piles?"

"They knew we built piles. They did not know that at last we had the certainty that a pile would work. The fact that a chain reaction was feasible remained classified material for a while. I could talk freely with Stearns because he was one of the leaders."

"If you were sure a larger pile would work, why didn't you start it at once?"

"We did not have enough materials, neither uranium nor graphite. Procurement of uranium metal was always an obstacle. It hampered progress."

While waiting for more materials, Herbert Anderson went to the Goodyear Tire and Rubber Company to place an order for a square balloon. The Goodyear people had never heard of square balloons, they did not think they could fly. At first they threw suspicious glances at Herbert. The young man, however, seemed to be in full possession of his wits. He talked earnestly, had figured out precise specifications, and knew exactly what he wanted. The Goodyear people promised to make a square balloon of rubberized cloth. They delivered it a couple of months later to the Squash Court. It came neatly folded, but, once unfolded, it was a huge thing that reached from floor to ceiling.

The Squash Court ceiling could not be pushed up as the physicists would have liked. They had calculated that their final pile ought to chain-react somewhat before it reached the ceiling. But not much margin was left, and calculations are never to be trusted entirely. Some impurities might go unnoticed, some unforeseen factor might upset theory. The critical size of the pile might not be reached at the ceiling. Since the physicists were compelled to stay within that very concrete limit, they thought of improving the performance of the pile by means other than size.

The experiment at Columbia with a canned pile had indicated that such an aim might be attained by removing the air from the pores of the graphite. To can as large a pile as they were to build now would be impracticable, but they could assemble it inside a square balloon and pump the air from it if necessary.

The Squash Court was not large. When the scientists opened the balloon and tried to haul it into place, they could not see its top from the floor. There was a movable elevator in the room, some sort of scaffolding on wheels that could raise a platform. Fermi climbed onto it, let himself be hoisted to a height that gave him a good view of the entire balloon, and from there he gave orders:

"All hands stand by!"

"Now haul the rope and heave her!"

"More to the right!"

"Brace the tackles to the left!"

To the people below he seemed an admiral on his bridge, and "Admiral" they called him for a while.

When the balloon was secured on five sides, with the flap that formed the sixth left down, the group began to assemble the pile inside it. Not all the material had arrived, but they trusted that it would come in time.

From the numerous experiments they had performed so far, they had an idea of what the pile should be, but they had not worked out the details, there were no drawings nor blueprints and no time to spare to make them. They planned their pile even as they built it. They were to give it the shape of a sphere of about 26 feet in diameter, supported by a square frame, hence the square balloon.

The pile supports consisted of blocks of wood. As a block was put in place inside the balloon, the size and shape of the next were figured. Between the Squash Court and the nearby carpenter's shop there was a steady flow of boys, who fetched finished blocks and brought specifications for more on bits of paper.

When the physicists started handling graphite bricks, everything became black. The walls of the Squash Court were black to start with. Now a huge black wall of graphite was going up fast. Graphite powder covered the floor and made it black and as slippery as a dance floor. Black figures skidded on it, figures in overalls and goggles under a layer of graphite dust. There was one woman among them, Leona Woods; she could not be distinguished from the men, and she got her share of cussing from the bosses.

The carpenters and the machinists who executed orders with no knowledge of their purpose and the high-school boys who helped lay bricks for the pile must have wondered at the black scene. Had they been aware that the ultimate result would be an atomic bomb, they might have renamed the court Pluto's Workshop or Hell's Kitchen.

To solve difficulties as one meets them is much faster than to try to foresee them all in detail. As the pile grew, measurements were taken and further construction adapted to results.

The pile never reached the ceiling. It was planned as a sphere 26 feet in diameter, but the last layers were never put into place. The sphere remained flattened at the top. To make a vacuum proved unnecessary, and the balloon was never sealed. The critical size of the pile was attained sooner than was anticipated.

Only six weeks had passed from the laying of the first graphite brick, and it was the morning of December 2.

Herbert Anderson was sleepy and grouchy. He had been up until two in the morning to give the pile its finishing touches. Had he pulled a control rod during the night, he could have operated the pile and have been the first man to achieve a chain reaction, at least in a material, mechanical sense. He had a moral duty not to pull that rod, despite the strong temptation. It would not be fair to Fermi. Fermi was the leader. He had directed research and worked out theories. His were the basic ideas. His were the privilege and the responsibility of conducting the final experiment and controlling the chain reaction.

"So the show was all Enrico's, and he had gone to bed early the night before," Herbert told me years later, and a bit of regret still lingered in his voice.

Walter Zinn also could have produced a chain reaction during the night. He, too, had been up and at work. But he did not care whether he operated the pile or not; he did not care in the least. It was not his job.

His task had been to smooth out difficulties during the pile construction. He had been some sort of general contractor: he had placed orders for material and made sure that they were delivered in time; he had supervised the machine shops where graphite was milled; he had spurred others to work faster, longer, more efficiently. He had become angry, had shouted, and had reached his goal. In six weeks the pile was assembled, and now he viewed it with relaxed nerves and with that vague feeling of emptiness, of slight disorientation, which never fails to follow completion of a purposeful task.

There is no record of what were the feelings of the three young men who crouched on top of the pile, under the ceiling of the square balloon. They were called the "suicide squad." It was a joke, but perhaps they were asking themselves whether the joke held some truth. They were like firemen alerted to the possibility of a fire, ready to extinguish it. If something unexpected were to happen, if the pile should get out of control, they would "extinguish" it by flooding it with a cadmium solution. Cadmium absorbs neutrons and prevents a chain reaction.

A sense of apprehension was in the air. Everyone felt it, but outwardly, at least, they were all calm and composed.

Among the persons who gathered in the Squash Court on that morning, one was not connected with the Met. Lab.—Mr. Crawford H. Greenewalt of E. I. duPont de Nemours, who later became the president of the company. Arthur Compton had led him there out of a near-by room where, on that day, he and other men from his company happened to be holding talks with top Army officers.

Mr. Greenewalt and the duPont people were in a difficult position, and they did not know how to reach a decision. The Army had taken over the Uranium Project on the previous August and renamed it Manhattan District. In September General Leslie R. Groves was placed in charge of it. General Groves must have been of a trusting nature: before a chain reaction was achieved, he was already urging the duPont de Nemours Company to build and operate piles on a production scale.

In a pile, Mr. Greenewalt was told, a new element, plutonium, is created during uranium fission. Plutonium would probably be suited for making atomic bombs. So Greenewalt and his group had been taken to Berkeley to see the work done on plutonium, and then flown to Chicago for more negotiations with the Army.

Mr. Greenewalt was hesitant. Of course his company would like to help win the war! But piles and plutonium!

With the Army's insistent voice in his ears, Compton, who had attended the conference, decided to break the rules and take Mr. Greenewalt to witness the first operation of a pile.

They all climbed onto the balcony at the north end of the Squash Court; all, except the three boys perched on top of the pile and except a young physicist, George Weil, who stood alone on the floor by a cadmium rod that he was to pull out of the pile when so instructed.

And so the show began.

There was utter silence in the audience, and only Fermi spoke. His gray eyes betrayed his intense thinking, and his hands moved along with his thoughts.

"The pile is not performing now because inside it there are rods of cadmium which absorb neutrons. One single rod is sufficient to prevent a chain reaction. So our first step will be to pull out of the pile all control rods, but the one that George Weil will man." As he spoke others acted. Each chore had been assigned in advance and rehearsed. So Fermi went on speaking, and his hands pointed out the things he mentioned.

"This rod, that we have pulled out with the others, is

automatically controlled. Should the intensity of the reaction become greater than a pre-set limit, this rod would go back inside the pile by itself.

"This pen will trace a line indicating the intensity of the radiation. When the pile chain-reacts, the pen will trace a line that will go up and up and that will not tend to level off. In other words, it will be an exponential line.

"Presently we shall begin our experiment. George will pull out his rod a little at a time. We shall take measurements and verify that the pile will keep on acting as we have calculated.

"Weil will first set the rod at thirteen feet. This means that thirteen feet of the rod will still be inside the pile. The counters will click faster and the pen will move up to this point, and then its trace will level off. Go ahead, George!"

Eyes turned to the graph pen. Breathing was suspended. Fermi grinned with confidence. The counters stepped up their clicking; the pen went up and then stopped where Fermi had said it would. Greenewalt gasped audibly. Fermi continued to grin.

He gave more orders. Each time Weil pulled the rod out some more, the counters increased the rate of their clicking, the pen raised to the point that Fermi predicted, then it leveled off.

The morning went by. Fermi was conscious that a new experiment of this kind, carried out in the heart of a big city, might become a potential hazard unless all precautions were taken to make sure that at all times the operation of the pile conformed closely with the results of the calculations. In his mind he was sure that if George Weil's rod had been pulled out all at once, the pile would have started reacting at a leisurely rate and could have been stopped at will by reinserting one of the rods. He chose, however, to take his time and be certain that no unforeseen phenomenon would disturb the experiment.

It is impossible to say how great a danger this unforeseen element constituted or what consequences it might have brought about. According to the theory, an explosion was out of the question. The release of lethal amounts of radiation through an uncontrolled reaction was improbable. Yet the men in the Squash Court were working with the unknown. They could not claim to know the answers to all the questions that were in their minds. Caution was welcome. Caution was essential. It would have been reckless to dispense with caution.

So it was lunch time, and, although nobody else had given signs of being hungry, Fermi, who is a man of habits, pronounced the now historical sentence:

"Let's go to lunch."

After lunch they all resumed their places, and now Mr. Greenewalt was decidedly excited, almost impatient.

But again the experiment proceeded by small steps, until it was 3:20.

Once more Fermi said to Weil:

"Pull it out another foot"; but this time he added, turning to the anxious group in the balcony: "This will do it. Now the pile will chain-react."

The counters stepped up; the pen started its upward rise. It showed no tendency to level off. A chain reaction was taking place in the pile.

In the back of everyone's mind was one unavoidable question: "When do we become scared?"

Under the ceiling of the balloon the suicide squad was alert, ready with their liquid cadmium: this was the moment. But nothing much happened. The group watched the recording instruments for 28 minutes. The pile behaved as it should, as they all had hoped it would, as they had feared it would not.

The rest of the story is well known. Eugene Wigner, the Hungarian-born physicist who in 1939 with Szilard and Einstein had alerted President Roosevelt to the importance of uranium fission, presented Fermi with a bottle of Chianti. According to an improbable legend, Wigner had concealed the bottle behind his back during the entire experiment.

All those present drank. From paper cups, in silence, with no toast. Then all signed the straw cover on the bottle of Chianti. It is the only record of the persons in the Squash Court on that day.

The group broke up. Some stayed to round up their measurements and put in order the data gathered from their instruments. Others went to duties elsewhere. Mr. Greenewalt hastened to the room where his colleagues were still in conference with the military. He announced, all in one breath, that Yes, it would be quite all right for their company to go along with the Army's request and start to build piles. Piles were wonderful objects that performed with the precision of a Swiss watch, and, provided that the advice of such competent scientists as Fermi and his group were available, the duPont company was certainly taking no undue risk.

Arthur Compton placed a long-distance call to Mr. Conant of the Office of Scientific Research and Development at Harvard.

"The Italian Navigator has reached the New World," said Compton as soon as he got Conant on the line.

"And how did he find the natives?"

"Very friendly."

Here the official story ends, but there is a sequel to it, which started on that same afternoon when a young physicist, Al Wattemberg, picked up the empty Chianti bottle from which all had drunk. With the signatures on its cover, it would make a nice souvenir. In subsequent years Al Wattemberg did his share of traveling, like any other physicist, and the bottle followed him. When big celebrations for the pile's tenth anniversary were planned at the University of Chicago, the bottle and Al Wattemberg were both in Cambridge, Massachusetts. Both, Al promised, would be in Chicago on December 2.

It so happened, however, that a little Wattemberg decided to come into this world at about that time, and Al could not attend the celebrations. So he shipped his bottle, and, because he wanted to make doubly sure that it would not be broken, he insured it for a thousand dollars. It is not often that an empty bottle is considered worth so much money, and newspaper men on the lookout for sensation gave the story a prominent position in the press.

A couple of months later the Fermis and a few other physicists received a present: a case of Chianti wine. An importer had wished to acknowledge his gratitude for the free advertisement that Chianti had received.

SAMUEL GOUDSMIT

During World War II belief in the superiority of German science was not confined to the German scientists themselves. American physicists, working at desperate speed on the atom bomb, were sure the Germans were at least one or two years in the lead. Elaborate plans were carried through by the allies for blowing up a Norwegian plant supplying German physicists with heavy water. When invasion troops landed in France they carried equipment for detecting uranium piles thought to be installed along the coast.

On the heels of the invasion troops went a top secret intelligence mission known by the code name ALSOS. Its purpose: to find out how far the Germans had progressed in their atomic efforts. Samuel Abraham Goudsmit (1902-1978), a distinguished nuclear physicist from Holland (he was co-discoverer of electron "spin"), headed the mission's scientific side. After the war he wrote a book called *ALSOS* in which he gives an exuberant account of the mission.

ALSOS is an hilarious book not only because of the author's sense of humor that sparkles on almost every page, but because of a basic comedy inherent in the circumstances. The mission engaged in a huge amount of expensive cloak and dagger work only to discover, to its vast astonishment, that the Germans had gotten almost nowhere. They had not even produced a workable atomic pile and their only concept of an atom bomb was to drop the entire pile itself! It never occurred to them that the pile might be used for producing plutonium which in turn could be used in a bomb. Dr. Goudsmit attributes their fantastic bungling to many causes—the exiling of important Jewish physicists; distrust of "non-Aryan" relativity theory; hero-worship of their leading atomic expert, Werner Heisenberg; wasted efforts on crack-pot theories (*e.g.*, the theory that two infra-red rays intersecting at the proper angle could explode the bomb load of an enemy plane); above all, the elevation of knuckleheaded Nazis to positions of scientific authority.

There are scenes of pathos as well as comedy in Dr. Goudsmit's remarkable book. When he and his old friend Heisenberg finally met face to face, the great German physicist condescendingly offered to explain to his captors the great results of his research. Goudsmit could not tell him then how supremely insignificant it was. And there is a heart-rending episode in which Goudsmit stands in the wreckage of his old home in The Hague, weeping for the past and for his aged father and blind mother who had met death in Hitler's gas chambers. These episodes give depth of meaning to what is otherwise a lively, sardonic account of scientific sleuthing. The following selection, apart from some grim reminders of Nazi sadism, is one of the funniest chapters in the book.

SAMUEL GOUDSMIT

The Gestapo in Science

WHEN GOERING took over the Reich's Research Council, a curious character named Osenberg was placed in charge of the newly instituted "Planning Office."

This Osenberg was an obscure Professor of Mechanical Engineering at the University of Hannover, but a good Party member. His technical and scientific knowledge were well below par, but he had supervised some work on torpedoes for the German Navy, which was reported as creditable. He was inspired by a mania for organization and a passion for card indexes.

Osenberg started his career as an organizer of war science with the German Navy. He impressed the authorities with the observation that most academic research facilities were not being used and that the Navy might well take over these places before anyone else got the same idea. With this in view he headed the "Osenberg Committee" which surveyed the various universities. But the Navy soon dropped him, when they found out that he wanted to run and re-organize everything.

The Reich's Research Council, which employed him on the rebound, seemed quite pleased with him. At least most scientists were willing to put in a good word for him even after V-E Day. The reasons for this are fairly clear. First of all, Ramsauer's talks and other information had given German scientists an idealized picture of our American organization. What they admired most of all was our much publicized

"Roster of Scientific Personnel" and here was this Osenberg, who wanted to set up something just like it for Germany, a complete card index of all German scientists and engineers, and a complete card index of all scientific war projects. There was a still more important reason why they liked him. Osenberg was convinced that scientists should be taken out of the Army and put back in the laboratories on war work. What no one else had been able to do, Osenberg did. He got a Hitler decree passed in December, 1943, dubbed "Osenberg Action," for the release of 5000 scientists from the Armed Forces. "He is the man who really saved German science," the professors said later whenever they were questioned about him.

Osenberg did have a lot of drive. He needed it to get the release decree really executed. He was in continuous quarrels with the Army leaders for his salvaged scientists and at the end of the war only about half of the 5000 had been sent back.

With his extensive personnel files, the boss of the "Planning Office" had a lot to say about the assignment of scientific personnel to the various projects. He had the power to transfer technicians and scientists from one place to another and if one wanted to expand a certain project one had to come to him. He even wanted to supervise the actual programs of research, but in that he was less successful.

From what mysterious source did Osenberg derive his great power? It was no mystery. He was a high member of the Gestapo, Himmler's secret police. The "Security Service of the Elite Guard" (SD der SS), commonly called Gestapo, also boasted of a "cultural" department, Section IIIc, headed by a Wilhelm Spengler. Osenberg was Spengler's right-hand man for the sciences. The function of this section was to enforce the Nazi doctrine at educational and cultural institutions. This was done by means of squealers and investigators who reported directly to Osenberg. All scientific conferences and all important meetings concerning war research and co-ordination were attended by Osenberg's spies. They were also present in all laboratories, whether in the person of a professor or a scrub woman. These spies reported about quarrels between the scientists, inefficiency of the research workers, causes for delays, and other supposed reasons for lack of progress of the war work. In addition, Osenberg collected data on their attitude towards the Nazi doctrine.

This character's Gestapo files were probably the most revealing documents in his possession. From them we learned who among the leading scientists were considered politically reliable and professionally competent. The physicist Walther Gerlach, the chemists Thiessen and Richard Kuhn are highly praised, but the famous medical scientist Sauerbruch is reported no good as a leader and politically unreliable. Schumann, the chief scientist for the Army, is severely criticized. The able young physicist Gentner, who had been sent to Paris to work in Joliot-Curie's laboratory, is accused of having democratic ideals, probably influenced by his Swiss wife. Indeed Gentner's exemplary behavior during the war at the risk of his own life and freedom completely confirms the poor opinion the Gestapo spies had of him.

Osenberg's agents would investigate research institutes and report on the value and progress of the work done. They gave in some cases more pertinent information than could have been obtained by Allied technical teams.

It was Osenberg's outfit, too, which tried to push Mentzel out of his post as head of the Research Council because of his incompetence and finally almost succeeded, near the end of 1944, after two years of intrigue. A secret report to Goering early in 1943, probably by Osenberg himself, states that "Mentzel is unfit for leadership" and that "a state of chaotic confusion exists in German universities without any coherent discipline."

Mentzel's lengthy reply was quite significant. This "loyal Nazi," as he called himself, suddenly discovered some of the flaws of the regime when he himself was attacked. His defense could almost have been written by an anti-Nazi. He referred openly to the "early lack of recognition by the Nazi Party of the universities, when scientists were obviously regarded as liberal, reactionary, Jewish, or Freemason—in any case, anti-Nazi. This belief was partly justified, and led to a purification which lasted until 1937. . . . Nearly 40% of all professors were dismissed which led to a serious lack of personnel. This could only be repaired slowly; only a limited number of Nazi lecturers and assistants were available to fill the vacancies, and they did not always satisfy the scientific requirements." Mentzel denies the "chaotic confusion" but stresses the indiscriminate drafting of students in the sciences.

A later report, dated August, 1944, and written by one of Osenberg's stooges, critically analyzes the projects sponsored by Mentzel's Research Council, and points out that practically none of them are related to the war effort. Of the 800 projects studied, forestry and agriculture accounted for 70%, physics, only 3%. The only essential problems were on guided missiles. The investigator also complained bitterly about the administration and office routine of the Berlin headquarters of the Reich's Research Council. The files are in disorder, keys are missing, reports look dirty, and indexing is full of fatal mistakes.

In addition to these undercover reports the "cultural" section of the Gestapo also solicited direct information from the scientists. A secret letter to physicist Von Weizsäcker at Strasbourg, in August, 1944, asks for his views on theoretical physics in relation to German physics and the rôle of theoretical physics in the German war effort. About the same time scientists at the University of Bonn were asked their opinion on "the disintegration of research in the sciences as a result of insufficient governmental guidance."

This letter starts out with the statement that "The advantage which German science and technology possessed before World War I has been wiped out by tremendous developments in this field especially in America." The letter further stresses the important rôle of the scientist in modern warfare and criticizes the Reich's Research Council and similar organizations for having failed to use the German scientific potential exhaustively. It promises a new plan, which intends to remove all obstacles which so far had prevented scientists from contributing effectively to the war effort.

In addition to this, Osenberg sent frequent "Denkschrifte," or memoranda, to the chief of the Nazi Party, Martin Bormann. These were queer looking, immaculately typed pamphlets embellished with underscorings in blue and red ink, beautifully executed, meaningless diagrams, numerous appendixes, cross references, altogether long-winded, pompous affairs in which he aired his complaints. As they referred to almost everything there was hardly a file folder in his office which did not contain one of these pamphlets. It is doubtful whether Bormann or anyone else who received copies ever read the stuff, for occasionally Osenberg complains bitterly about not having gotten any reaction out of them.

One of these memos complains to Bormann that no one in the entourage of Hitler had the courage to tell him that one of his favorite "revenge weapons" against London was a total flop and should be discontinued. The weapon referred to bore the code name "high pressure pump" and consisted of a hundred yard long gun-barrel into which the explosive was fed at intervals along the barrel. Although tests had shown that the thing would not work, thousands of workers were still constructing such installations along the French coast in order not to disappoint the Führer.

Finally Osenberg's dream came true. Late in 1944, Goering was talked into adopting Osenberg's plan. Based on a Hitler decree of June, 1944, ordering the concentration of scientific research towards the war effort, Goering created a super research council, called the "War Research Pool" (Wehrforschungsgemeinschaft) with Osenberg as the leader directly responsible to Goering, but also keeping his advantageously powerful job in the planning and personnel bureau.

Goering's decree was intended merely as a strengthening of the old Reich's Research Council by putting energetic Osenberg at the top. However, Osenberg extended the interpretation; the new organization was to include also all research facilities of the Army, Navy, Air-Force and industry. He distributed a high-sounding, secret circular on the organization. This included a most complex organization chart, which the recipients promptly dubbed the "railroad switchyard" (Rangier Bahnhof). It looks more difficult than a radio circuit diagram.

It is, of course, superfluous to mention that the research establishments of the Armed Forces, including Goering's own Air Force, completely ignored Osenberg's attempt. The electrical industry was the only one willing to co-operate, but then it had been doing so unofficially for quite some time.

It was November, 1944. Bombing and the advancing Allied troops had increased the chaos inside Germany. Decidedly, this was not a good time for a new organization to get started. No wonder it never left the paper stage.

Documents found by the Alsos Mission in Strasbourg in November, 1944, had put us on the trail of Osenberg. We found that his office was evacuated to a little town near Hannover and thought that his files, if found intact, might give us all we ever wanted to know about German war research. We had given the capture of this office a high priority in our plans.

When early in April, 1945, the place was taken, a small group of Alsos military, led by the physicist Major R. A. Fisher and accompanied by physicist Walter Colby of Michigan and chemist C. P. Smyth of Princeton, moved in and captured the whole Osenberg outfit.

As was usual for these Nazis, Osenberg surrendered with all his papers and personnel intact, and offered us his services. The few normal German scientists we encountered always refused to reveal their war work and had hidden or destroyed their secret papers. Not so the Nazis. One reason for their easy surrender was, of course, to save their skins, but this was not the principal motive in a case such as Osenberg's. The truth was that he was so convinced of his own greatness, his indispensability to German science, he was sure the Allies could not govern an occupied Germany without putting him at the head of science. He was greatly impressed by the attention we paid him and even more so when he was taken to Paris.

While the Alsos members were occupied with a near-by secret nuclear laboratory, some colleagues from Supreme

Headquarters moved in and hijacked Osenberg and his menagerie, including all documents. They were put on planes and interned in the previously mentioned "Dustbin" in Versailles. Here Osenberg set up business as usual; he merely had his secretary change the address on his letterhead to "z.Zt.Paris"—"at present in Paris." He was, indeed, very helpful. Various officials asked him for information on technical and scientific programs and he would order his staff to write a very exhaustive report, excellently executed, containing all information on the required subject available in his extensive files and usually ready in an incredibly short time. This strengthened his belief in his indispensability.

A rather stoutish bachelor in his forties, Osenberg was always pleased with himself. People who wanted to get information from him were invariably forced to listen to lengthy talks of his own crackpot ideas on anti-aircraft rockets. It was amusing to observe how he tried to maintain decorum; one of his staff always had to announce his visitors to him. Alsos members felt they could dispense with this rule of etiquette.

Osenberg ruled his staff in a typical German way, by fear. During their internment a revolt broke out. He complained bitterly because his staff had lost respect for Germany's greatness; they would laugh with ridicule when they saw distinguished German internees walk past the office in the château garden on their daily airing. This, he said, was a change in his men which one could not tolerate. He must have sensed that their lack of respect included himself.

One exception was his seemingly sexless secretary, who turned out the best work in the shortest time. She gave the impression of a nervous, mechanical attachment to a typewriter and was under almost hypnotic influence of the "Herr Professor." His male employees, however, who were in many cases more able than Osenberg himself, began to disobey him. They told us how employees who had displeased him had had their draft deferment revoked and had been sent to the front. One sure way to displease Osenberg, it seemed, was to be seen in a movie theater with a girl. They produced the list of former employees and told the insufficient reasons why each one was fired. Even if the details of their tales were not true, they clearly reflected the abnormal relations between Osenberg and his people.

My friends at Supreme Headquarters, who had hijacked our quarry and who were praised in reports for the discovery of this most significant scientific Intelligence objective, had failed to make a preparatory study of their treasure. They were, therefore, unaware of the fact that some of the most important papers were still missing, namely, Osenberg's Gestapo files and the principal files of the Reich's Research Council which had been sent from Berlin to Osenberg's village for safe keeping. I had questioned Osenberg's men about these papers. They readily confirmed his relations with the Gestapo, but claimed that he had burned the papers.

One day, when Osenberg was again pestering me with his apologies and swearing his loyalty to the Allies, I became impatient. "I am not interested in your political views," I said, "but only in the technical information you have. At any rate, one cannot trust you. You were in charge of the scientific section of the Gestapo, which you never revealed to us and you burned all the relating papers." This unexpected outburst took him by surprise and he put up a defense by blurting out, "No, I did not burn those papers, I buried them and, moreover, I was not the chief of the scientific section of the Gestapo, I was merely the second in command!" After that it was a very simple matter to make him tell where those papers were buried and where the missing Berlin papers were stored.

Osenberg's signature would be worth a study by psychiatric graphologists if there are such experts. Many Nazis seemed to imitate Hitler and made their official signature into a hieroglyph, utterly unintelligible, but easy to fake and conveying an idea of pathological pomposity. This habit was especially widespread among Gestapo officials, although Himmler himself signed his name very clearly. Compared to their Teutonic calligraphy, an intricate oriental tughra is a thing of beauty and clarity.

I don't know what became of Osenberg. His Gestapo connections probably put him in the automatic arrest category. At any rate, the revolt of his underlings broke up his dream of future power. He was interned somewhere else and his papers left in "Dustbin" in the care of one of his former slaves.

If Himmler's Gestapo sported a cultural department, his all-embracing SS, or Elite Guard, boasted a whole academy. The SS was a state within a state, with its own government, its own army and, what interests us here, its own science. It was avowedly the last word in Nazi ideals. Its members were supposed to measure up to the ultimates of "pure" Aryanism, fertility, and other ill-digested dogmas, just as its philosophical and religious doctrines were supposed to derive from ancient Teutonic lore. The symbol of the organization—ᛋᛋ—was twice the ancient runic letter S, and not two lightning strokes as has often been incorrectly stated.

During the war the SS had a few technical research laboratories of its own, under the direction of an SS-General Schwab, but these did not amount to anything. They tried some work on heavy water, but soon gave up and sent their "expert" on this subject to the University of Hamburg to continue his work with the legitimate physicists.

The principal "scientific" interest of the SS was ancient Germanic history, with a view to proving the greatness of their Teutonic ancestry. It was for this purpose that Himmler created his own "scientific academy" in 1935, Das Ahnenerbe, or Academy of Ancestral Heritage. Because some of the activities of this strange academy were shrouded in mystery that might just possibly have concealed something really important, we assigned Carl Baumann to make a thorough investigation of the organization for Alsos.

Except for Himmler's letter to hangman Heydrich about the physicist Heisenberg, mentioned in Chapter IX, Baumann did not discover anything connected with atomic research in the Ahnenerbe material. But his report on this academy was most instructive.

In the beginning, the Ahnenerbe was merely a cultural-propaganda section of the SS. But Himmler could never be content with anything so modest. He wanted a full-fledged academy with himself as president. If, as it happened, his academy was duplicating in part the functions of the "culture" ministry of Rosenberg and the propaganda ministry of Goebbels, all the better. This fitted in very well with his method of muscling in wherever possible with a view to eventual control of everything.

Director of Scholarship in the Ahnenerbe was Dr. Walther Wüst, President of the University of Munich, and Professor of Sanskrit and Persian. His great qualification for his high post in Himmler's academy was that in the early days of Nazism he had defended the "positive" view of Aryan culture in controversies with other professors.

The administrative head was SS Colonel Wolfram Sievers. This psychopathic gentleman was so happy that his name began and ended with an S, he always signed it **ϟieverϟ**. He was steeped in Teutonic lore, and while silent about most things he was always willing to talk at length on the subject of runic symbols. Sievers was directly responsible to Himmler and kept him well informed about the activities of his academy. He was also in charge of the organization's publications, books as well as magazines. In addition, he held an important post in the Reich's Research Council. Here he was the understudy of our man Mentzel and had the right to sign all papers. It was another case of penetration on the part of the wily Himmler.

Although, as has been said, the "scientific" work of the Ahnenerbe was mainly historical research to prove that the Nazi ideology was directly descended from ancient Teutonic culture and was therefore superior to all other ideologies, the pseudo-sciences were not neglected. There were divisions for "Genealogy," "Research on the Origin of Proper Names," "Research on Family Symbols (Sippenzeichen) and House Markings," "Spelaeology" and "Folk Lore," not to mention divining rods, and the mysteries of the occult.

Himmler himself was a graduate of an agricultural college and perhaps it was due to this background that he occasionally suggested a sensible research program. Thus he planned an entomological division to study all aspects of insect life and its effect on man. But every so often he could be counted on to come up with something really extraordinary, as may be seen from the following letter he wrote to Sievers from his field headquarters in March, 1944:

> "In future weather researches, which we expect to carry out after the war by systematic organization of an immense number of single observations, I request you to take note of the following:
>
> "The roots, or onions, of the meadow saffron are located at depths that vary from year to year. The deeper they are, the more severe the winter will be; the nearer they are to the surface, the milder the winter.
>
> "This fact was called to my attention by the Führer."

The academy had a few divisions on natural science, although their work was frowned upon by the scientists in the universities. Thus, there was a botany division under Von Luetzelburg, a cousin of Himmler's, who had spent some twenty-seven years in Brazil studying jungle plants and their medicinal properties. There was a section on applied geology which did secret work on the location of oil, minerals and water. Its chief, a Professor Wimmer, spent considerable time with the Army to help them find water in occupied territories. Wimmer is said to have done this by using a divining rod in combination with studies of the slope of the ground.

Among the Ahnenerbe publications was a "Journal for All the Natural Sciences," in which Nazi sympathizers wrote about their "scientific" work. For instance, they had their own pet theory about the structure of the universe, which they called the "Welteislehre," or world ice theory. According to this theory, the inner core of all the planets and all the stars consisted of ice. Not any fancy kind of ice. Just ordinary ice.

In his letter to Heydrich about Heisenberg, Himmler had written: "It would be advisable to bring Professor Heisenberg together with Professor Wüst. . . . Wüst must then try to make contact with Heisenberg, because we might be able to use him in the Ahnenerbe, when eventually it becomes a complete academy, for he is a good scientist and we might make him co-operate with our people of the Welteislehre." It was a suggestion that might well have made Heisenberg shiver.

The Ahnenerbe sponsored archaeological and historical expeditions (Ur-, Vor- und Frühgeschichte!) to foreign countries. These, with true German efficiency, could also serve as bases for military and spy activities. In occupied Russia, the academy's "experts" were on hand to loot the museums of their ancient Gothic art. The only trouble was that the gangs of "culture" minister Rosenberg had been there first. Sievers protested violently against this outrage. He could not see, he wrote, "how these art objects contributed anything to Rosenberg's assignment, which was to collect material for the spiritual fight against Jews and Masons and related world philosophical opponents of National Socialism."

Another example of Sievers' "scientific" interest may be seen in the following letter, written to a Fräulein Erna Piffl, in March, 1943, when the war was at its height.

> "Dear Fräulein Piffl:
>
> "There was a recent report in the press that there is an old woman living in Ribe in Jutland [Denmark], who still possesses knowledge of the knitting methods of the Vikings.
>
> "The Reichsleader [Himmler] desires that we send someone to Jutland immediately to visit this old woman and learn these knitting methods.
>
> "Heil Hitler!
>
> "ϟievers"

Unfortunately for the future of science, the records fail to reveal if Miss Piffl's mission was successful.

During the war it was found necessary to add an important new department to the Ahnenerbe—the division for

"Applied War Research." This division was responsible for all experimentation on human beings. Since Himmler was in charge of all concentration camps, it followed that any "scientific" work involving their inmates had to be relegated to his academy. When there was a shortage of mathematicians to do computing work in connection with the V-1 and V-2 weapons, a "mathematical section" was made up of concentration camp prisoners who had had mathematical training. They were reported to have done very good work.

But it was rarely that the "Applied War Research" division of the Ahnenerbe operated so humanely. There was, for instance, the notorious "Section H," under Professor August Hirt and Dr. E. Haagen at Strasbourg, who worked on prisoners in the Natzweiler camp. Worst of all was "Section R" at Dachau, where the cruelest experiments were performed by a Dr. Rascher and his very pretty and elegant wife, Nini Rascher, née Diehl. These experiments, which were requested by the Air Force, comprised such investigations as survival after long exposure to extreme cold, and the effects of exposure to extremely low pressures. A thorough study of the complete files of these activities, which were found intact, was made by the well-known Boston physician, Major Leo Alexander of the Medical Corps, and his reports were available to the Alsos Mission.

Aside from the inhuman cruelty of the Rascher operations at Dachau, the work stands out for its absurd attempt at perfection which amounts to a kind of parody of the scientific method. Thus one set of experiments consisted of immersing victims for several hours in ice-cold water until they were almost dead. (Only the hardiest satisfied this particular; most of them succumbed.) Then various ways of revival of the almost dead were tried and their results compared in order to discover the best.

One method was to put the frozen victim in bed with a young woman. With typical Teutonic thoroughness they then tried revival by putting the victim in bed with two women. If time had permitted, they would no doubt have experimented with three, four, and more women, and plotted a learned graph of the results. During the whole experiment the victim's temperature was recorded electrically by means of a thermocouple in the rectum. Major Alexander even found graphs, showing the temperature change until death or revival, marked "rewarming by one woman," "rewarming by two women," and "rewarming by women after coitus."

Although Sievers was in direct charge of all the war research sections of the Ahnenerbe, Himmler himself seems to have taken a personal interest. Most of the letters and reports from Rascher were directed to him. In one of those Rascher requests that he be transferred to Concentration Camp Auschwitz, because it was much colder there and he could cool the victims by leaving them out in the open, naked. Dachau, he wrote, was also too small; his experiments caused some trouble among the other inmates, because "his patients roared while being frozen."

When we interrogated Sievers, he first denied all knowledge of human experiments until we confronted him with some conclusive evidence that he was lying. Even then he seemed only mildly interested and preferred to discuss the prehistoric glory of the Teutonic peoples. He did, however, tell us that his close friends, the Raschers, had ended up as concentration camp prisoners themselves. There were probably several reasons for this, but the one stressed by Sievers was that they had violated the SS code of honor. Nini had had a miscarriage and had substituted another child as her own.

As we have seen, the thoroughness which we admire so much in German science can at times become a parody of science. In the professional library of the Gestapo in Berlin we found a book on "Germanic Symbols." It shows hundreds of runes and other emblems for which it supplies nauseating explanations. For example: "The dumbbell is the symbol for oppositions, counterpart. Birth and death, life and death, old and new year, winter and summer, heaven and earth, the receiving, the conserving, the generality, etc." Eminently logical and thorough, this book naturally starts with the beginning of everything, the sacred dot, "der Punkt." The dot, we are informed, is "the symbol of all symbols, meaning the beginning and end of all life, the innermost core and source of power of all formations. It is the symbol of the germ, but also of the remainder of all life. . . ."

This type of exaggerated nonsense flourished especially during the Nazi regime, but it had always been taken more seriously in Germany than anywhere else because of the pretentious and pompous way the pseudo sciences were presented. The very style of German books often made it impossible for a non-expert to judge the soundness of their contents; compendiums of utter nonsense were written like learned texts with numerous footnotes and references, tables and illustrations. Sometimes good books were spoiled by this overthoroughness.

There is an old anecdote that goes the rounds every five years or so in slightly revised form. It tells about a group of learned men of different nationalities who meet in the zoo and are greatly impressed by the camel. They decide that each shall write a book about it. The Englishman is first with a book titled "Camel Hunting in the Colonies." The Frenchman writes about "Le Chameau et Ses Amours"; the American on "Bigger and Better Camels." The German, after two years, announces a "Handbook on Camels"; Volume I—"The Camel in the Middle Ages," Volume II—"The Camel in Modern German Civilization."

There is more than just an element of truth in the last part of this. With some modification, we found just such a German book. It is not about camels but about dogs. Not about all dogs, of course, but only about German dogs, and specifically the German shepherd. The book I am referring to is "The German Shepherd Dog in Words and Pictures" by Captain Von Stephanitz. It was first published in 1901 and I have the sixth edition published in 1921, long before the advent of Hitler. It is definitely one of the best books on the care and breeding of dogs, but of interest to us here is the way its informative material is smothered under a weight of pretentious nonsense.

Almost eight hundred pages long, this tome starts, like so

many German books, at the very beginning of things—the creation of the world. To make it more impressive the creation is introduced by a quotation of the "Vendidad, the oldest book of the Zend-Avesta." No doubt every German dog breeder has this ancient Persian work on his bookshelf. The first two hundred pages then deal with the origin of the shepherd dog and its occurrence in all periods and all over the world—the dog in China, the dog in ancient Greece, the dog in the Bible, the dog in Egypt. There is also an interesting section about "the dog and the Jews."

We learn that the ancient Jews despised dogs and this explains partially "the present contempt for dogs even among Aryan people, which can be blamed on the great influence of Jewish notions, which smuggled themselves in, hidden under the Christian religion." Furthermore, "the attitude of the Jew to the dog is still the same nowadays. . . . Never can the dog have any emotional value to him, never can he devote himself unselfishly. . . . That only a German can do, for 'being a German means to do a thing for its own sake.' [Wagner]"

Among the illustrations in this by no means singular book is a drawing of a dog with the caption "Friendly greeting; after Professor B. Schmid." Of course, a professor was needed to analyze the dog's mood; otherwise the readers could not accept the statement as authoritative.

Finally, German books always have such an excellent index. "The German Shepherd Dog in Words and Pictures" is no exception. I shall list just a few consecutive items from the index of this book to indicate to what extremes German thoroughness can go.

Hund und andere Tiere	Dog and other animals
" und Dienstboten	" and servants
" und Frau	" and mistress
" und Herr	" and master
" und Hund	" and dog
" und Kinder	" and children

Hund und Spielzeug	Dog and toys
" und Werkseug	" and tools

As I have said, Captain Stephanitz' classic on the German shepherd dog was written long before Hitler and Himmler. It helps us to understand that not all the farcical elements in the SS Academy were supplied by Nazism.

ROBERT LOUIS STEVENSON

EVERYDAY LIFE is a stimulating mixture of order and haphazardry. The sun rises and sets on schedule but the wind bloweth where it listeth. Science is a perpetual search for underlying order, and so successful has it been that many scientists suppose all nature must blend together in one ultimate harmonic melody as pure as the notes of Apollo's lyre. On the other hand, the more order science uncovers, the more hitherto unsuspected disorder it brings to light. Galileo's opponents were so certain the moon was a perfect sphere that they could not believe the lunar surface was ragged with mountains. Is there in nature an inescapable element of carelessness—a wanton swerving from the perfectly plotted curve, like the "sweet disorder" of a woman's dress that the poet Robert Herrick found more bewitching than precision?

This is the refreshing theme of the little essay to follow by Robert Louis Stevenson (1850-1894). He has chosen Pan, the merry, goat-footed demi-god as the symbol of nature's rowdy randomness. It is, Stevenson tells us, a shaggy world. One must add that even science itself cannot escape from the careless tunes of Pan's pipe. The interior of the atom gets shaggier every day. What could be neater than the deductive webs of mathematics? They are heavy with shagginess. The ratio of a circle's diameter to the circumference is a precise one, but express it in any number system and it becomes a series of unending digits that satisfy every test of randomness. Mathematicians used to dream of one vast deductive system that would embrace all the theorems of logic and mathematics. Then in 1931 mathematician Kurt Gödel, to the panic of certain mathematicians but to the great delight of all Pan worshippers, discovered an elegant proof that this was a hopeless dream.

Wherever science turns she seems to come upon a cloven hoof print or hear a note or two of wild music. And for this all higher Pan-theists are duly grateful, for a completely tidy universe would be to them as unbearable as a completely tidy life or home.

ROBERT LOUIS STEVENSON

Pan's Pipes

THE WORLD in which we live has been variously said and sung by the most ingenious poets and philosophers: these reducing it to formulæ and chemical ingredients, those striking the lyre in high-sounding measures for the handiwork of God. What experience supplies is of a mingled tissue, and the choosing mind has much to reject before it can get together the materials of a theory. Dew and thunder, destroying Attila and the Spring lambkins, belong to an order of contrasts which no repetition can assimilate. There is an uncouth, outlandish strain throughout the web of the world, as from a vexatious planet in the house of life. Things are not congruous and wear strange disguises: the consummate flower is fostered out of dung, and after nourishing itself awhile with heaven's delicate distillations, decays again into indistinguishable soil; and with Cæsar's ashes, Hamlet tells us, the urchins make dirt pies and filthily besmear their countenance. Nay, the kindly shine of summer, when tracked home with the scientific spyglass, is found to issue from the most portentous nightmare of the universe—the great, conflagrant sun: a world of hell's squibs, tumultuary, roaring aloud, inimical to life. The sun itself is enough to disgust a human being of the scene which he inhabits; and you would not fancy there was a green or habitable spot in a universe thus awfully lighted up. And yet it is by the blaze of such a conflagration, to which the fire of Rome was but a spark, that we do all our fiddling, and hold domestic tea-parties at the arbour door.

The Greeks figured Pan, the god of Nature, now terribly stamping his foot, so that armies were dispersed; now by the woodside on a summer noon trolling on his pipe until he charmed the hearts of upland ploughmen. And the Greeks, in so figuring, uttered the last word of human experience. To certain smoke-dried spirits matter and motion and elastic aethers, and the hypothesis of this or that other spectacled professor, tell a speaking story; but for youth and all ductile and congenial minds, Pan is not dead, but of all the classic hierarchy alone survives in triumph; goat-footed, with a gleeful and an angry look, the type of the shaggy world: and in every wood, if you go with a spirit properly prepared, you shall hear the note of his pipe.

For it is a shaggy world, and yet studded with gardens; where the salt and tumbling sea receives clear rivers running from among reeds and lilies; fruitful and austere; a rustic world; sunshiny, lewd, and cruel. What is it the birds sing among the trees in pairing-time? What means the sound of the rain falling far and wide upon the leafy forest? To what tune does the fisherman whistle, as he hauls in his net at morning, and the bright fish are heaped inside the boat? These are all airs upon Pan's pipe; he it was who gave them breath in the exultation of his heart, and gleefully modulated their outflow with his lips and fingers. The coarse mirth of herdsmen, shaking the dells with laughter and striking out high echoes from the rock; the tune of moving feet in the lamplit city, or on the smooth ballroom floor; the hooves of many horses, beating the wide pastures in alarm; the song of hurrying rivers; the colour of clear skies; and smiles and the live touch of hands; and the voice of things, and their significant look, and the renovating influence they breathe forth—these are his joyful measures, to which the whole earth treads in choral harmony. To this music the young lambs bound as to a tabor, and the London shop-girl skips rudely in the dance. For it puts a spirit of gladness in all hearts; and to look on the happy side of nature is common, in their hours, to all created things. Some are vocal under a good influence, are pleasing whenever they are pleased, and hand on their happiness to others, as a child who, looking upon lovely things, looks lovely. Some leap to the strains with unapt foot, and make a halting figure in the universal dance. And some, like sour spectators at the play, receive the music into their hearts with an unmoved countenance, and walk like strangers through the general rejoicing. But let him feign never so carefully, there is not a man but has his pulses shaken when Pan trolls out a stave of ecstasy and sets the world a-singing.

Alas if that were all! But oftentimes the air is changed; and in the screech of the night wind, chasing navies, subverting the tall ships and the rooted cedar of the hills; in the random deadly levin or the fury of headlong floods, we recognize the "dread foundation" of life and the anger in Pan's heart. Earth wages open war against her children, and under her softest touch hides treacherous claws. The cool waters invite us in to drown; the domestic hearth burns up in the hour of sleep, and makes an end of all. Everything is good or bad, helpful or deadly, not in itself, but by its circumstances. For a few bright days in England the hurricane must break forth and the North Sea pay a toll of populous ships. And when the universal music has led lovers into the paths of dalliance, confident of Nature's sympathy, suddenly the air shifts into a minor, and death makes a clutch from his ambuscade below the bed of marriage. For death is given in a kiss; the dearest kindnesses are fatal; and into this life, where one thing preys upon another, the child too often makes its entrance from the mother's corpse. It is no wonder, with so traitorous a scheme of things, if the wise people who created for us the idea of Pan thought that of all fears the fear of him was the most terrible, since it embraces all. And still we preserve the phrase: a panic terror. To reckon dangers too curiously, to hearken too intently for the threat that runs through all the winning music of the world, to hold back the hand from the rose because of the thorn, and from life because of death: this it is to be afraid of Pan. Highly respectable citizens who flee life's pleasures and responsibilities and keep, with upright hat, upon the midway of custom, avoiding the right hand and the left, the ecstasies and the agonies, how surprised they would be if they could hear their attitude mythologically expressed, and knew themselves as tooth-chattering ones, who flee from Nature because they fear the hand of Nature's God! Shrilly sound Pan's pipes; and behold the banker instantly concealed in the bank parlour! For to distrust one's impulses is to be recreant to Pan.

There are moments when the mind refuses to be satisfied with evolution, and demands a ruddier presentation of the sum of man's experience. Sometimes the mood is brought about by laughter at the humorous side of life, as when, abstracting ourselves from earth, we imagine people plodding on foot, or seated in ships and speedy trains, with the planet all the while whirling in the opposite direction, so that, for all their hurry, they travel back-foremost through the universe of space. Sometimes it comes by the spirit of delight, and sometimes by the spirit of terror. At least, there will always be hours when we refuse to be put off by the feint of explanation, nicknamed science; and demand instead some palpitating image of our estate, that shall represent the troubled and uncertain element in which we dwell, and satisfy reason by the means of art. Science writes of the world as if with the cold finger of a starfish; it is all true; but what is it when compared to the reality of which it discourses? where hearts beat high in April, and death strikes, and hills totter in the earthquake, and there is a glamour over all the objects of sight, and a thrill in all noises for the ear, and Romance herself has made her dwelling among men? So we come back to the old myth, and hear the goat-footed piper making the music which is itself the charm and terror of things; and when a glen invites our visiting footsteps, fancy that Pan leads us thither with a gracious tremolo; or when our hearts quail at the thunder of the cataract, tell ourselves that he has stamped his hoof in the nigh thicket.

SIGMUND FREUD

SHORTLY AFTER the Reverend Mr. Davidson, in Somerset Maugham's classic story "Rain," succeeds in reducing Miss Sadie Thompson to a state of sniffling repentance, the minister dreams of the mountains of Nebraska—"like huge molehills, rounded and smooth, and they rose from the plain abruptly." Maugham would never have included this episode if Freud had not published, in 1899, his startling work on *The Interpretation of Dreams*. Dreams were then regarded by psychologists as a meaningless play of ideas, like the notes struck by a man ignorant of music who lets his fingers wander idly over the keys. In Freud's view they were symbols of repressed wishes, appearing in distorted, censored forms so as not to shock and wake the sleeper. Freud always considered this his most valuable discovery, and in the history of the analytic movement his book on dreams has become one of its greatest milestones.

Sigmund Freud (1856-1939) concludes his *Autobiographical Study* with these modest words: "Looking back, then, over the patchwork of my life's labors, I can say that I have made many beginnings and thrown out many suggestions. Something will come of them in the future, though I cannot myself tell whether it will be much or little. I can, however, express a hope that I have opened up a pathway for an important advance in our knowledge."

Freud himself did not always live up to this humility, and many of his followers have abandoned it entirely, elaborating the master's views into those "self-sealing systems" of which Oppenheimer complains in an essay earlier in this volume. More open-minded analysts have not hesitated to modify, add to, and even discard Freud's suggestions when they thought the evidence warranted. As this revisionary work proceeds, it should eventually become clear how many of his ideas were audacious insights based on sound interpretations of clinical data, and how many were simply bad guesses or projections of his own neuroses. Whatever the outcome, Freud is certain to remain a towering figure in the history of psychology and a man whose views had an incalculable effect on twentieth century life and thought.

The selection reprinted here, from James Strachey's magnificent new translation of *The Interpretation of Dreams,* contains Freud's first allusions to the "Oedipus complex." He found in this concept a key that he believed would explain the compelling power of Sophocles' famous tragedy as well as the baffling indecision of Shakespeare's Hamlet. Thus this brief selection, written with Freud's usual brilliance and persuasive fire, introduces us to many of his major contributions, not the least of which was his recognition of the role played by repressed wishes in artistic creation.

The footnotes on the pages to follow are from various editions of Freud's book. Comments by the translator are within square brackets.

SIGMUND FREUD

Dreams of the Death of Beloved Persons

Another group of dreams which may be described as typical are those containing the death of some loved relative—for instance, of a parent, of a brother or sister, or of a child. Two classes of such dreams must at once be distinguished: those in which the dreamer is unaffected by grief, so that on awakening he is astonished at his lack of feeling, and those in which the dreamer feels deeply pained by the death and may even weep bitterly in his sleep.

We need not consider the dreams of the first of these classes, for they have no claim to be regarded as 'typical'. If we analyse them, we find that they have some meaning other than their apparent one, and that they are intended to conceal some other wish. Such was the dream of the aunt who saw her sister's only son lying in his coffin. It did not mean that she wished her little nephew dead; as we have seen, it merely concealed a wish to see a particular person of whom she was fond and whom she had not met for a long time—a person whom she had once before met after a similarly long interval beside the coffin of another nephew. This wish, which was the true content of the dream, gave no occasion for grief, and no grief, therefore, was felt in the dream. It will be noticed that the affect felt in the dream belongs to its latent and not to its manifest content, and that the dream's *affective* content has remained untouched by the distortion which has overtaken its *ideational* content.

Very different are the dreams of the other class—those in which the dreamer imagines the death of a loved relative and is at the same time painfully affected. The meaning of such dreams, as their content indicates, is a wish that the person in question may die. And since I must expect that the feelings of all of my readers and any others who have experienced similar dreams will rebel against my assertion, I must try to base my evidence for it on the broadest possible foundation.

I have already discussed a dream which taught us that the wishes which are represented in dreams as fulfilled are not always present-day wishes. They may also be wishes of the past which have been abandoned, overlaid and repressed, and to which we have to attribute some sort of continued existence only because of their re-emergence in a dream. They are not dead in our sense of the word but only like the shades in the Odyssey, which awoke to some sort of life as soon as they had tasted blood. In the dream of the dead child in the 'case' what was involved was a wish which had been an immediate one fifteen years earlier and was frankly admitted as having existed at that time. I may add—and this may not be without its bearing upon the theory of dreams—that even behind this wish there lay a memory from the dreamer's earliest childhood. When she was a small child—the exact date could not be fixed with certainty—she had heard that her mother had fallen into a deep depression during the pregnancy of which she had been the fruit and had passionately wished that the child she was bearing might die. When the dreamer herself was grown-up and pregnant, she merely followed her mother's example.

If anyone dreams, with every sign of pain, that his father or mother or brother or sister has died, I should never use the dream as evidence that he wishes for that person's death *at the present time*. The theory of dreams does not require as much as that; it is satisfied with the inference that this death has been wished for at some time or other during the dreamer's childhood. I fear, however, that this reservation will not appease the objectors; they will deny the possibility of their *ever* having had such a thought with just as much energy as they insist that they harbour no such wishes now. I must therefore reconstruct a portion of the vanished mental life of children on the basis of the evidence of the present.[1]

Let us first consider the relation of children to their brothers and sisters. I do not know why we presuppose that that relation must be a loving one; for instances of hostility between adult brothers and sisters force themselves upon everyone's experience and we can often establish the fact that the disunity originated in childhood or has always existed. But it is further true that a great many adults, who are on affectionate terms with their brothers and sisters and are ready to stand by them to-day, passed their childhood on almost unbroken terms of enmity with them. The elder child ill-treats the younger, maligns him and robs him of his toys; while the younger is consumed with impotent rage against the elder, envies and fears him, or meets his oppressor with the first stirrings of a love of liberty and a sense of justice. Their parents complain that the children do not get on with one another, but cannot discover why. It is easy to see that the character of even a good child is not what we should wish to find it in an adult. Children are completely egoistic; they feel their needs intensely and strive ruthlessly to satisfy them —especially as against the rivals, other children, and first and foremost as against their brothers and sisters. But we do not on that account call a child 'bad', we call him 'naughty'; he is no more answerable for his evil deeds in our judgement than in the eyes of the law. And it is right that this should be so; for we may expect that, before the end of the period which we count as childhood, altruistic impulses and morality will awaken in the little egoist and (to use Meynert's terms [e.g. 1892, 169 ff.]) a secondary ego will overlay and inhibit the primary one. It is true, no doubt, that morality

[1] [*Footnote added* 1909:] Cf. my 'Analysis of a Phobia in a Five-Year-Old Boy' (1909*b*) and my paper 'On the Sexual Theories of Children' (1908*c*).

does not set in simultaneously all along the line and that the length of non-moral childhood varies in different individuals. If this morality fails to develop, we like to talk of 'degeneracy', though what in fact faces us is an inhibition in development. After the primary character has already been overlaid by later development, it can still be laid bare again, at all events in part, in cases of hysterical illness. There is a really striking resemblance between what is known as the hysterical character and that of a naughty child. Obsessional neurosis, on the contrary, corresponds to a super-morality imposed as a reinforcing weight upon fresh stirrings of the primary character.

Many people, therefore, who love their brothers and sisters and would feel bereaved if they were to die, harbour evil wishes against them in their unconscious, dating from earlier times; and these are capable of being realized in dreams.

It is of quite particular interest, however, to observe the behaviour of small children up to the age of two or three or a little older towards their younger brothers and sisters. Here, for instance, was a child who had so far been the only one; and now he was told that the stork had brought a new baby. He looked the new arrival up and down and then declared decisively: 'The stork can take him away again!'[1] I am quite seriously of the opinion that a child can form a just estimate of the set-back he has to expect at the hands of the little stranger. A lady of my acquaintance, who is on very good terms to-day with a sister four years her junior, tells me that she greeted the news of her first arrival with this qualification: 'But all the same I shan't give her my red cap!' Even if a child only comes to realize the situation later on, his hostility will date from that moment. I know of a case in which a little

[1] [*Footnote added* 1909:] The three-and-a-half-year-old Hans (whose phobia was the subject of the analysis mentioned in the preceding footnote) exclaimed shortly after the birth of a sister, while he was suffering from a feverish sore throat: 'I don't *want* a baby sister!' [Freud, 1909*b*, Section I.] During his neurosis eighteen months later he frankly confessed to a wish that his mother might drop the baby into the bath so that she would die. [Ibid., Section II (April 11).] At the same time, Hans was a good-natured and affectionate child, who soon grew fond of this same sister and particularly enjoyed taking her under his wing.

girl of less than three tried to strangle an infant in its cradle because she felt that its continued presence boded her no good. Children at that time of life are capable of jealousy of any degree of intensity and obviousness. Again, if it should happen that the baby sister does in fact disappear after a short while, the elder child will find the whole affection of the household once more concentrated upon himself. If after that the stork should bring yet another baby, it seems only logical that the little favourite should nourish a wish that his new competitor may meet with the same fate as the earlier one, so that he himself may be as happy as he was originally and during the interval.[1] Normally, of course, this attitude of a child towards a younger brother or sister is a simple function of the difference between their ages. Where the gap in time is sufficiently long, an elder girl will already begin to feel the stirring of her maternal instincts towards the helpless newborn baby.

Hostile feelings towards brothers and sisters must be far more frequent in childhood than the unseeing eye of the adult observer can perceive.[2]

In the case of my own children, who followed each other in rapid succession, I neglected the opportunity of carrying out observations of this kind; but I am now making up for this

[1] [*Footnote added* 1914:] Deaths that are experienced in this way in childhood may quickly be forgotten in the family; but psycho-analytic research shows that they have a very important influence on subsequent neuroses.

[2] [*Footnote added* 1914:] Since this was written, a large number of observations have been made and recorded in the literature of psycho-analysis upon the originally hostile attitude of children towards their brothers and sisters and one of their parents. The [Swiss] author and poet Spitteler has given us a particularly genuine and naïve account of this childish attitude, derived from his own childhood [1914, 40]: 'Moreover there was a second Adolf there: a little creature who they alleged was my brother, though I could not see what use he was and still less why they made as much fuss of him as of me myself. I was sufficient so far as I was concerned; why should I want a brother? And he was not merely useless, he was positively in the way. When I pestered my grandmother, he wanted to pester her too. When I was taken out in the perambulator, he sat opposite to me and took up half the space, so that we were bound to kick each other with our feet.'

neglect by observing a small nephew, whose autocratic rule was upset, after lasting for fifteen months, by the appearance of a female rival. I am told, it is true, that the young man behaves in the most chivalrous manner to his little sister, that he kisses her hand and strokes her; but I have been able to convince myself that even before the end of his second year he made use of his powers of speech for the purpose of criticizing someone whom he could not fail to regard as superfluous. Whenever the conversation touched upon her he used to intervene in it and exclaim petulantly: 'Too 'ickle! too 'ickle!' During the last few months the baby's growth has made enough progress to place her beyond this particular ground for contempt, and the little boy has found a different basis for his assertion that she does not deserve so much attention: at every suitable opportunity he draws attention to the fact that she has no teeth.[1] We all of us recollect how the eldest girl of another of my sisters, who was then a child of six, spent half-an-hour in insisting upon each of her aunts in succession agreeing with her: 'Lucie can't understand that yet, can she?' she kept asking. Lucie was her rival—two and a half years her junior.

In none of my women patients, to take an example, have I failed to come upon this dream of the death of a brother or sister, which tallies with an increase in hostility. I have only found a single exception; and it was easy to interpret this as a confirmation of the rule. On one occasion during an analytic session I was explaining this subject to a lady, since in view of her symptom its discussion seemed to me relevant. To my astonishment she replied that she had never had such a dream. Another dream, however, occurred to her, which ostensibly had no connection with the topic—a dream which she had first dreamt when she was four years old and at that time the youngest of the family, and which she had dreamt repeatedly since: *A whole crowd of children—all her brothers, sisters and cousins of both sexes—were romping in a field.*

---

[1] [*Footnote added* 1909:] Little Hans, when he was three and a half, gave vent to a crushing criticism of his sister in the same words. It was because of her lack of teeth, he supposed, that she was unable to talk. [Freud, 1909*b*, Section I.]

Suddenly they all grew wings, flew away and disappeared. She had no idea what this dream meant; but it is not hard to recognize that in its original form it had been a dream of the death of all her brothers and sisters, and had been only slightly influenced by the censorship. I may venture to suggest the following analysis. On the occasion of the death of one of this crowd of children (in this instance the children of two brothers had been brought up together as a single family) the dreamer, not yet four years old at the time, must have asked some wise grown-up person what became of children when they were dead. The reply must have been: 'They grow wings and turn into little angels.' In the dream which followed upon this piece of information all the dreamer's brothers and sisters had wings like angels and—which is the main point—flew away. Our little baby-killer was left alone, strange to say: the only survivor of the whole crowd! We can hardly be wrong in supposing that the fact of the children romping in a *field* before flying away points to butterflies. It is as though the child was led by the same chain of thought as the peoples of antiquity to picture the soul as having a butterfly's wings.

At this point someone will perhaps interrupt: 'Granted that children have hostile impulses towards their brothers and sisters, how can a child's mind reach such a pitch of depravity as to wish for the *death* of his rivals or of playmates stronger than himself, as though the death penalty were the only punishment for every crime?' Anyone who talks like this has failed to bear in mind that a child's idea of being 'dead' has nothing much in common with ours apart from the word. Children know nothing of the horrors of corruption, of freezing in the ice-cold grave, of the terrors of eternal nothingness—ideas which grown-up people find it so hard to tolerate, as is proved by all the myths of a future life. The fear of death has no meaning to a child; hence it is that he will play with the dreadful word and use it as a threat against a playmate: 'If you do that again, you'll die, like Franz!' Meanwhile the poor mother gives a shudder and remembers, perhaps, that the greater half of the human race fail to survive their childhood years. It was actually possible for a child, who was over eight years old at the time, coming home from a visit to the Natural History Museum, to say to his mother: 'I'm so fond of you, Mummy: when you die I'll have you stuffed and I'll keep you in this room, so that I can see you *all* the time.' So little resemblance is there between a child's idea of being dead and our own![1]

To children, who, moreover, are spared the sight of the scenes of suffering which precede death, being 'dead' means approximately the same as being 'gone'—not troubling the survivors any longer. A child makes no distinction as to how this absence is brought about: whether it is due to a journey, to a dismissal, to an estrangement, or to death.[2] If, during a child's prehistoric epoch, his nurse has been dismissed, and if soon afterwards his mother has died, the two events are superimposed on each other in a single series in his memory as revealed in analysis. When people are absent, children do not miss them with any great intensity; many mothers have learnt this to their sorrow when, after being away from home for some weeks on a summer holiday, they are met on their return by the news that the children have not once asked after their mummy. If their mother does actually make the

---

[1] [*Footnote added* 1909:] I was astonished to hear a highly intelligent boy of ten remark after the sudden death of his father: 'I know father's dead, but what I can't understand is why he doesn't come home to supper.' —[*Added* 1919:] Further material on this subject will be found in the first [seven] volumes of the periodical *Imago* [1912-21], under the standing rubric of '*Vom wahren Wesen der Kinderseele*' ['The True Nature of the Child Mind'], edited by Frau Dr. H. von Hug-Hellmuth.

[2] [*Footnote added* 1919:] An observation made by a parent who had a knowledge of psycho-analysis caught the actual moment at which his highly intelligent four-year-old daughter perceived the distinction between being 'gone' and being 'dead'. The little girl had been troublesome at meal-time and noticed that one of the maids at the pension where they were staying was looking at her askance. 'I wish Josefine was dead,' was the child's comment to her father. 'Why dead?' enquired her father soothingly; 'wouldn't it do if she went away?' 'No,' replied the child; 'then she'd come back again.' The unbounded self-love (the narcissism) of children regards any interference as an act of *lèse majesté;* and their feelings demand (like the Draconian code) that any such crime shall receive the one form of punishment which admits of no degrees.

journey to that 'undiscover'd country, from whose bourn no traveller returns', children seem at first to have forgotten her, and it is only later on that they begin to call their dead mother to mind.

Thus if a child has reasons for wishing the absence of another, there is nothing to restrain him from giving his wish the form of the other child being dead. And the psychical reaction to dreams containing death-wishes proves that, in spite of the different content of these wishes in the cases of children, they are nevertheless in some way or other the same as wishes expressed in the same terms by adults.[1]

If, then, a child's death-wishes against his brothers and sisters are explained by the childish egoism which makes him regard them as his rivals, how are we to explain his death-wishes against his parents, who surround him with love and fulfil his needs and whose preservation that same egoism should lead him to desire?

A solution of this difficulty is afforded by the observation that dreams of the death of parents apply with preponderant frequency to the parent who is of the same sex as the dreamer: that men, that is, dream mostly of their father's death and women of their mother's. I cannot pretend that this is universally so, but the preponderance in the direction I have indicated is so evident that it requires to be explained by a factor of general importance.[2] It is as though—to put it bluntly—a sexual preference were making itself felt at an early age: as though boys regarded their fathers and girls their mothers as their rivals in love, whose elimination could not fail to be to their advantage.

Before this idea is rejected as a monstrous one, it is as well

[1] [The adult attitude to death is discussed by Freud more particularly in the second essay of his *Totem and Taboo* (1912-13), Section 3 *(c)*, in his paper on 'The Three Caskets' (1913 *f*) and in the second part of his 'Thoughts on War and Death' (1915*b*).]

[2] [*Footnote added* 1925:] The situation is often obscured by the emergence of a self-punitive impulse, which threatens the dreamer, by way of a moral reaction, with the loss of the parent whom he loves.

in this case, too, to consider the real relations obtaining—this time between parents and children. We must distinguish between what the cultural standards of filial piety demand of this relation and what everyday observation shows it in fact to be. More than one occasion for hostility lies concealed in the relation between parents and children—a relation which affords the most ample opportunities for wishes to arise which cannot pass the censorship.

Let us consider first the relation between father and son. The sanctity which we attribute to the rules laid down in the Decalogue has, I think, blunted our powers of perceiving the real facts. We seem scarcely to venture to observe that the majority of mankind disobey the Fifth Commandment. Alike in the lowest and in the highest strata of human society filial piety is wont to give way to other interests. The obscure information which is brought to us by mythology and legend from the primaeval ages of human society gives an unpleasing picture of the father's despotic power and of the ruthlessness with which he made use of it. Kronos devoured his children, just as the wild boar devours the sow's litter; while Zeus emasculated his father[1] and made himself ruler in his place. The more unrestricted was the rule of the father in the ancient family, the more must the son, as his destined successor, have found himself in the position of an enemy, and the more impatient must he have been to become ruler himself through his father's death. Even in our middle-class families fathers are as a rule inclined to refuse their sons independence and the means necessary to secure it and thus to foster the growth of the germ of hostility which is inherent in their relation. A physician will often be in a position to notice how a son's grief at the loss of his father cannot suppress

---

[1] [*Footnote added* 1909:] Or so he is reported to have done according to some myths. According to others, emasculation was only carried out by Kronos on his father Uranus. [This passage is discussed in Chapter X (3) of *The Psychopathology of Everyday Life* (Freud, 1901*b*).] For the mythological significance of this theme, cf. Rank, 1909, [*added* 1914:] and Rank, 1912*c*, Chapter IX, Section 2.—[These sentences in the text are, of course, an early hint at the line of thought developed later by Freud in his *Totem and Taboo* (1912–13).]

his satisfaction at having at length won his freedom. In our society to-day fathers are apt to cling desperately to what is left of a now sadly antiquated *potestas partis familias;* and an author who, like Ibsen, brings the immemorial struggle between fathers and sons into prominence in his writings may be certain of producing his effect.

Occasions for conflict between a daughter and her mother arise when the daughter begins to grow up and long for sexual liberty, but finds herself under her mother's tutelage; while the mother, on the other hand, is warned by her daughter's growth that the time has come when she herself must abandon her claims to sexual satisfaction.

All of this is patent to the eyes of everyone. But it does not help us in our endeavour to explain dreams of a parent's death in people whose piety towards their parents has long been unimpeachably established. Previous discussions, moreover, will have prepared us to learn that the death-wish against parents dates back to earliest childhood.

This supposition is confirmed with a certainty beyond all doubt in the case of psychoneurotics when they are subjected to analysis. We learn from them that a child's sexual wishes —if in their embryonic stage they deserve to be so described— awaken very early, and that a girl's first affection is for her father[1] and a boy's first childish desires are for his mother. Accordingly, the father becomes a disturbing rival to the boy and the mother to the girl; and I have already shown in the case of brothers and sisters how easily such feelings can lead to a death-wish. The parents too give evidence as a rule of sexual partiality: a natural predilection usually sees to it that a man tends to spoil his little daughters, while his wife takes her sons' part; though both of them, where their judgement is not disturbed by the magic of sex, keep a strict eye upon their children's education. The child is very well aware of this partiality and turns against that one of his parents who is opposed to showing it. Being loved by an adult does not merely bring a child the satisfaction of a special need; it also

[1] [Freud's views on this point were later modified. Cf. Freud, 1925*j* and 1931*b*.]

means that he will get what he wants in every other respect as well. Thus he will be following his own sexual instinct and at the same time giving fresh strength to the inclination shown by his parents if his choice between them falls in with theirs.

The signs of these infantile preferences are for the most part overlooked; yet some of them are to be observed even after the first years of childhood. An eight-year-old girl of my acquaintance, if her mother is called away from the table, makes use of the occasion to proclaim herself her successor: '*I*'m going to be Mummy now. Do you want some more greens, Karl? Well, help yourself, then!' and so on. A particularly gifted and lively girl of four, in whom this piece of child psychology is especially transparent, declared quite openly: 'Mummy can go away now. Then Daddy must marry me and I'll be his wife.' Such a wish occurring in a child is not in the least inconsistent with her being tenderly attached to her mother. If a little boy is allowed to sleep beside his mother when his father is away from home, but has to go back to the nursery and to someone of whom he is far less fond as soon as his father returns, he may easily begin to form a wish that his father should *always* be away, so that he himself could keep his place beside his dear, lovely Mummy. One obvious way of attaining this wish would be if his father were dead; for the child has learnt one thing by experience—namely that 'dead' people, such as Granddaddy, are always away and never come back.

Though observations of this kind on small children fit in perfectly with the interpretation I have proposed, they do not carry such complete conviction as is forced upon the physician by psycho-analyses of adult neurotics. In the latter case dreams of the sort we are considering are introduced into the analysis in such a context that it is impossible to avoid interpreting them as *wishful* dreams.

One day one of my women patients was in a distressed and tearful mood. 'I don't want ever to see my relations again,' she said, 'they must think me horrible.' She then went on, with almost no transition, to say that she remembered a dream, though of course she had no idea what it meant. When she was four years old she had a dream that *a lynx or fox*[1] *was walking on the roof; then something had fallen down or she had fallen down; and then her mother was carried out of the house dead*—and she wept bitterly. I told her that this dream must mean that when she was a child she had wished she could see her mother dead, and that it must be on account of the dream that she felt her relations must think her horrible. I had scarcely said this when she produced some material which threw light on the dream. 'Lynx-eye' was a term of abuse that had been thrown at her by a street-urchin when she was a very small child. When she was three years old, a tile off the roof had fallen on her mother's head and made it bleed violently.

I once had an opportunity of making a detailed study of a young woman who passed through a variety of psychical conditions. Her illness began with a state of confusional excitement during which she displayed a quite special aversion to her mother, hitting and abusing her whenever she came near her bed, while at the same period she was docile and affectionate towards a sister who was many years her senior. This was followed by a state in which she was lucid but somewhat apathetic and suffered from badly disturbed sleep. It was during this phase that I began treating her and analysing her dreams. An immense number of these dreams were concerned, with a greater or less degree of disguise, with the death of her mother: at one time she would be attending an old woman's funeral, at another she and her sister would be sitting at table dressed in mourning. There could be no question as to the meaning of these dreams. As her condition improved still further, hysterical phobias developed. The most tormenting of these was a fear that something might have happened to her mother. She was obliged to hurry home, wherever she might be, to convince herself that her mother

[1] [The German names for these animals are very much alike: '*Luchs*' and '*Fuchs*'.]

was still alive. This case, taken in conjunction with what I had learnt from other sources, was highly instructive: it exhibited, translated as it were into different languages, the various ways in which the psychical apparatus reacted to one and the same exciting idea. In the confusional state, in which, as I believe, the second psychical agency was overwhelmed by the normally suppressed first one, her unconscious hostility to her mother found a powerful *motor* expression. When the calmer condition set in, when the rebellion was suppressed and the domination of the censorship re-established, the only region left open in which her hostility could realize the wish for her mother's death was that of dreaming. When a normal state was still more firmly established, it led to the production of her exaggerated worry about her mother as a hysterical counter-reaction and defensive phenomenon. In view of this it is no longer hard to understand why hysterical girls are so often attached to their mothers with such exaggerated affection.

On another occasion I had an opportunity of obtaining a deep insight into the unconscious mind of a young man whose life was made almost impossible by an obsessional neurosis. He was unable to go out into the street because he was tortured by the fear that he would kill everyone he met. He spent his days in preparing his alibi in case he might be charged with one of the murders committed in the town. It is unnecessary to add that he was a man of equally high morals and education. The analysis (which, incidentally, led to his recovery) showed that the basis of this distressing obsession was an impulse to murder his somewhat over-severe father. This impulse, to his astonishment, had been consciously expressed when he was seven years old, but it had, of course, originated much earlier in his childhood. After his father's painful illness and death, the patient's obsessional self-reproaches appeared—he was in his thirty-first year at the time—taking the shape of a phobia transferred on to strangers. A person, he felt, who was capable of wanting to push his own father over a precipice from the top of a mountain was not to be trusted to respect the lives of those less closely related to him; he was quite right to shut himself up in his room.

In my experience, which is already extensive, the chief part in the mental lives of all children who later become psychoneurotics is played by their parents. Being in love with the one parent and hating the other are among the essential constituents of the stock of psychical impulses which is formed at that time and which is of such importance in determining the symptoms of the later neurosis. It is not my belief, however, that psychoneurotics differ sharply in this respect from other human beings who remain normal—that they are able, that is, to create something absolutely new and peculiar to themselves. It is far more probable—and this is confirmed by occasional observations on normal children—that they are only distinguished by exhibiting on a magnified scale feelings of love and hatred to their parents which occur less obviously and less intensely in the minds of most children.

This discovery is confirmed by a legend that has come down to us from classical antiquity: a legend whose profound and universal power to move can only be understood if the hypothesis I have put forward in regard to the psychology of children has an equally universal validity. What I have in mind is the legend of King Oedipus and Sophocles' drama which bears his name.

Oedipus, son of Laïus, King of Thebes, and of Jocasta, was exposed as an infant because an oracle had warned Laïus that the still unborn child would be his father's murderer. The child was rescued, and grew up as a prince in an alien court, until, in doubts as to his origin, he too questioned the oracle and was warned to avoid his home since he was destined to murder his father and take his mother in marriage. On the road leading away from what he believed was his home, he met King Laïus and slew him in a sudden quarrel. He came next to Thebes and solved the riddle set him by the Sphinx who barred his way. Out of gratitude the Thebans made him their king and gave him Jocasta's hand in marriage. He reigned long in peace and honour, and she who, unknown to him, was his mother bore him two sons and two daughters. Then at last a plague broke out and the Thebans made enquiry once more of the oracle. It is at this point that Sophocles' tragedy opens. The messengers bring back the reply that the plague will cease when the murderer of Laïus has been driven from the land.

> But he, where is he? Where shall now be read
> The fading record of this ancient guilt?[1]

The action of the play consists in nothing other than the process of revealing, with cunning delays and ever-mounting excitement—a process that can be likened to the work of a psychoanalysis—that Oedipus himself is the murderer of Laïus, but further that he is the son of the murdered man and of Jocasta. Appalled at the abomination which he has unwittingly perpetrated, Oedipus blinds himself and forsakes his home. The oracle has been fulfilled.

Oedipus Rex is what is known as a tragedy of destiny. Its tragic effect is said to lie in the contrast between the supreme will of the gods and the vain attempts of mankind to escape the evil that threatens them. The lesson which, it is said, the deeply moved spectator should learn from the tragedy is submission to the divine will and realization of his own impotence. Modern dramatists have accordingly tried to achieve a similar tragic effect by weaving the same contrast into a plot invented by themselves. But the spectators have looked on unmoved while a curse or an oracle was fulfilled in spite of all the efforts of some innocent man: later tragedies of destiny have failed in their effect.

If *Oedipus Rex* moves a modern audience no less than it did the contemporary Greek one, the explanation can only be that its effect does not lie in the contrast between destiny and human will, but is to be looked for in the particular nature of the material on which that contrast is exemplified. There must be something which makes a voice within us ready to recognize the compelling force of destiny in the *Oedipus*, while we can

---

[1] [Lewis Campbell's translation (1883), line 108 f.]

dismiss as merely arbitrary such dispositions as are laid down in [Grillparzer's] *Die Ahnfrau* or other modern tragedies of destiny. And a factor of this kind is in fact involved in the story of King Oedipus. His destiny moves us only because it might have been ours—because the oracle laid the same curse upon us before our birth as upon him. It is the fate of all of us, perhaps, to direct our first sexual impulse towards our mother and our first hatred and our first murderous wish against our father. Our dreams convince us that that is so. King Oedipus, who slew his father Laïus and married his mother Jocasta, merely shows us the fulfilment of our own childhood wishes. But, more fortunate than he, we have meanwhile succeeded, in so far as we have not become psychoneurotics, in detaching our sexual impulses from our mothers and in forgetting our jealousy of our fathers. Here is one in whom these primaeval wishes of our childhood have been fulfilled, and we shrink back from him with the whole force of the repression by which those wishes have since that time been held down within us. While the poet, as he unravels the past, brings to light the guilt of Oedipus, he is at the same time compelling us to recognize our own inner minds, in which those same impulses, though suppressed, are still to be found. The contrast with which the closing Chorus leaves us confronted—

> . . . Fix on Oedipus your eyes,
> Who resolved the dark enigma, noblest champion and most wise.
> Like a star his envied fortune mounted beaming far and wide:
> Now he sinks in seas of anguish, whelmed beneath a raging tide . . .[1]

—strikes as a warning at ourselves and our pride, at us who since our childhood have grown so wise and so mighty in our own eyes. Like Oedipus, we live in ignorance of these wishes, repugnant to morality, which have been forced upon us by

---

[1] [Lewis Campbell's translation, line 1524 ff.]

Nature, and after their revelation we may all of us well seek to close our eyes to the scenes of our childhood.[1]

There is an unmistakable indication in the text of Sophocles' tragedy itself that the legend of Oedipus sprang from some primaeval dream-material which had as its content the distressing disturbance of a child's relation to his parents owing to the first stirrings of sexuality. At a point when Oedipus, though he is not yet enlightened, has begun to feel troubled by his recollection of the oracle, Jocasta consoles him by referring to a dream which many people dream, though, as she thinks, it has no meaning:

> Many a man ere now in dreams hath lain
> With her who bare him. He hath least annoy
> Who with such omens troubleth not his mind.[2]

To-day, just as then, many men dream of having sexual relations with their mothers, and speak of the fact with indignation and astonishment. It is clearly the key to the tragedy and the complement to the dream of the dreamer's father being dead. The story of Oedipus is the reaction of the imagination to these

---

[1] [*Footnote added* 1914:] None of the findings of psycho-analytic research has provoked such embittered denials, such fierce opposition—or such amusing contortions—on the part of critics as this indication of the childhood impulses towards incest which persist in the unconscious. An attempt has even been made recently to make out, in the face of all experience, that the incest should only be taken as 'symbolic'.—Ferenczi (1912) has proposed an ingenious 'over-interpretation' of the Oedipus myth, based on a passage in one of Schopenhauer's letters.—[*Added* 1919:] Later studies have shown that the 'Oedipus complex' which was touched upon for the first time in the above paragraphs in the *Interpretation of Dreams,* throws a light of undreamt-of importance on the history of the human race and the evolution of religion and morality. (See my *Totem and Taboo,* 1912–13 [Essay IV].)—[Actually the gist of this discussion of the Oedipus complex and of the *Oedipus Rex,* as well as of what follows on the subject of *Hamlet,* had already been put forward by Freud in a letter to Fliess as early as October 15th, 1897. (See Freud, 1950*a*, Letter 71.) A still earlier hint at the discovery of the Oedipus complex was included in a letter of May 31st, 1897. (Ibid., Draft N.)—The actual term 'Oedipus complex' seems to have been first used by Freud in his published writings in the first of his 'Contributions to the Psychology of Love' (1910*h*).]

[2] [Lewis Campbell's translation, line 982 ff.]

two typical dreams. And just as these dreams, when dreamt by adults, are accompanied by feelings of repulsion, so too the legend must include horror and self-punishment. Its further modification originates once again in a misconceived secondary revision of the material, which has sought to exploit it for theological purposes. The attempt to harmonize divine omnipotence with human responsibility must naturally fail in connection with this subject-matter just as with any other.

Another of the great creations of tragic poetry, Shakespeare's *Hamlet*, has its roots in the same soil as *Oedipus Rex*.[1] But the changed treatment of the same material reveals the whole difference in the mental life of these two widely separated epochs of civilization: the secular advance of repression in the emotional life of mankind. In the *Oedipus* the child's wishful phantasy that underlies it is brought into the open and realized as it would be in a dream. In *Hamlet* it remains repressed; and —just as in the case of a neurosis—we only learn of its existence from its inhibiting consequences. Strangely enough, the overwhelming effect produced by the more modern tragedy has turned out to be compatible with the fact that people have remained completely in the dark as to the hero's character. The play is built up on Hamlet's hesitations over fulfilling the task of revenge that is assigned to him; but its text offers no reasons or motives for these hesitations and an immense variety of attempts at interpreting them have failed to produce a result. According to the view which was originated by Goethe and is still the prevailing one to-day, Hamlet represents the type of man whose power of direct action is paralysed by an excessive development of his intellect. (He is 'sicklied o'er with the pale cast of thought'.) According to another view, the dramatist has tried to portray a pathologically irresolute character which might be classed as neurasthenic. The plot of the drama shows us, however, that Hamlet is far from being represented as a person incapable of taking any action. We see him doing so on two occasions: first in a sudden outburst

[1] [This paragraph was printed as a footnote in the first edition (1900) and included in the text from 1914 onward.]

of temper, when he runs his sword through the eavesdropper behind the arras, and secondly in a premeditated and even crafty fashion, when, with all the callousness of a Renaissance prince, he sends the two courtiers to the death that had been planned for himself. What is it, then, that inhibits him in fulfilling the task set him by his father's ghost? The answer, once again, is that it is the peculiar nature of the task. Hamlet is able to do anything—except take vengeance on the man who did away with his father and took that father's place with his mother, the man who shows him the repressed wishes of his own childhood realized. Thus the loathing which should drive him on to revenge is replaced in him by self-reproaches, by scruples of conscience, which remind him that he himself is literally no better than the sinner whom he is to punish. Here I have translated into conscious terms what was bound to remain unconscious in Hamlet's mind; and if anyone is inclined to call him a hysteric, I can only accept the fact as one that is implied by my interpretation. The distaste for sexuality expressed by Hamlet in his conversation with Ophelia fits in very well with this: the same distaste which was destined to take possession of the poet's mind more and more during the years that followed, and which reached its extreme expression in *Timon of Athens*. For it can of course only be the poet's own mind which confronts us in Hamlet. I observe in a book on Shakespeare by Georg Brandes (1896) a statement that *Hamlet* was written immediately after the death of Shakespeare's father (in 1601), that is, under the immediate impact of his bereavement and, as we may well assume, while his childhood feelings about his father had been freshly revived. It is known, too, that Shakespeare's own son who died at an early age bore the name of 'Hamnet', which is identical with 'Hamlet'. Just as *Hamlet* deals with the relation of a son to his parents, so *Macbeth* (written at approximately the same period) is concerned with the subject of childlessness. But just as all neurotic symptoms, and, for that matter, dreams, are capable of being 'over-interpreted' and indeed need to be, if they are to be fully understood, so all genuinely creative writings are the product of more than a single motive and more than a single impulse in the poet's mind, and are open to more than a single interpretation. In what I have written I have only attempted to interpret the deepest layer of impulses in the mind of the creative writer.[1]

---

[1] [*Footnote added* 1919:] The above indications of a psycho-analytic explanation of *Hamlet* have since been amplified by Ernest Jones and defended against the alternative views put forward in the literature of the subject. (See Jones, 1910*a* [and, in a completer form, 1949].)—[*Added* 1930:] Incidentally, I have in the meantime ceased to believe that the author of Shakespeare's works was the man from Stratford. [See Freud, 1930*e*.]—[*Added* 1919:] Further attempts at an analysis of *Macbeth* will be found in a paper of mine [Freud, 1916*d*] and in one by Jekels (1917).—[The first part of this footnote was included in a different form in the edition of 1911 but omitted from 1914 onwards: 'The views on the problem of *Hamlet* contained in the above passage have since been confirmed and supported with fresh arguments in an extensive study by Dr. Ernest Jones of Toronto (1910*a*). He has also pointed out the relation between the material in *Hamlet* and the myths of the birth of heroes discussed by Rank (1909).'—Freud further discussed *Hamlet* in a posthumously published sketch dealing with 'Psychopathic Characters on the Stage' (1942*b*), probably written in 1905 or 1906.]

BERTRAND RUSSELL

LORD BERTRAND ARTHUR WILLIAM RUSSELL (1872-1970), recipient of a Nobel Prize and the Order of Merit, in his time was the most distinguished philosopher of the English-speaking world, perhaps of the world at large. His technical contributions to symbolic logic and the philosophy of science were of the highest order. In addition, he wrote an astounding number of popular books, their subjects ranging from relativity and quantum theory to marriage, happiness, education, and politics.

Popular or technical, Russell's style is never dull or opaque, and his dry humor often catches the reader by surprise. The most notorious of his flippant footnotes occurs in a discussion of Köhler's well-known experiments in gestalt learning. The experiments suggested that when German apes are faced with the problem of obtaining a banana, they sit down to meditate until they have a sudden flash of insight, whereas American apes, Russell observes, rush about frantically trying to solve the problem by trial and error. A footnote on the word "banana" reads: "Called by Köhler 'the objective' because the word 'banana' is too humble for a learned work. The pictures disclose the fact that 'the objective' was a mere banana."

Russell's long life was Byronic and colorful. His *Introduction to Mathematical Philosophy* was written during the six months he spent in prison for his pacifist views at the time of the First World War. In the twenties and thirties his "premature anti-Communism" caused him to be snubbed by friends on the left until they eventually discovered that Bolshevism was just as bad as Russell had always said it was. In this country he was ousted from two institutions: from the College of the City of New York after a public ruckus over his atheism and views on romantic love, and from the Barnes Foundation in Philadelphia where the terrible tempered Mr. Barnes objected to the clicking of young Mrs. Russell's knitting needles as she sat in the back of her husband's classes. An airplane in which Russell was traveling once crashed off the coast into any icy sea. The 77-year-old Earl was picked up later, still wearing his heavy overcoat and swimming calmly toward the shore.

Any of several dozen essays by Russell would have been appropriate for this volume. The two reprinted here were selected partly because they have not yet appeared in book form. The first was published by *The New York Times Magazine*. The second is from *The New Leader*, an American liberal weekly to which Russell often contributed.

BERTRAND RUSSELL

The Science to Save Us from Science

SINCE THE BEGINNING of the seventeenth century scientific discovery and invention have advanced at a continually increasing rate. This fact has made the last three hundred and fifty years profoundly different from all previous ages. The gulf separating man from his past has widened from generation to generation, and finally from decade to decade. A reflective person, meditating on the extinction of trilobites, dinosaurs and mammoths, is driven to ask himself some very disquieting questions. Can our species endure so rapid a change? Can the habits which insured survival in a comparatively stable past still suffice amid the kaleidoscopic scenery of our time? And, if not, will it be possible to change ancient patterns of behavior as quickly as the inventors change our material environment? No one knows the answer, but it is possible to survey probabilities, and to form some hypotheses as to the alternative directions that human development may take.

The first question is: Will scientific advance continue to grow more and more rapid, or will it reach a maximum speed and then begin to slow down?

The discovery of scientific method required genius, but its utilization requires only talent. An intelligent young scientist, if he gets a job giving access to a good laboratory, can be pretty certain of finding out something of interest, and may stumble upon some new fact of immense importance. Science, which was still a rebellious force in the early seventeenth century, is now integrated with the life of the community by the support of governments and universities. And as its importance becomes more evident, the number of people employed in scientific research continually increases. It would seem to follow that, so long as social and economic conditions do not become adverse, we may expect the rate of scientific advance to be maintained, and even increased, until some new limiting factor intervenes.

It might be suggested that, in time, the amount of knowledge needed before a new discovery could be made would become so great as to absorb all the best years of a scientist's life, so that by the time he reached the frontier of knowledge he would be senile. I suppose this may happen some day, but that day is certainly very distant. In the first place, methods of teaching improve. Plato thought that students in his academy would have to spend ten years learning what was then known of mathematics; nowadays any mathematically minded schoolboy learns much more mathematics in a year.

In the second place, with increasing specialization, it is possible to reach the frontier of knowledge along a narrow path, involving much less labor than a broad highway. In the third place, the frontier is not a circle but an irregular contour, in some places not far from the center. Mendel's epoch-making discovery required little previous knowledge; what it needed was a life of elegant leisure spent in a garden. Radio-activity was discovered by the fact that some specimens of pitchblende were unexpectedly found to have photographed themselves in the dark. I do not think, therefore, that purely intellectual reasons will slow up scientific advances for a very long time to come.

There is another reason for expecting scientific advance to continue, and that is that it increasingly attracts the best brains. Leonardo da Vinci was equally pre-eminent in art and science, but it was from art that he derived his greatest fame. A man of similar endowments living at the present day would almost certainly hold some post which would require his giving all his time to science; if his politics were orthodox, he would probably be engaged in devising the hydrogen bomb, which our age would consider more useful than his pictures. The artist, alas, has not the status that he once had. Renaissance princes might compete for Michelangelo; modern states compete for nuclear physicists.

There are considerations of quite a different sort which might lead to an expectation of scientific retrogression. It may be held that science itself generates explosive forces which will, sooner or later, make it impossible to preserve the kind of society in which science can flourish. This is a large and different question, to which no confident answer can be given. It is a very important question, which deserves to be examined. Let us therefore see what is to be said about it.

Industrialism, which is in the main a product of science, has provided a certain way of life and a certain outlook on the world. In America and Britain, the oldest industrial countries, this outlook and this way of life have come gradually, and the population has been able to adjust itself to them without any violent breach of continuity. These countries, accordingly, did not develop dangerous psychological stresses. Those who preferred the old ways could remain on the land, while the more adventurous could migrate to the new centers of industry. There they found pioneers who were compatriots, who shared in the main the general outlook of their neighbors. The only protests came from men like Carlyle and Ruskin, whom everybody at once praised and disregarded.

It was a very different matter when industrialism and science, as well-developed systems, burst violently upon countries hitherto ignorant of both, especially since they came as something foreign, demanding imitation of enemies and disruption of ancient national habits. In varying degrees this shock has been endured by Germany, Russia, Japan, India and the natives of Africa. Everywhere it has caused and is causing upheavals of one sort or another, of which as yet no one can foresee the end.

The earliest important result of the impact of industrialism on Germans was the Communist Manifesto. We think of this now as the Bible of one of the two powerful groups into which the world is divided, but it is worth while to think back to its origin in 1848. It then shows itself as an expression of admiring horror by two young university students from a pleasant and peaceful cathedral city, brought roughly and without intellectual preparation into the hurly-burly of Manchester competition.

Germany, before Bismarck had "educated" it, was a deeply religious country, with a quiet, exceptional sense of public duty. Competition, which the British regarded as essential to efficiency, and which Darwin elevated to an almost cosmic dignity, shocked the Germans, to whom service to the state seemed the obviously right moral ideal. It was therefore natural that they should fit industrialism into a framework of nationalism or socialism. The Nazis combined both. The somewhat insane and frantic character of German industrialism and the policies it inspired is due to its foreign origin and its sudden advent.

Marx's doctrine was suited to countries where industrialism was new. The German Social Democrats abandoned his dogmas when their country became industrially adult. But by that time Russia was where Germany had been in 1848, and it was natural that Marxism should find a new home. Stalin, with great skill, has combined the new revolutionary creed with the traditional belief in "Holy Russia" and the "Little Father." This is as yet the most notable example of the arrival of science in an environment that is not ripe for it. China bids fair to follow suit.

Japan, like Germany, combined modern technique with worship of the state. Educated Japanese abandoned as much of their ancient way of life as was necessary in order to secure industrial and military efficiency. Sudden change produced collective hysteria, leading to insane visions of world power unrestrained by traditional pieties.

These various forms of madness—communism, nazism, Japanese imperialism—are the natural result of the impact of science on nations with a strong pre-scientific culture. The effects in Asia are still at an early stage. The effects upon the native races of Africa have hardly begun. It is therefore unlikely that the world will recover sanity in the near future.

The future of science—nay more, the future of mankind—depends upon whether it will be possible to restrain these various collective hysterias until the populations concerned have had time to adjust themselves to the new scientific environment. If such adjustment proves impossible, civilized society will disappear, and science will be only a dim memory. In the Dark Ages science was not distinguished from sorcery, and it is not impossible that a new Dark Ages may revive this point of view.

The danger is not remote; it threatens within the next few years. But I am not now concerned with such immediate issues. I am concerned with the wider question: Can a society based, as ours is, on science and scientific technique, have the sort of stability that many societies had in the past, or is it bound to develop explosive forces that will destroy it? This question takes us beyond the sphere of science into that of ethics and moral codes and the imaginative understanding of mass psychology. This last is a matter which political theorists have quite unduly neglected.

Let us begin with moral codes. I will illustrate the problem by a somewhat trivial illustration. There are those who think it wicked to smoke tobacco, but they are mostly people untouched by science. Those whose outlook has been strongly influenced by science usually take the view that smoking is neither a vice nor a virtue. But when I visited a Nobel works, where rivers of nitro-glycerine flowed like water, I had to leave all matches at the entrance, and it was obvious that to smoke inside the works would be an act of appalling wickedness.

This instance illustrates two points: first, that a scientific outlook tends to make some parts of traditional moral codes appear superstitious and irrational; second, that by creating a new environment science creates new duties, which may happen to coincide with those that have been discarded. A world containing hydrogen bombs is like one containing rivers of nitro-glycerine; actions elsewhere harmless may become dangerous in the highest degree. We need therefore, in a scientific world, a somewhat different moral code from the one inherited from the past. But to give to a new moral code sufficient compulsive force to restrain actions formerly considered harmless is not easy, and cannot possibly be achieved in a day.

As regards ethics, what is important is to realize the new dangers and to consider what ethical outlook will do most to diminish them. The most important new facts are that the world is more unified than it used to be, and that communities at war with each other have more power of inflicting mutual disaster than at any former time. The question of power has a new importance. Science has enormously increased human power, but has not increased it without limit. The increase of power brings an increase of responsibility; it brings also a danger of arrogant self-assertion, which can only be averted by continuing to remember that man is not omnipotent.

The most influential sciences, hitherto, have been physics and chemistry; biology is just beginning to rival them. But before very long psychology and especially mass psychology, will be recognized as the most important of all sciences from the standpoint of human welfare. It is obvious that populations have dominant moods, which change from time to time according to their circumstances. Each mood has a corresponding ethic. Nelson inculcated these ethical principles on midshipmen: to tell the truth, to shoot straight, and to hate a Frenchman as you would the devil. This last was chiefly because the English were angry with France for intervening on the side of America. Shakespeare's Henry V says:

If it be a sin to covet honor,
I am the most offending soul alive.

This is the ethical sentiment that goes with aggressive imperialism: "honor" is proportional to the number of harmless people you slaughter. A great many sins may be excused under the name of "patriotism." On the other hand complete powerlessness suggests humility and submission as the greatest virtues; hence the vogue of stoicism in the Roman Empire and of Methodism among the English poor in the early nineteenth century. When, however, there is a chance of successful revolt, fierce vindictive justice suddenly becomes the dominant ethical principle.

In the past, the only recognized way of inculcating moral precepts has been by preaching. But this method has very definite limitations: it is notorious that, on the average, sons of clergy are not morally superior to other people. When science has mastered this field, quite different methods will be adopted. It will be known what circumstances generate what moods, and what moods incline men to what ethical systems. Governments will then decide what sort of morality their subjects are to have and their subjects will adopt what the Government favors, but will do so under the impression that they are exercising free will. This may sound unduly cynical, but that is only because we are not yet accustomed to applying science to the human mind. Science has powers for evil, not only physically, but mentally: the hydrogen bomb can kill the body, and government propaganda (as in Russia) can kill the mind.

In view of the terrifying power that science is conferring on governments, it is necessary that those who control governments should have enlightened and intelligent ideals, since otherwise they can lead mankind to disaster.

I call an ideal "intelligent" when it is possible to approximate to it by pursuing it. This is by no means sufficient as an ethical criterion, but it is a test by which many aims can be condemned. It cannot be supposed that Hitler desired the fate which he brought upon his country and himself, and yet it was pretty certain that this would be the result of his arrogance. Therefore the ideal of "Deutschland ueber Alles" can be condemned as unintelligent. (I do not mean to suggest that this is its only defect.) Spain, France, Germany and Russia have successively sought world dominion: three of them have endured defeat in consequence, but their fate has not inspired wisdom.

Whether science—and indeed civilization in general—can long survive depends upon psychology, that is to say, it depends upon what human beings desire. The human beings concerned are rulers in totalitarian countries, and the mass of men and women in democracies. Political passions determine political conduct much more directly than is often supposed. If men desire victory more than cooperation, they will think victory possible.

But if hatred so dominates them that they are more anxious to see their enemies killed than to keep their own children alive, they will discover all kinds of "noble" reasons in favor of war. If they resent inferiority or wish to preserve superiority, they will have the sentiments that promote the class war. If they are bored beyond a point, they will welcome excitement even of a painful kind.

Such sentiments, when widespread, determine the policies and decisions of nations. Science can, if rulers so desire, create sentiments which will avert disaster and facilitate cooperation. At present there are powerful rulers who have no such wish. But the possibility exists, and science can be just as potent for good as for evil. It is not science, however, which will determine how science is used.

Science, by itself, cannot supply us with an ethic. It can show us how to achieve a given end, and it may show us that some ends cannot be achieved. But among ends that can be achieved our choice must be decided by other than purely scientific considerations. If a man were to say, "I hate the human race, and I think it would be a good thing if it were exterminated," we could say, "Well, my dear sir, let us begin the process with you." But this is hardly argument, and no amount of science could prove such a man mistaken.

But all who are not lunatics are agreed about certain things: That it is better to be alive than dead, better to be adequately fed than starved, better to be free than a slave. Many people desire those things only for themselves and their friends; they are quite content that their enemies should suffer. These people can be refuted by science: Mankind has become so much one family that we cannot insure our own prosperity except by insuring that of everyone else. If you wish to be happy yourself, you must resign yourself to seeing others also happy.

Whether science can continue, and whether, while it continues, it can do more good than harm, depends upon the capacity of mankind to learn this simple lesson. Perhaps it is necessary that all should learn it, but it must be learned by all who have great power, and among those some still have a long way to go.

BERTRAND RUSSELL

The Greatness of Albert Einstein

EINSTEIN WAS indisputably one of the greatest men of our time. He had in a high degree the simplicity characteristic of the best men of science—a simplicity which comes of a single-minded desire to know and understand things that are completely impersonal. He had also the faculty of not taking familiar things for granted. Newton wondered why apples fall; Einstein expressed "surprised thankfulness" that four equal rods can make a square, since, in most of the universes that he could imagine, there would be no such things as squares.

He showed greatness also in his moral qualities. In private, he was kindly and unassuming; toward colleagues he was (so far as I could see) completely free from jealousy, which is more than can be said of Newton or Leibniz. In his later years, relativity was more or less eclipsed, in scientific interest, by quantum theory, but I never discovered any sign that this vexed him. He was profoundly interested in world affairs. At the end of the First World War, when I first came in contact with him, he was a pacifist, but Hitler led him (as he led me) to abandon this point of view. Having previously thought of himself as a citizen of the world, he found that the Nazis compelled him to think of himself as a Jew, and to take up the cause of the Jews throughout the world. After the end of the Second World War, he joined the group of American scientists who were attempting to find a way of avoiding the disasters to mankind that are threatened as a result of the atomic bomb.

After Congressional committees in America began their inquisitorial investigations into supposed subversive activities, Einstein wrote a well publicized letter urging that all men in academic posts should refuse to testify before these committees or before the almost equally tyrannical boards set up by some universities. His argument for this advice was that, under the Fifth Amendment, no man is obliged to answer a question if the answer will incriminate him, but that the purpose of this Amendment had been defeated by the inquisitors, since they held that refusal to answer may be taken as evidence of guilt. If Einstein's policy had been followed even in cases where it was absurd to presume guilt, academic freedom would have greatly profited. But, in the general *sauve qui peut*, none of the "innocent" listened to him. In these various public activities, he has been completely self-effacing and only anxious to find ways of saving the human race from the misfortunes brought about by its own follies. But while the world applauded him as a man of science, in practical affairs his wisdom was so simple and so profound as to seem to the sophisticated like mere foolishness.

Although Einstein has done much important work outside the theory of relativity, it is by this theory that he is most famous—and rightly, for it is of fundamental significance both for science and for philosophy. Many people (including myself) have attempted popular accounts of this theory, and I will not add to their number on this occasion. But I will try to say a few words as to how the theory affects our view of the universe. The theory, as everyone knows, appeared in two stages: the special theory in 1905, and the general theory in 1915. The special theory was important both in science and philosophy—first, because it accounted for the result of the Michelson-Morley experiment, which had puzzled the world for thirty years; secondly, because it explained the increase of mass with velocity, which had been observed in electrons; thirdly, because it led to the interchangeability of mass and energy, which has become an essential principle in physics. These are only some of the ways in which it was scientifically important.

Philosophically, the special theory demanded a revolution in deeply rooted ways of thought, since it compelled a change in our conception of the spatio-temporal structure of the world. Structure is what is most significant in our knowledge of the physical world, and for ages structure had been conceived as depending upon two different manifolds, one of space, the other of time. Einstein showed that, for reasons partly experimental and partly logical, the two must be replaced by one which he called "space-time." If two events happen in different places, you cannot say, as was formerly supposed, that they are separated by so many miles and minutes, because different observers, all equally careful, will make different estimates of the miles and minutes, all equally legitimate. The only thing that is the same for all observers is what is called "interval," which is a sort of combination of space-distance and time-distance as previously estimated.

The general theory has a wider sweep than the special theory, and is scientifically more important. It is primarily a theory of gravitation. No advance whatever had been made in explaining gravitation during the 230 years since Newton, although the action at a distance that it seems to demand had always been repugnant. Einstein made gravitation part of geometry; he said that it was due to the character of space-time. There is a law called the "Principle of Least Action," according to which a body, in going from one place to another, chooses always the easiest route, which may not be a straight line: It may pay you to avoid mountain-tops and deep valleys. According to Einstein (to use crude language, misleading if taken literally), space-time is full of mountains and valleys, and that is why planets do not move in straight lines. The sun is at the top of a hill, and a lazy planet prefers going round the hill to climbing up to the summit. There were some very delicate experimental tests by which it could be decided whether Einstein or Newton fitted the facts more accurately. The observations came out on Einstein's side, and almost everybody except the Nazis accepted his theory.

Some odd things have emerged as a consequence of the general theory of relativity. It appears that the universe is of finite size, although unbounded. (Do not attempt to understand this unless you have studied non-Euclidean geometry.) It appears also that the universe is continually getting bigger. Theory shows that it must be always getting bigger, or always getting smaller; observation of distant nebulae shows that it is getting bigger. Our present universe seems to have begun about 2 billion years ago; what, if anything, there was before that, it is impossible to conjecture.

I suppose that, in the estimation of the general public, Einstein is still reckoned as a revolutionary innovator. Among physicists, however, he has become the leader of the Old Guard. This is due to his refusal to accept some of the innovations of quantum theory. Heisenberg's principle of indeterminacy, along with other principles of that theory, has had very curious results. It seems that individual occurrences in atoms do not obey strict laws, and that the observed regularities in the world are only statistical. What we know about the behavior of matter, according to this view, is like what insurance companies know about mortality. Insurance companies do not know and do not care which of the individuals who insured their lives will die in any given year. All that matters to them is the statistical average of mortality. The regularities to which classical physics accustomed us are, we are now told, of this merely statistical sort. Einstein never accepted this view. He continued to believe that there are laws, though as yet they have not been ascertained, which determine the behavior of individual atoms. It would be exceedingly rash for anyone who is not a professional physicist to allow himself an opinion on this matter until the physicists are all agreed, but I think it must be conceded that on this matter the bulk of competent opinion was opposed to Einstein. This is all the more remarkable in view of the fact that he had done epoch-making work in quantum theory, which would have put him in the first rank among physicists even if he had never thought of the theory of relativity.

Quantum theory is more revolutionary than the theory of relativity, and I do not think that its power of revolutionizing our conceptions of the physical world is yet completed.

Its imaginative effects are very curious. Although it has given us new powers of manipulating matter, including the sinister powers displayed in the atom and hydrogen bombs, it has shown us that we are ignorant of many things which we thought we knew. Nobody before quantum theory doubted that at any given moment a particle is at some definite place and moving with some definite velocity. This is no longer the case. The more accurately you determine the place of a particle, the less accurate will be its velocity; the more accurately you determine the velocity, the less accurate will be its position. And the particle itself has become something quite vague, not a nice little billiard ball as it used to be. When you think you have caught it, it produces a convincing alibi as a wave and not a particle. In fact, you know only certain equations of which the interpretation is obscure.

This point of view was distasteful to Einstein, who struggled to remain nearer to classical physics. In spite of this, he was the first to open the imaginative vistas which have revolutionized science during the present century. I will end as I began: He was a great man, perhaps the greatest of our time.

ALBERT EINSTEIN

IT IS FITTING that this anthology, opening with a selection by Darwin, should almost close with a piece by Albert Einstein (1879-1955), for the theories of evolution and relativity mark the two greatest turning points in the history of modern science. The brief article selected here was written by Einstein in 1946 for a popular science magazine. It explains clearly and simply, without a wasted word, his famous formula for the equivalence of mass and energy. It is of course the equation that measures the energy released by the loss of mass in an atomic or hydrogen bomb. Einstein arrived at this formula long before the German discovery of uranium fission that prompted his famous letter to President Roosevelt urging federal backing of research on nuclear weapons. The formula was not the product of a laboratory. It came from a giant electronic brain that functioned behind a pair of kindly, humorous eyes.

Henry Miller once suggested in a misanthropic novel that our planet, which he believed to be, like all of us, rotting with the cancer of time, is in need of a fitting *coup de grâce*. Why, Miller wondered, did not someone insert a bomb into a crevice of the world and blow it to smithereens? In view of the fact that mankind now possesses the means—if not to do precisely what Miller suggested, at least to destroy all living things—Einstein's final sentence comes as something of an understatement. Yet it carries within it the double trumpet note of doom and triumph that is now sounding in our ears.

ALBERT EINSTEIN

$E = mc^2$

IN ORDER TO UNDERSTAND the law of the equivalence of mass and energy, we must go back to two conservation or "balance" principles which, independent of each other, held a high place in pre-relativity physics. These were the principle of the conservation of energy and the principle of the conservation of mass. The first of these, advanced by Leibnitz as long ago as the seventeenth century, was developed in the nineteenth century essentially as a corollary of a principle of mechanics.

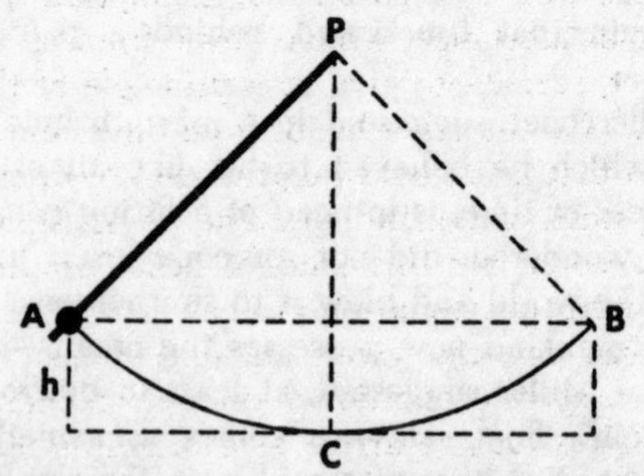

Drawing from Dr. Einstein's manuscript.

Consider, for example, a pendulum whose mass swings back and forth between the points A and B. At these points the mass m is higher by the amount h than it is at C, the lowest point of the path (see drawing). At C, on the other hand, the lifting height has disappeared and instead of it the mass has a velocity v. It is as though the lifting height could be converted entirely into velocity, and vice versa. The exact relation would be expressed as $mgh = \frac{m}{2} v^2$, with g representing the acceleration of gravity. What is interesting here is that this relation is independent of both the length of the pendulum and the form of the path through which the mass moves.

The significance is that something remains constant throughout the process, and that something is energy. At A and at B it is an energy of position, or "potential" energy; at C it is an energy of motion, or "kinetic" energy. If this concept is correct, then the sum $mgh + m \frac{v^2}{2}$ must have the same value for any position of the pendulum, if h is understood to represent the height above C, and v the velocity at that point in the pendulum's path. And such is found to be actually the case. The generalization of this principle gives us the law of the conservation of mechanical energy. But what happens when friction stops the pendulum?

The answer to that was found in the study of heat phenomena. This study, based on the assumption that heat is an indestructible substance which flows from a warmer to a colder object, seemed to give us a principle of the "conservation of heat." On the other hand, from time immemorial it has been known that heat could be produced by friction, as in the fire-making drills of the Indians. The physicists were for long unable to account for this kind of heat "production." Their difficulties were overcome only when it was successfully established that, for any given amount of heat produced by friction, an exactly proportional amount of energy had to be expended. Thus did we arrive at a principle of the "equivalence of work and heat." With our pendulum, for example, mechanical energy is gradually converted by friction into heat.

In such fashion the principles of the conservation of mechanical and thermal energies were merged into one. The physicists were thereupon persuaded that the conservation principle could be further extended to take in chemical and electromagnetic processes—in short, could be applied to all fields. It appeared that in our physical system there was a sum total of energies that remained constant through all changes that might occur.

Now for the principle of the conservation of mass. Mass is defined by the resistance that a body opposes to its acceleration (inert mass). It is also measured by the weight of the body (heavy mass). That these two radically different definitions lead to the same value for the mass of a body is, in itself, an astonishing fact. According to the principle—namely, that masses remain unchanged under any physical or chemical changes—the mass appeared to be the essential (because unvarying) quality of matter. Heating, melting, vaporization, or combining into chemical compounds would not change the total mass.

Physicists accepted this principle up to a few decades ago. But it proved inadequate in the face of the special theory of relativity. It was therefore merged with the energy principle—just as, about 60 years before, the principle of the conservation of mechanical energy had been combined with the principle of the conservation of heat. We might say that the principle of the conservation of energy, having previously swallowed up that of the conservation of heat, now proceeded to swallow that of the conservation of mass—and holds the field alone.

It is customary to express the equivalence of mass and energy (though somewhat inexactly) by the formula $E = mc^2$, in which c represents the velocity of light, about 186,000 miles per second. E is the energy that is contained in a stationary body; m is its mass. The energy that belongs to the mass m is equal to this mass, multiplied by the square of the enormous speed of light—which is to say, a vast amount of energy for every unit of mass.

But if every gram of material contains this tremendous energy, why did it go so long unnoticed? The answer is simple enough: so long as none of the energy is given off externally, it cannot be observed. It is as though a man who is fabulously rich should never spend or give away a cent; no one could tell how rich he was.

Now we can reverse the relation and say that an increase of E in the amount of energy must be accompanied by an increase of $\frac{E}{c^2}$ in the mass. I can easily supply energy to the mass—for instance, if I heat it by 10 degrees. So why not measure the mass increase, or weight increase, connected with this change? The trouble here is that in the mass increase the enormous factor c^2 occurs in the denominator of the fraction. In such a case the increase is too small to be measured directly; even with the most sensitive balance.

For a mass increase to be measurable, the change of energy per mass unit must be enormously large. We know of only one sphere in which such amounts of energy per mass unit are released: namely, radioactive disintegration. Schematically, the process goes like this: An atom of the mass M splits into two atoms of the mass M′ and M″, which separate with tremendous kinetic energy. If we imagine these two masses as brought to rest—that is, if we take this energy of motion from them—then, considered together, they are essentially poorer in energy than was the original atom. According to the equivalence principle, the mass sum $M' + M''$ of the disintegration products must also be somewhat smaller than the original mass M of the disintegrating atom—in contradiction to the old principle of the conservation of mass. The relative difference of the two is on the order of 1/10 of one percent.

Now, we cannot actually weigh the atoms individually. However, there are indirect methods for measuring their weights exactly. We can likewise determine the kinetic energies that are transferred to the disintegration products M′ and M″. Thus it has become possible to test and confirm the equivalence formula. Also, the law permits us to calculate in advance, from precisely determined atom weights, just how much energy will be released with any atom disintegration we have in mind. The law says nothing, of course, as to whether—or how—the disintegration reaction can be brought about.

What takes place can be illustrated with the help of our

rich man. The atom M is a rich miser who, during his life, gives away no money (*energy*). But in his will he bequeaths his fortune to his sons M′ and M″, on condition that they give to the community a small amount, less than one thousandth of the whole estate (*energy or mass*). The sons together have somewhat less than the father had (*the mass sum M′ + M″ is somewhat smaller than the mass M of the radioactive atom*). But the part given to the community, though relatively small, is still so enormously large (*considered as kinetic energy*) that it brings with it a great threat of evil. Averting that threat has become the most urgent problem of our time.

LEWIS THOMAS

LEWIS THOMAS is a research pathologist who has held a variety of administrative posts at hospitals and medical schools around the nation. For seven years he was president of New York's Memorial Sloan-Kettering Cancer Center before he became its chancellor, the position he still holds.

The public became aware of a new luminary on the science-writing scene when Thomas's first book, *The Lives of a Cell,* became a best-seller and won a National Book Award. Other volumes soon followed, based on the short essays he contributes regularly to the *New England Journal of Medicine.* Behind the modest, unpretentious prose of these books is a quiet wisdom, a deep understanding of science, an intense enthusiasm for research, and a keen awareness of the unpredictable discoveries that lie ahead.

Thomas's autobiography, *The Youngest Science: Notes of a Medicine Watcher,* tells of his boyhood in Flushing, New York, where his father was a compassionate general practitioner. Why does Thomas call medicine the youngest science? Because, he reminds us, only in very recent times have doctors learned enough to be able to treat diseases with something more than placebos—those harmless substances made effective by the confidence patients have in doctors who were once willing to take the time to listen to them, who wrote mysterious prescriptions in Latin, and who kept them cheerful until a disease had run its natural course.

The essay I have chosen was originally a commencement address in 1983. After its appearance in the *New York Times,* it became a chapter in Thomas's *Late Night Thoughts on Listen-*

ing to Mahler's Ninth Symphony. This delightful essay, like so many others by Thomas, is suffused with Chestertonian humility and wonder, yet touched by a strong Wellsian sense of foreboding—a fear that the glorious march of science may finally bring to an inglorious end the not-so-intelligent life on that complicated stone ball that Thomas lists as the first of his seven wonders.

LEWIS THOMAS

Seven Wonders

A WHILE AGO I received a letter from a magazine editor inviting me to join six other people at dinner to make a list of the Seven Wonders of the Modern World, to replace the seven old, out-of-date Wonders. I replied that I couldn't manage it, not on short order anyway, but still the question keeps hanging around in the lobby of my mind. I had to look up the old biodegradable Wonders, the Hanging Gardens of Babylon and all the rest, and then I had to look up that word "wonder" to make sure I understood what it meant. It occurred to me that if the magazine could get any seven people to agree on a list of any such seven things you'd have the modern Seven Wonders right there at the dinner table.

Wonder is a word to wonder about. It contains a mixture of messages: something marvelous and miraculous, suprising, raising unanswerable questions about itself, making the observer wonder, even raising skeptical questions like, "I *wonder* about that." Miraculous and marvelous are clues; both words come from an ancient Indo-European root meaning simply to smile or to laugh. Anything wonderful is something to smile in the presence of, in admiration (which, by the way, comes from the same root, along with, of all telling words, "mirror").

I decided to try making a list, not for the magazine's dinner party but for this occasion: seven things I wonder about the most.

I shall hold the first for the last, and move along.

My Number Two Wonder is a bacterial species never seen on

the face of the earth until 1982, creatures never dreamed of before, living violation of what we used to regard as the laws of nature, things literally straight out of Hell. Or anyway what we used to think of as Hell, the hot unlivable interior of the earth. Such regions have recently come into scientific view from the research submarines designed to descend twenty-five hundred meters or more to the edge of deep holes in the sea bottom, where open vents spew superheated seawater in plumes from chimneys in the earth's crust, known to oceanographic scientists as "black smokers." This is not just hot water, or steam, or even steam under pressure as exists in a laboratory autoclave (which we have relied upon for decades as the surest way to destroy all microbial life). This is extremely hot water under extremely high pressure, with temperatures in excess of 300 degrees centigrade. At such heat, the existence of life as we know it would be simply inconceivable. Proteins and DNA would fall apart, enzymes would melt away, anything alive would die instantaneously. We have long since ruled out the possibility of life on Venus because of that planet's comparable temperature; we have ruled out the possibility of life in the earliest years of this planet, four billion or so years ago, on the same ground.

B. J. A. Baross and J. W. Deming have recently discovered the presence of thriving colonies of bacteria in water fished directly from these deep-sea vents. Moreover, when brought to the surface, encased in titanium syringes and sealed in pressurized chambers heated to 250 degrees centigrade, the bacteria not only survive but reproduce themselves enthusiastically. They can be killed only by chilling them down in boiling water.

And yet they look just like ordinary bacteria. Under the electron microscope they have the same essential structure—cell walls, ribosomes, and all. If they were, as is now being suggested, the original archebacteria, ancestors of us all, how did they or their progeny ever learn to cool down? I cannot think of a more wonderful trick.

My Number Three Wonder is *oncideres,* a species of beetle encountered by a pathologist friend of mine who lives in Houston and has a lot of mimosa trees in his backyard. This beetle is not new, but it qualifies as a Modern Wonder because of the exceedingly modern questions raised for evolutionary biologists about the three consecutive things on the mind of the female of the species. Her first thought is for a mimosa tree, which she finds and climbs, ignoring all other kinds of trees in the vicinity. Her second thought is for the laying of eggs, which she does by crawling out on a limb, cutting a longitudinal slit with her mandible and depositing her eggs beneath the slit. Her third and last thought concerns the welfare of her offspring; beetle larvae cannot survive in live wood, so she backs up a foot or so and cuts a neat circular girdle all around the limb, through the bark and down into the cambium. It takes her eight hours to finish this cabinetwork. Then she leaves and where she goes I do not know. The limb dies from the girdling, falls to the ground in the next breeze, the larvae feed and grow into the next generation, and the questions lie there unanswered. How on earth did these three linked thoughts in her mind evolve together in evolution? How could any one of the three become fixed as beetle behavior by itself, without the other two? What are the odds favoring three totally separate bits of behavior—liking a particular tree, cutting a slit for eggs, and then girdling the limb—happening together by random chance among a beetle's genes? Does this smart beetle know what she is doing? And how did the mimosa tree enter the picture in its evolution? Left to themselves, unpruned, mimosa trees have a life expectancy of twenty-five to thirty years. Pruned each year, which is what the beetle's girdling labor accomplishes, the tree can flourish for a century. The mimosa-beetle relationship is an elegant example of symbiotic partnership, a phenomenon now recognized as pervasive in nature. It is good for us to have around on our intellectual mantelpiece such creatures as this insect and its friend the tree, for they keep reminding us how little we know about nature.

The Fourth Wonder on my list is an infectious agent known as the scrapie virus, which causes a fatal disease of the brain in sheep, goats, and several laboratory animals. A close cousin of scrapie is the C-J virus, the cause of some cases of senile dementia in human beings. These are called "slow viruses," for the excellent reason that an animal exposed to infection today will not become ill until a year and a half or two years from today. The agent, whatever it is, can propagate itself in abundance from a few infectious units today to more than a billion next year. I use the phrase "whatever it is" advisedly. Nobody has yet been able to find any DNA or RNA in the scrapie or C-J viruses. It may be there, but if so it exists in amounts too small to detect. Meanwhile, there is plenty of protein, leading to a serious proposal that the virus may indeed be *all* protein. But protein, so far as we know, does not replicate itself all by itself, not on this planet anyway. Looked at this way, the scrapie agent seems the strangest thing in all biology and, until someone in some laboratory figures out what it is, a candidate for Modern Wonder.

My Fifth Wonder is the olfactory receptor cell, located in the epithelial tissue high in the nose, sniffing the air for clues to the environment, the fragrance of friends, the smell of leaf smoke, breakfast, nighttime and bedtime, and a rose, even, it is said, the odor of sanctity. The cell that does all these things, firing off urgent messages into the deepest parts of the brain, switching on one strange unaccountable memory after another, is itself a proper brain cell, a certified neuron belonging to the brain but miles away out in the open air, nosing around the world. How it manages to make sense of what it senses, discriminating between jasmine and anything else non-jasmine with infallibility, is one of the deep secrets of neurobiology. This would be wonder enough, but there is more. This population of brain cells, unlike any other neurons of the vertebrate central nervous system, turns itself over every few weeks; cells wear out, die, and are replaced by brand-new cells rewired to the same deep centers miles back in the brain, sensing and remembering the same wonderful smells. If and when we reach an understanding of these cells and their functions, including the moods and whims under their governance, we will know a lot more about the mind than we do now, a world away.

Sixth on my list is, I hesitate to say, another insect, the

termite. This time, though, it is not the single insect that is the Wonder, it is the collectivity. There is nothing at all wonderful about a single, solitary termite, indeed there is really no such creature, functionally speaking, as a lone termite, any more than we can imagine a genuinely solitary human being; no such thing. Two or three termites gathered together on a dish are not much better; they may move about and touch each other nervously, but nothing happens. But keep adding more termites until they reach a critical mass, and then the miracle begins. As though they had suddenly received a piece of extraordinary news, they organize in platoons and begin stacking up pellets to precisely the right height, then turning the arches to connect the columns, constructing the cathedral and its chambers in which the colony will live out its life for the decades ahead, air-conditioned and humidity-controlled, following the chemical blueprint coded in their genes, flawlessly, stoneblind. They are not the dense mass of individual insects they appear to be, they are an organism, a thoughtful, meditative brain on a million legs. All we really know about this new thing is that it does its architecture and engineering by a complex system of chemical signals.

The Seventh Wonder of the modern world is a human child, any child. I used to wonder about childhood and the evolution of our species. It seemed to me unparsimonious to keep expending all that energy on such a long period of vulnerability and defenselessness, with nothing to show for it, in biological terms, beyond the feckless, irresponsible pleasure of childhood. After all, I used to think, it is one sixth of a whole human life span! Why didn't our evolution take care of that, allowing us to jump catlike from our juvenile to our adult (and, as I thought) productive stage of life? I had forgotten about language, the single human trait that marks us out as specifically human, the property that enables our survival as the most compulsively, biologically, obsessively social of all creatures on earth, more interdependent and interconnected even than the famous social insects. I had forgotten that, and forgotten that children *do* that in childhood. Language is what childhood is for.

There is another related but different creature, nothing like so wonderful as a human child, nothing like so hopeful, something to worry about all day and all night. It is *us,* aggregated together in our collective, critical masses. So far, we have learned how to be useful to each other only when we collect in small groups—families, circles of friends, once in a while (although still rarely) committees. The drive to be useful is encoded in our genes. But when we gather in very large numbers, as in the modern nation-state, we seem capable of levels of folly and self-destruction to be found nowhere else in all of Nature.

As a species, taking all in all, we are still too young, too juvenile, to be trusted. We have spread across the face of the earth in just a few thousand years, no time at all as evolution clocks time, covering all livable parts of the planet, endangering other forms of life, and now threatening ourselves. As a species, we have everything in the world to learn about living, but we may be running out of time. Provisionally, but only provisionally, we are a Wonder.

And now the first on my list, the one I put off at the beginning of making a list, the first of all Wonders of the modern world. To name this one, you have to redefine the world as it has indeed been redefined in this most scientific of all centuries. We named the place we live in the *world* long ago, from the Indo-European root *wiros,* which meant man. We now live in the whole universe, that stupefying piece of expanding geometry. Our suburbs are the local solar system, into which, sooner or later, we will spread life, and then, likely, beyond into the galaxy. Of all celestial bodies within reach or view, as far as we can see, out to the edge, the most wonderful and marvelous and mysterious is turning out to be our own planet earth. There is nothing to match it anywhere, not yet anyway.

It is a living system, an immense organism, still developing, regulating itself, making its own oxygen, maintaining its own temperature, keeping all its infinite living parts connected and interdependent, including us. It is the strangest of all places, and there is everything in the world to learn about it. It can keep us awake and jubilant with questions for millennia ahead, if we can learn not to meddle and not to destroy. Our great hope is in being such a young species, thinking in language only a short while, still learning, still growing up.

We are not like the social insects. They have only the one way of doing things and they will do it forever, coded for that way. We are coded differently, not just for binary choices, *go* or *no-go*. We can go four ways at once, depending on how the air feels: *go, no-go,* but also *maybe,* plus *what the hell let's give it a try.* We are in for one surprise after another if we keep at it and keep alive. We can build structures for human society never seen before, thoughts never thought before, music never heard before.

Provided we do not kill ourselves off, and provided we can connect ourselves by the affection and respect for which I believe our genes are also coded, there is no end to what we might do on or off this planet.

At this early stage in our evolution, now through our infancy and into our childhood and then, with luck, our growing up, what our species needs most of all, right now, is simply a future.

Index

Alexander, Leo, 363
Alien, 35
amoeba, 117
ancient world, the, 154–5
Anderson, Herbert, 338
animals, 116; behaviour of, 235–6
anthills, 118
ants, 193
applied science, 145–6
Archimedes, 143
argon, 222
Ariosto, Ludovico, 66
Aristotle, 22, 48, 143, 331
Arnold, Matthew, 131, 137, 149, 287
arthropods, 305
Asimov, Isaac, 167–72
Aston, F. W., 222
astronomy, 143, 169–72
atom bomb, 336–48, 349, 412, 413
atomic energy, 323–9
atomic physics, 201–5, 251, 262–4
atoms, 333–4
Audubon, John J., 158, 163
Augustus Cæsar, 3–4
automation, 174–80

Bacon, Francis, 1–4
bacteria, 184, 422–3
Barton, Otis, 287–8
Baudelaire, Charles, 71
beauty, 54–71, 169–72
Beebe, William, 287, 294
bees, 307–20
behaviourism, 235–6
being, problem of, 48–52, 331–2
Bergson, Henri, 122
biology, 143
birds, 233–50
black holes, 170
Blake, William, 234
Blanchard, Émile, 82–3, 90
body, the, 54–71,
Bohr, Niels, 197, 253; Correspondence Principle, 259
Born, Max, 253, 263
brain, the, 106–7
Brantôme, Pierre de Bourdeille, Seigneur de, 67
Brave New World, 273–85
Bridgeman, Prof., 211
Bridgewater, Earl of, 34–5
Broad, C. D., 256
Buckland, William, 34–5
Burroughs, John, 149–66, 286
Burton, Robert, 68

cactus, Christmas, 115
carbon, 185, 186, 187
Carlyle, Thomas, 46, 181, 401
Carson, Rachel, 286–304
Causality, Principle of, 266
cells, human, 194
Cervantes Saavedra, Miguel de, 67–8
chance, 251–2, 368–73; design and, 25–8
Chance, Edgar, 236–7
change, 23; scientific, 399–407
Chaucer, Geoffrey, 54
Chesterton, G. K., 47, 95–101
children, 425; dreams, 425
classical education, 136–7, 140–2, 152
classification, biological, 14–15
clothing, beauty and, 55–6
coelacanths, 303–4
colloids, 112–20
Communist Manifesto, 402
complementarity, principle of, 197, 199
Compton, Arthur, 338, 343
Conant, James, 210, 211
concentration camps, 363–4
conditioning, 119
consciousness, human, 271, 331
conservation of heat, 415

control, automatic, 174–80
co-operation, scientific, 211–12
Copernicus, Nicolaus, 24
cosmic rays, 201
Cosmas, 217
creatures, things and, 116
crested grebe, 240–4
croakers, 301
crystals, 112–20
cuckoos, 236–7
culture, science and, 132–48

d'Annunzio, Gabriele, 70
Dante, 141, 152
darters, 247–8
Darwin, Charles, 5–17, 44, 155–6, 157, 158–9, *and see* evolution
death, dreams of, 376–96
democracy, 121, 179–80
Democritus, 143
design in nature, 9, 25–7, 114
destruction, 112
determinism, 251–72, 331
Dewey, John, 18–31, 110, 197
Divers, 246
Dos Passos, John, 229–32
dreams, 374–96
dung beetle, 73–94

Earth, planet, 426
Eastman, Max, 19
Eddington, Arthur Stanley, 251–72
education, science in, 132–48
egrets, 237–9, 249
Einstein, Albert, 108, 128, 252, 408–418
electrons, 201, 262–3
elementarity, 200
Ellis, Havelock, 53–71
Emerson, Ralph Waldo, 104, 156
emotion in animals, 236
empiricism, 50
energy, 409, 414–18
error, 213
eternity, 48–52
ethics, *see* morality
eugenics, 45
evil, 34–45
evolution, 5–17, 130–1, 191–3, 218, 249–50, 373; influence of, on philosophy, 20–31
existence, 48–52
existentialism, 47
extinction, 17
extraterrestrial life, 182–95

Fabre, Jean Henri, 72–94, 305
fairy tales, 96–101
faith, 120
falcons, 245
Fermi, Enrico, 322, 336–48
Fermi, Laura, 336–48
Firenzuola, Agnolo, 66
fish, 292, 297–300, *and see* croakers
Fleming, Richard, 293
Forbes, Edward, 288
fossils, 15–16
Franklin, Benjamin, 112
Frazer, James, G., 56
freedom, 268–71, 331
Freud, Sigmund, 213, 374–96
Frisch, Karl von, 306

galaxies, 171
Galen, Claudius, 143
Galileo, 24, 25, 125, 218, 219
genes, 191–2
geology, 15–16
German shepherd (dog), 367
Germany, Nazi, 349–67, 402
Gestapo, the, 351–67
Gladstone, W. E., 32
glass sponge, 297
God, 29, 33, 120, 227, 228; evil and, 34–7; existence of, 50–1
Gödel, Kurt, 368–9
Goering, Hermann, 355–6
Goethe, J. W. von, 150, 153, 160, 161–2
Goudsmit, Samuel, 349–67
Gould, Stephen Jay, 32–45
government regulation, 179–80
gravity, 410
Gray, Asa, 12, 27
Greeks, the, 21, 140, 371
Greenewalt, Crawford, 343–6
Gregory the Great, Pope, 225–6
Grimm's Fairy Tales, 99
Groves, Leslie, 344

Haber, Heinz, 188, 189
Haldane, J. B. S., 183
Hamlet, 394–6
Heaven, 225
Hegel, G. W. F., 50–1
Heisenberg, Werner, 350, 361–2, 411
Hell, 225
herons, 237–9
Herrick, Robert, 368
herring gulls, 248
Heyerdahl, Thor, 293
Hilbert, David, 203–4
Himmler, Heinrich, 359–65
Hipparchus, 143
Hitler, Adolf, 274, 406
Hjort, Johan, 293
Hooker, Joseph, 21, 43
Houdoy, Jules, 66–7
Howard, Eliot, 236, 246
Hudson, W. H., 234
humanism, 141
Humboldt, Alexander von, 64, 160–1, 163
Hume, David, 95
Hutchins, Robert, 149
Huxley, Aldous, 273–85
Huxley, Julian, 45, 233–50
Huxley, Thomas H., 21, 32, 43–4, 99, 102, 130–48, 149, 197
Huyghens, Christian, 219–20

ichneumon fly, 35–45
Illiger, Johann, 83, 90
imperialism, 403, 405
indeterminism, *see* determinism
industrialism, 401
industry, 147; automation of, 174–80
insects, social groups of, 192–3, 305, 427
integration, 115
intellectual progress, 31
isotopes, 222

James, William, 46–52, 122
Japan, 402–3
Jeans, Sir James, 331
jellyfish, 297
Johnson, Martin W., 290
Joliot-Curie, Frédéric, 321–2
Jupiter, 194

Kant, Immanuel, 128, 181
Kepler, Johannes, 24
kestrels, 244–5
Klein, Felix, 203
knowledge, 400; object of, 28; of the universe, 104–9
Köhler, Wolfgang, 397
Krutch, Joseph Wood, 110–20

Lamarck, Jean, 6
Lambda, Prof., 265
language, 425
languages, science and, 144–5
Laplace, Pierre, 115, 256, 257–9, 262
laws of nature, *see* natural laws
Leibnitz, G. W. von, 8, 408, 414
Lemaire, Jehan, 67
Leonard, Jonathan Norton, 181–95
Leonardo da Vinci, 400
Lewes, G. H., 12
liberal education, 136–8
life: definition of, 110, 183; origin of, 12, 184–6
light, 219–20
Lippman, Walter, 121
liquid hydrocarbons, 188
literature, 150–66
logic, classic type of, 29
Lucian, 181
Luetzelberg, Herr von, 361
Luigini, Federico, 66
Lyell, Charles, 9, 21, 166
Lysenko, T. D., 274

Mach, Ernst, 128
McIlhenney, E. A., 237
Maeterlinck, Maurice, 305–20
Magnell's Questions, 333
mammals, 189
man: determinism and, 267; nature of, 3; socialization of, 192–3
Mars, 190–1, 194
Marx, Karl, 213, 251
Marxism, 402
mass, 409, 414–18
materialism, 27, 100
mathematics, 143, 148
Maugham, Somerset, 374
meaning, 114
mediæval philosophy, 139

Mencken, H. L., 121
Mendel, Gregor, 400
Mercury, 190
metaphysics, 215
Milky Way, 171
Miller, Henry, 413
Milton, John, 181
Mivart, St. George, 42–3
molecules, 191, 192
morality, 403–6; nature and, 34–45; religion and, 227–8
Morrison, Philip, 102
mutation of species, 8–9

Nagel, Ernest, 173–80
natural laws, 107–9, 115, 116, 118–19, 260–2
natural selection, 6, 7, 17
nature, description of, 99–100
Nazi Germany, 349–67, 402
neurosis, 387–90
Newman, Cardinal, 218
Newton, Isaac, 8, 98, 125, 128, 218, 219, 408
Niphus, Augustinus, 66
nitrogen, 221–2
Noyes, Alfred, 130

oceans, 287–304
Oedipus, 1–3, 375, 390–4
olfactory receptor cell, 424
Omar Khayyam, 256–7
oncideres, 422–3
Oppenheimer, J. Robert, 196–214, 374
order, 142, 368
Ortega y Gasset, José, 121–9
Orwell, George, 273
Osenberg, Prof. 351–9
oxygen, 186, 187

Pan, 370–3
pantheism, 117
parasites, 35–45, 184, 192
particle accelerators, 201, 202
particles, 251, 412; elementary, 200
patriotism, 405
Peirce, Charles S., 118
penguins, 245–6
perception, 23
Petavius, Father, 218
Petrarch, Francesco, 66
philosophy: evolution and, 20; science and, 206–7
physics, 143, 198–214
Planck, Max, 253
planets, 170, 190
plankton, 291–2, 294
plants, 116
Plato, 121, 206–7, 400
play, animal, 247–9
Plutarch, 154
politics, 147–8
Pontecorvo, Bruno, 337
porpoises, 302
power, human, 404, 407
Priestley, Joseph, 132–3
psychoanalysis, 376–96
psychology, 16, 404–6
Ptolemy, 143
Pubilius Syrus, 71
purpose, *see* design in nature

quantum theory, 108, 197, 258–9, 271, 411–12
questions, asking, 105

radar, 203
radio-activity, 400
Ramsay, Sir William, 221–2
randomness, *see* chance
Rascher, Dr., 363
Ravens, 248–9
Rayleigh, Lord, 221–2
reality, 52
relativity theory, 108, 271, 409–418
religion, evolution and, 5, 8, 20–1, 26; science and, 215–28; sexuality and, 56
Renaissance, the, 143
responsibility, 270
Rollefson, Gunnar, 293–4
Ross, Sir John, 288–9
Ruskin, John, 401
Russell, Bertrand, 110–11, 181, 214, 215, 397–412

sacred beetle, 73–94
Sagan, Carl, 102–9
St. Francis of Assisi, 220–1

salt crystal, 106–7
Santayana, George, 46
Sartre, Jean Paul, 47
Schopenhauer, Arthur, 48, 63
scientific advance, 399–407
scientists: responsibility of, 205–6; specialization of, 123–9
scrapie virus, 423–4
sea, the, 287–304
seals, 296
sedge warblers, 246
self-understanding, 270
sentimentalism, 100
sexual attraction, 54–71
sexual reproduction, 12, 17
Shakespeare, William, 152, 405
sharks, 304
shrimps, 297, 302
Sidgwick, Henry, 28
Sievers, Col. Wolfram, 360, 362
silicones, 188
Smith, J. L. B, 304
social Darwinism, 45
socialization, 191–3
society, 119, 193–4; stability of, 403
sociology, 147–8
Soviet science, 251–2
space-time, 410
specialization in science, 123–9, 182, 400
species, 16, 22–3; defining, 13–14
Spencer, Herbert, 16, 29, 50, 153
Sphinx, the, 1–4
squid, 292–3, 295–6, 299–300
Stalin, Joseph, 274
stars, 170
Steinmetz, Charles Proteus, 229–32
Stephanitz, Capt. Von, 365–7
Stevenson, Robert Louis, 368–73
Stratz, Carl H., 59–60, 70
sun, the, 185
symbols, 269

Taine, Hippolyte, 156
teaching science, 209–10
technicism, 123
technology, 204
termites, 425
theology, *see* religion
Theophrastus, 143
thermodynamics, 257
things: creatures and, 116; nature of, 3
Thomas, Lewis, 419–27
Thoreau, Henry, 149
toucans, 248
truth, 18, 218–19; ultimate, 330–5

unemployment, 177–8
universe, the, 142; beauty and, 169–172; knowledge of, 104–9; size of, 410–11
uranium fission, 339–48, 413

Venus, 190, 194
vertebrates, 305
viruses, 192
vitalism, 122

wasps, 34–45
water, 187
water turkeys, 247–8
Wattemberg, Al, 347
wave mechanics, 271
weapons, 205
Weil, George, 344–5
Wells, H. G., 273, 321–35
Wesley, John, 220–1
Weyl, Hermann, 254
whales, 294–7
Whitehead, Alfred North, 110, 214, 215–28
Whitman, Walt, 105, 149, 169
Wigner, Eugene, 346
Wilberforce, Bishop, 32, 130–1
Wilson, Alexander, 163, 166
Wimmer, Prof., 361
women, beauty and, 54–71
Wonders of the World, 421–7
Wordsworth, William, 151, 234
world, beginning and end of, 142
World War II, 199–200
Wüst, Walther, 360, 362

Zinn, Walter, 343